Human Protein Metabolism

Springer
New York
Berlin
Heidelberg
Barcelona
Hong Kong
London
Milan
Paris
Singapore
Tokyo

Stephen Welle

Human Protein Metabolism

With 39 Figures

Stephen Welle
University of Rochester Medical Center
Endocrine Metabolism Unit
Monroe Community Hospital
Rochester, NY 14620, USA

Library of Congress Cataloging-in-Publication Data
Welle, Stephen.
Human protein metabolism / Stephen Welle.
p. cm.
Includes bibliographical references and index.
ISBN 0-387-98750-9 (alk. paper)
1. Proteins—Metabolism. I. Title.
QP551.W43 1999
612.3′98—dc21 98-53841

Printed on acid-free paper.

Production coordinated by Chernow Editorial Services, Inc., and managed by Lesley Poliner; manufacturing supervised by Nancy Wu.
Typeset by KP Company, Brooklyn, NY.
Printed and bound by Braun-Brumfield, Inc., Ann Arbor, MI.
Printed in the United States of America.

9 8 7 6 5 4 3 2 1

ISBN 0-387-98750-9 Springer-Verlag New York Berlin Heidelberg SPIN 10710160

To
Linda, Megan, and Kevin

Preface

The last comprehensive book on the topic of mammalian protein metabolism was published in 1978 by Waterlow, Garlick, and Millward. At that time, few studies of protein dynamics in human subjects had been performed because of limitations in the available methods. Since then, hundreds of studies have been published in biomedical journals that describe protein metabolism in humans. The advent of better mass spectrometers and the availability of reasonably-priced stable isotope tracers allowed the field to grow tremendously. The results of these studies have not, until now, been synthesized in a single book. Researchers in this field are often frustrated in trying to keep track of all the key findings over the last decade and often wish for a single source for this information. Wolfe (1992) published a book that focused on the some of the isotopic methods to study protein metabolism, but did not review the results of the studies that employed these methods. Therefore, the goal of this book is to provide a thorough source of information on the methods used to study protein metabolism in humans, and what we have learned by using these methods. The book focuses on understanding the principles of the methods, and their limitations, rather than giving "cookbook" recipes for performing research. It tries to give a balanced interpretation of the research in the field, pointing out areas of disagreement and why some conclusions are more likely to be valid than others.

The book is limited to human protein metabolism. There is a huge literature on protein metabolism in mammals, in cell lines, and in cell-free systems, but a comprehensive analysis of all of these areas would require several volumes. Some animal studies have been cited to elucidate closely related human studies. No attempt was made to describe every study done in humans. The goal was to describe succinctly what is known (or in some cases, what is not known) in most of the areas relevant to the study of human protein metabolism. Although not every study of human protein metabolism is cited, the literature citation is thorough and points readers to the most relevant primary work.

Readers should have some understanding of biochemistry and physiology, but undergraduate courses should be sufficient for understanding the principles described in this book. Some algebraic formulas are given, primarily in the chapter on methods. However, every effort has been made to write so that the basic principles can be understood even if the reader ignores the formulas. The audience for this book will include nutritionists, physiologists, biochemists, and clinicians. It is my hope that it will be a helpful resource for anyone interested in performing research in this field. For those already doing research on human protein metabolism, it should be a useful review of work outside their particular focus. For those more interested in the basic biochemistry of protein metabolism, it shows how the more basic research has been applied to humans.

I wish to thank my mentors and collaborators who stimulated my interest in this field, including Sree Nair, Robert Campbell, Berch Griggs, Gilbert Forbes, and Dwight Matthews. I am indebted to the National Institutes of Health, which has supported my research for many years.

STEPHEN WELLE
October 1998

Contents

1 The Importance of Protein Dynamics

Protein Structure and the Concept of Protein Turnover

The *Webster's College Dictionary* defines protein as "any of numerous organic molecules constituting a large portion of the mass of every life form, composed of 20 or more amino acids linked in one or more long chains, the final shape and other properties of each protein being determined by the side chains of the amino acids and their chemical attachments." The term protein derives from the Greek prōte, or primary. The primary importance of proteins to tissue structure and function and as a component of the diet has been recognized for more than a century (see Munro 1964a) for a review of nineteenth-century developments that led to modern research in protein metabolism).

The primary structural feature of all proteins is that they are chains of amino acid residues (the part of the amino acid remaining after removal of an oxygen and two hydrogen atoms to form a peptide bond). Figure 1.1 shows the chemical structures of the 20 amino acids[1] that are incorporated into the primary amino acid chains, as described in Chapter 2. The amino acid residues are held together by peptide bonds (Figure 1.2). The amino acid chains, especially the shorter ones, are often called peptides or polypeptides, although there is no formal rule for when to call the amino acid chain a peptide rather than a protein. Peptide bonds are very stable and are broken only by specific enzymes or harsh chemical treatment, such as heating in strong acid. Although the genetic code specifies incorporation of only 20 amino acids into the peptide chain, numerous other amino acids derive from these 20 that are found in proteins (Chapter 2).

The amino acid sequence of a protein is its primary structure. Its secondary structure is the 3-dimensional structure of the peptide chain in specific regions of the protein, including helical structures and pleated sheets (Stryer 1995; Schultz and Liebman 1997). Its tertiary structure is determined by the folding of the protein because of covalent and noncovalent interactions

[1] Proline actually is an imino acid rather than an amino acid, but will be referred to as an amino acid for simplicity.

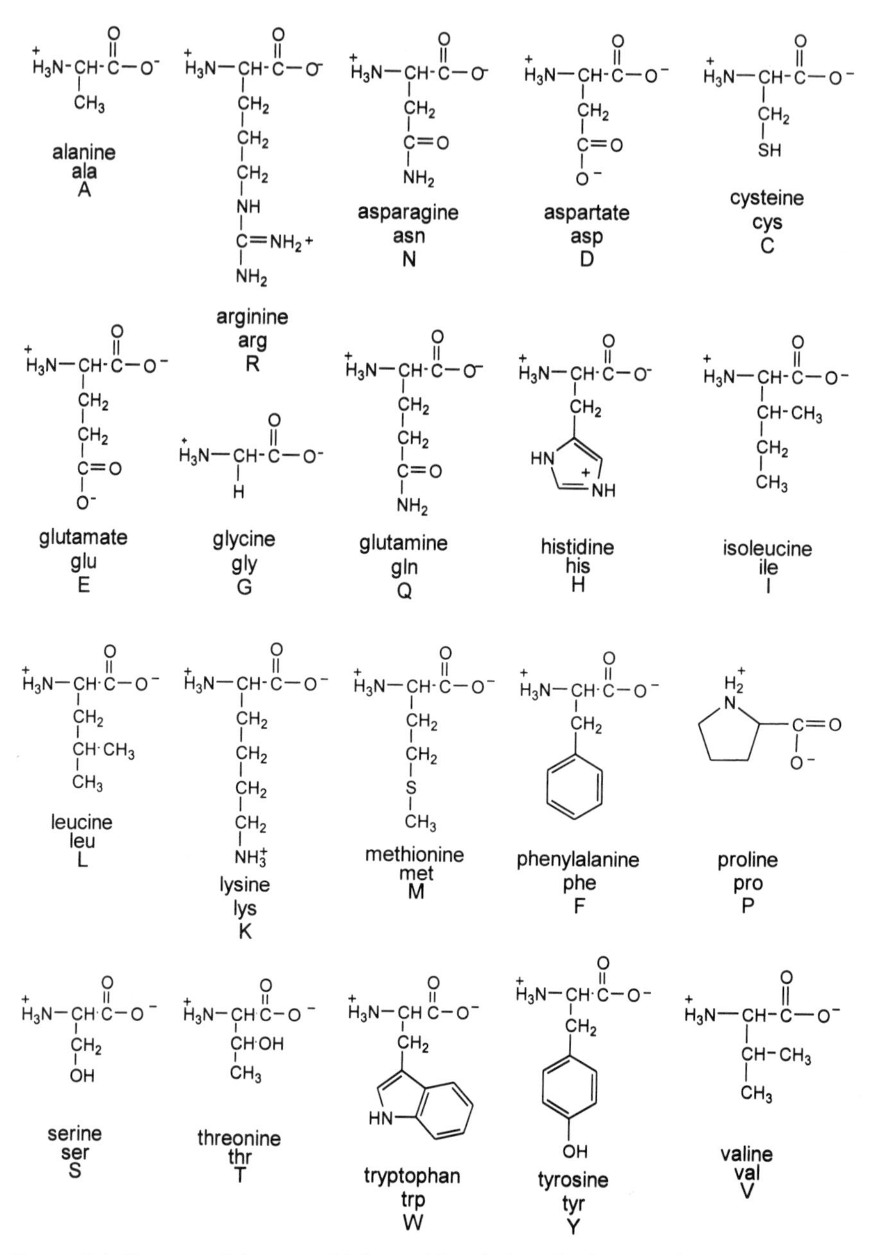

FIGURE 1.1. Structure, 3-letter, and 1-letter abbreviations for the 20 amino acids incorporated into proteins.

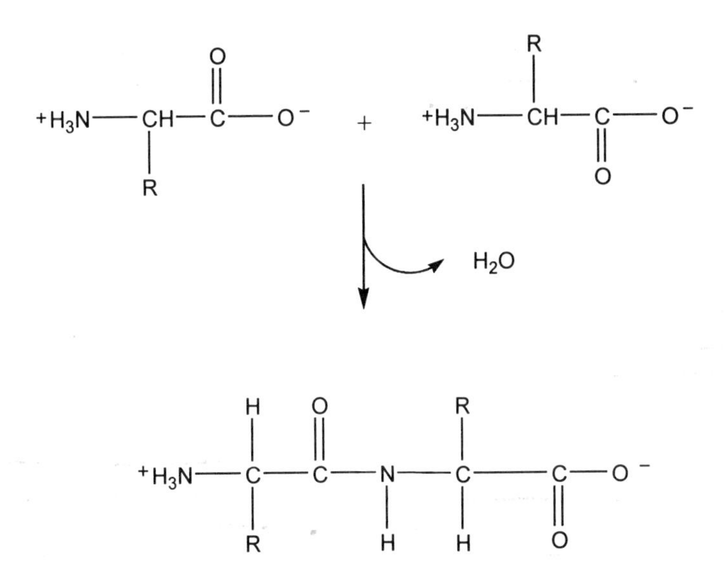

FIGURE 1.2. Formation of a peptide bond. *R* is side chain of any amino acid. *Trans* configuration of amino acid side chains is shown. This configuration separates adjacent side chains maximally and is more stable than the *cis* configuration. *Cis* configuration occurs with proline.

between amino acid residues that are farther apart than those that determine secondary structure. In some cases, proteins have a quaternary structure, which is determined by the binding together of two or more separate peptide chains. In such cases, the individual peptide chains are called subunits. Each subunit can be considered an individual protein, but it may lack biological activity when separated from the other subunits. Biological activity is determined by the tertiary and quaternary structures, but the primary structure is what determines how the peptide chain is folded into these structures. Various characteristics of the amino acid side chains, including charge, polarity, and size, ultimately determine how a protein folds into its final form. Although there are many different ways that a particular peptide chain could fold into different structures, the most thermodynamically stable conformation is the one that prevails. Higher order structure is unstable compared with primary structure, and can be disrupted by mild heating or small changes in ionic strength or pH. A protein is said to be denatured when higher order structure is disrupted. Elaborate physiological mechanisms have evolved to maintain homeostasis of temperature, pH, and ionic environment, largely to ensure that the activity and structural integrity of the proteins are preserved.

Much of our dry weight consists of proteins. A typical 70 kg man contains about 44 kg of water, 11 kg of proteins, and 10 kg of fats (Forbes 1987a). The protein mass consists of thousands of different proteins, ranging in size from small peptides with molecular weights less than a thousand daltons to huge proteins with molecular weights over a million daltons. They are involved in all aspects life. They provide the structural framework for cells and organs, catalyze biochemical reactions, and regulate intracellular and intercellular communication. Although most of the research discussed refers to the metabolism of total body or tissue proteins in many instances, it should be emphasized that this bulk protein metabolism is merely the sum of the metabolism of the many individual proteins. The responses of some of the individual proteins to various stimuli may be very different from the overall response of protein metabolism to these stimuli.

Although the total bulk of proteins appears to remain fairly constant throughout adult life, the dynamic nature of protein homeostasis was suggested in the nineteenth century (Munro 1964a) and was proven with isotopic tracers several decades ago (Schoenheimer, Rutner, and Rittenberg 1939). Proteins are constantly degraded into their constituent amino acids, a process termed proteolysis. For protein mass to remain constant, new proteins must be synthesized to replace the degraded proteins. Of the 11 kg of protein in our typical 70 kg man, approximately 0.3 kg are degraded and replaced each day.

The term "turnover" often is used to describe simultaneous degradation and synthesis. This term can be somewhat confusing when the rate of degradation does not equal the rate of synthesis, for it is unclear whether it then refers to the degradation or synthesis rate. Thus, the term turnover is not quite accurate if the protein mass is changing. The most common meaning of turnover in such instances is in reference to the rate of protein degradation. Fractional turnover refers to the rate of turnover relative to the amount of protein already present. For example, if you have 2 kg of myosin (the most abundant muscle protein) in your body and you synthesize and degrade 40 g of myosin each day, your fractional myosin turnover rate is 2%/d.

Protein turnover accounts for a significant portion of our basal energy expenditure. We have estimated that approximately 20% of energy expenditure at rest after overnight fasting is devoted to protein turnover (Welle and Nair 1990b). Most of this energy is devoted to protein synthesis rather than proteolysis. Protein turnover must have considerable biological importance for so much energy to be expended for it. Obviously, cell replication and growth require synthesis of new proteins, but can account for only a small proportion of human protein metabolism. As discussed below, the main reasons that humans expend so much energy for protein turnover are the need to rapidly alter concentrations of certain proteins and the need to replace damaged proteins.

Rapid Turnover of Proteins Facilitates Rapid Changes in Protein Concentrations

There are many seemingly "futile" substrate cycles—those that use a lot of energy but do not appear to do much other than eliminate useful molecules and then resynthesize them. Such cycles are not really futile, but have the important function of allowing changes in either the synthesis or the removal pathway to result in rapid changes in the concentrations of important molecules (Newsholme and Crabtree 1976). For example, suppose that Protein X is present at a concentration of 1 pg/cell. It is being synthesized and degraded at a rate of 0.1 pg/h per cell. If some stimulus for Protein X synthesis were to double the synthesis rate to 0.2 pg/h, and the degradation rate were to remain constant, then the amount of Protein X would double in 10 h (Figure 1.3). If the initial turnover rate had been only 0.01 pg/h, then doubling the synthesis rate would lead to only a 10% increase in Protein X concentration after 10 h (Figure 1.3). A greater turnover rate allows a much more rapid adaptation, with the same relative change in synthesis.

Usually, if rapid increases in levels of a protein are beneficial to the organism, it also is beneficial to rapidly remove the excess protein when higher levels are no longer needed. In the example above, if synthesis of Protein X were to return to the initial level after its concentration had doubled, again with no change in degradation, the Protein X concentration would remain at twice the baseline level (Figure 1.3). The only way to revert back to the initial

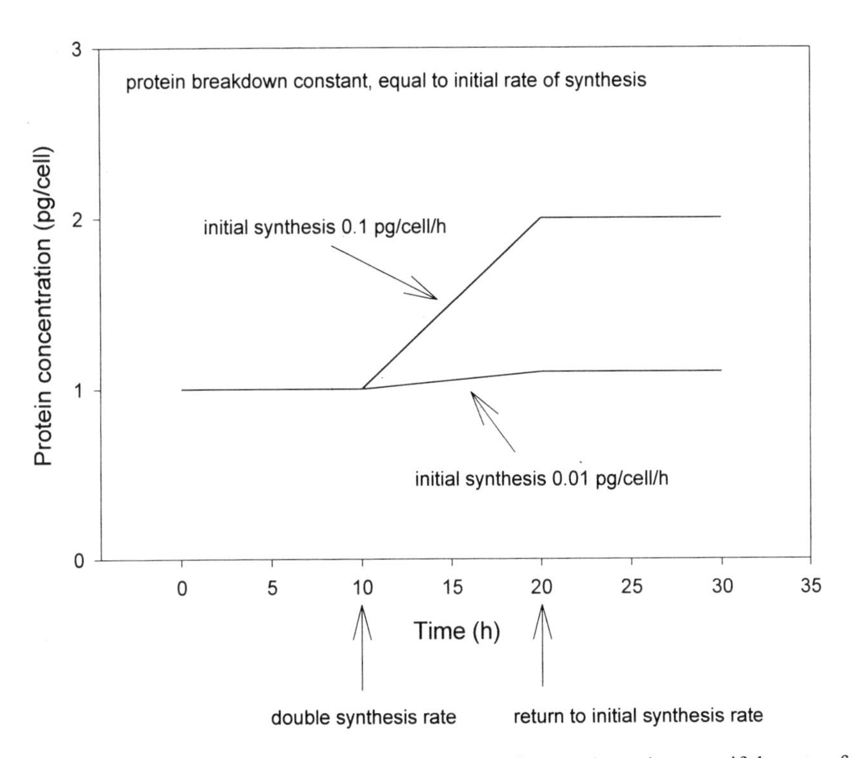

FIGURE 1.3. Effect of doubling the rate of synthesis of a protein on its mass if the rate of breakdown is constant and independent of the current mass ("zero order" breakdown). Mass would continue to increase as long as synthesis is faster than original degradation rate. Note that protein mass remains elevated even after synthesis rate returns to original value.

protein concentration would be either to reduce synthesis below the basal rate, increase degradation above the initial rate, or both. A simple system to ensure that protein concentrations are proportional to the current synthesis rate is one in which the degradation rate depends on the protein mass, such that fractional degradation is constant regardless of synthesis rate. In such a system, the equilibrium protein mass is directly proportional to the synthesis rate (Figure 1.4). This is a very simple system for regulating protein concentration. When Protein X synthesis increases from 0.1 to 0.2 pg/h, only the *initial* rate of increase in the concentration will be 0.1 pg/h per cell. When the protein level reaches 1.1 pg/cell, the degradation rate would be 0.11 pg/h rather than 0.1 pg/h if the fractional degradation were constant at 10%/h. As long as degradation is less than 0.2 pg/h, there is an increasing concentration of Protein X. When the concentration reaches 2 pg/cell, then synthesis would equal degradation and a new equilibrium concentration would be maintained as long as synthesis remains at 0.2 pg/h. If the stimulus for increased protein

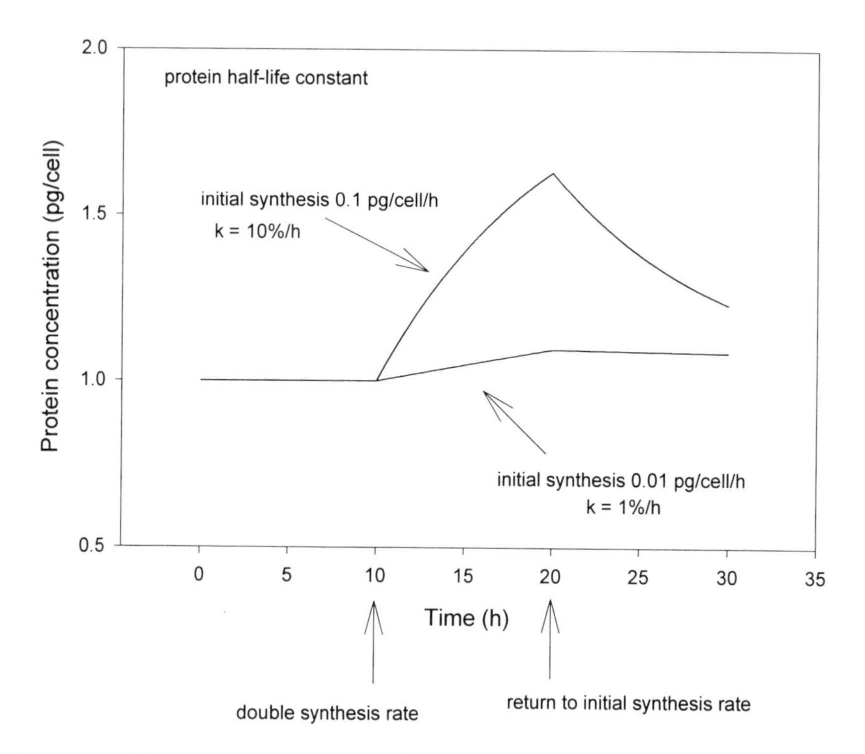

FIGURE 1.4. Effect of doubling the rate of synthesis of a protein on its mass if the rate of breakdown is a constant fraction of the current mass ("first order" breakdown). Mass eventually would stop increasing, after doubling, if synthesis rate were to remain at twice the original value. Note that protein mass declines after synthesis rate returns to original value, even though fractional degradation does not increase.

synthesis were removed, and the synthesis rate reverted to its original 0.1 pg/h, then protein concentrations would begin to decline immediately (Figure 1.4). Eventually the absolute degradation rate would slow to 0.1 pg/h, at which point the synthesis and degradation would again be equal and the original equilibrium concentration would be restored. Maintaining a constant fractional degradation slows the rate of accumulation of protein when synthesis is stimulated, compared with maintaining the same absolute rate of protein degradation (compare Figures 1.3 and 1.4). However, this disadvantage is minor compared with the complexity that would be required for Protein X to have a mechanism for matching its synthesis and degradation rates independently.

Although there are undoubtedly important exceptions in which degradation of specific proteins has a regulatory function, protein degradation processes (Chapter 2) are not specific enough to prevent an increase in the absolute degradation rate of a protein as its concentration increases. Degradation may not always be an ideal first order process (i.e., fractional rate of degradation is *always* the same), but absolute rates of degradation usually increase when protein concentration increases. This system ensures that an equilibrium between protein synthesis and degradation can be achieved without needing some more complicated process to match protein breakdown to protein synthesis. (The appendix to this chapter describes the mathematics of a simple system in which equilibrium protein mass is always proportional to protein synthesis because fractional degradation is always the same.) Altering the synthesis rate is the major mechanism for regulating concentrations of individual proteins. Synthesis of specific proteins can be regulated by altering the transcription and translation of specific genes (Chapter 2).

The term half-life often is used to describe the turnover rate of proteins. Half-life ($T_{1/2}$) is the time required for 50% of the mass of a protein to be degraded and replaced. It is calculated from the fractional degradation rate, as described in the appendix. Protein mass at equilibrium is directly proportional to both the protein synthesis rate and the half-life. Therefore, protein mass could be changed just as easily by changing the fractional degradation rate as it can be by changing the synthesis rate. When a general anabolic state of protein metabolism is needed, inhibiting the proteolytic pathways is a more energy-efficient approach than increasing overall protein synthesis. When a more general catabolic state is required, for example to derive free amino acids as a fuel source and for gluconeogenesis or to provide amino acids for the local protein synthesis required by wound healing, an overall increase in proteolysis is needed. However, proteolysis cannot be the major regulator of the relative concentrations of the proteins within a cell, because such a system would require an elaborate mechanism to selectively regulate degradation of each protein.

Knowledge about protein half-life is useful for estimating the time course of the response to a physiological challenge. For example, suppose that the rate of protein turnover in skeletal muscle is 1.5% per day in a sedentary person. Now suppose this person starts a weightlifting program and stimulates muscle protein synthesis by one third, while fractional degradation is unchanged. According to the concept that protein mass is proportional to synthesis, the final muscle protein mass will be one third more. The time required to achieve 50% of the maximal response would be 46 days (see appendix for derivation). The time required to achieve 95% of the final response would be 200 days. Any study to examine changes in muscle mass after only a few weeks would underestimate the effect of the intervention. This example also reveals other complications to the general model. Usually, changes in protein mass are adaptive. The muscle hypertrophy associated with weightlifting makes it easier to lift weights. When the lifting becomes easier, the stimulus for increasing protein synthesis is reduced, in relative terms. Unless the lifter adds more weight, the rate of protein synthesis might decline over time, and the increase in muscle mass could be less than one third of the initial muscle mass. In more general terms, the rate of protein synthesis can be influenced by the protein mass. There also is some evidence that weightlifting exercises can increase muscle protein degradation. The point here is that the final effect of some intervention on protein mass cannot always be predicted quantitatively based on the initial effect on protein synthesis or degradation. Nevertheless, information about half-life provides an important starting point for estimating how long it will take for protein mass to restabilize after some event that changes protein synthesis or half-life.

Protein Turnover Is Necessary to Maintain Protein Quality

Regulatory proteins such as hormones and rate-limiting enzymes account for a small proportion of the total protein mass of the human body. Most of the protein mass comes from constitutively expressed proteins, which are required for tissue structure and function, but which normally do not require significant alterations in abundance over time (except during growth). These proteins tend to have longer half-lives than regulatory proteins. Nevertheless, because of their large mass, a considerable portion of our energy expenditure is devoted to the turnover of these proteins. This turnover process is necessary to maintain optimal function because of the constant chemical damage to which proteins are subjected.

Oxidative damage to amino acid side chains is a major source of chemical damage to proteins (Stadtman 1988; Stadtman 1992). In the process of normal cellular respiration, there is always some production of free radicals of oxygen (i.e., oxygen atoms with an unpaired electron) or other reactive oxygen species such as hydrogen peroxide. The most damaging oxygen free radical appears to be the hydroxyl radical (·OH) (not to be confused with the hydroxide ion, OH^-). Although there are various cellular defense mechanisms to protect macromolecules from oxidative damage, the system is not perfect. These free radicals can oxidize the amino acid side chains, leading to a number of products that cause cause cross-linking, fragmentation, or loss or reduction of protein activity. Some amino acid side chains are oxidized to carbonyl derivatives, and the carbonyl content of proteins has been used to estimate the extent of oxidative damage to protein (Stadtman 1992). In young adults, approximately 10% of the protein molecules have a carbonyl modification, whereas in older individuals as many as 20% to 30% of the protein molecules may have carbonyl modifications. Because some forms of oxidation do not lead to carbonyl formation, an even greater portion of the protein molecules may have some type of oxidative damage. Although oxidative damage to one or a few amino acid side chains does not necessarily inactivate a protein molecule, accumulation of random damage could certainly have a major impact on the specific activity of the protein. Removal and replacement of proteins therefore is essential for maintaining cellular integrity.

Oxidation is not the only type of damage to proteins (Stadtman 1988). Side chains of asparagine and glutamine are subject to spontaneous deamidation. Aspartyl residues are subject to racemization and isomerization, and prolyl residues to isomerization. Perhaps a quantitatively more important form of protein modification is glycation (Kristal and Yu 1992). Glucose, fructose, other monosaccharides, and any compound containing an α-hydroxy aldo or keto group can react with the free amino groups of proteins through a series of reactions to yield various compounds known as advanced glycosylation end products (AGEs), otherwise known as Maillard products. Although these products are more abundant under hyperglycemic conditions, they also are present in nondiabetic individuals. Protein damage by glycation is more prevalent for extracellular proteins than for intracellular proteins, except for erythrocyte proteins and lens crystallins (Monnier et al. 1991).

Thus, the continuous turnover of proteins is essential for maintaining protein quality and ensuring that abundance of specific proteins can adapt to physiological challenges within an appropriate time frame. The next chapter will review the molecular machinery involved in this process.

Appendix

When there is a constant fractional rate of degradation, the change in protein mass associated with a change in protein synthesis is given by the expression:

$$\Delta P_t = \Delta P_f \times (1 - e^{-kt})$$

where ΔP_t is the change in protein mass at time t, ΔP_f is the change in protein mass at equilibrium, k is the fractional degradation rate, and t is time elapsed since the change in synthesis rate. The ultimate protein mass achieved will be directly proportional to the synthesis rate and fractional degradation:

$$P = S/k \text{ (at equilibrium, i.e. when protein mass is constant)}$$

where P is protein mass and S is the rate of protein synthesis. Hence,

$$\Delta P_f = (S_f - S_i)/k$$

where S_f is the final rate of protein synthesis and S_i is the initial rate.

Half-life ($T_{1/2}$)is the time required for 50% of the mass of a protein to be degraded and replaced. $T_{1/2}$ is easily derived from k:

$$T_{1/2} = ln2/k = 0.693/k$$

Thus protein mass at equilibrium can also be derived from S and $T_{1/2}$:

$$P = S \times 1.44 \times T_{1/2}$$

The formulas above are simplifications, because fractional degradation is not always constant from minute to minute or hour to hour. However, over long periods of time the integrated fractional degradation rate is usually fairly constant as long as nutritional, metabolic, and physiological status is constant. Even if degradation is not a true first order process (degradation is not the same at all protein concentrations), *P at equilibrium* is determined by S and k as described above.

2
Basic Mechanisms of Protein Turnover

Although this book is about human protein metabolism, very little of the information on the molecular mechanisms of protein turnover comes from studies of human cells or tissues. Nevertheless, the basic mechanisms are generally the same in all eukaryotic cells, so that the processes described in this chapter are applicable to human tissues.

Protein Degradation

There is great diversity in the half-lives of the various proteins residing in the same cell, even though only a few different pathways of proteolysis have been identified. Although it is possible that there are many more proteolytic systems that have not yet been discovered, a more likely explanation for the diversity of protein half-lives is that multiple properties of a protein molecule affect the probability that it will enter a degradative pathway. Structure (from primary amino acid sequence to quaternary structure), chemical properties determined by this structure (charge, hydrophobicity), susceptibility to oxidative damage, posttranslational modifications, and subcellular or extracellular location are potential factors that can interact to give a particular protein its characteristic half-life. This chapter discusses only processes that hydrolyze proteins to their constituent amino acids and not selective cleavage of specific proteins from inactive precursors to active forms. The latter process will be considered as posttranslational processing rather than degradation (see discussion of posttranslational processing later in this chapter).

This section describes the intracellular degradation of proteins produced by the body and not the gastrointestinal digestion of proteins in food. The combined action of gastric acid and digestive enzymes in the lumen of the gut and peptidases in the intestinal cells results in entry of dietary protein into the body mostly as free amino acids. Once these free amino acids enter the systemic circulation, they are indistinguishable from the free amino acids in the circulation derived from intracellular proteolysis throughout the body.

Protein degradation is accomplished by numerous enzymes that are called proteases. Proteases have been classified as exopeptidases, which cleave peptide bonds adjacent to free amino or carboxyl groups at the ends of peptides, and endopeptidases, which cleave the interior peptide bonds of proteins or smaller peptides. The term proteinase is synonymous with endopeptidase. The term cathepsin often is used for lysosomal or other intracellular proteases. These enzymes all cleave peptide bonds at specific sites on amino acid chains, and many only act on short peptides. The combined action of many proteases is required to completely digest a protein to its constituent amino acids. Proteases are widely distributed throughout the cells and extracellular matrix. Overall proteolysis usually is measured in humans by measuring the rate of appearance of amino acids (Chapter 3), but hydrolysis of a single peptide bond, without release of an amino acid, can inactivate a protein.

Degradation of extracellular proteins and some intracellular proteins occurs in specialized vacuoles (Figure 2.1). Vacuolar protein degradation can be either heterophagic (degradation of proteins derived from the extracellular matrix or from other cells) or autophagic (degradation of intracellular proteins). The primary organelle involved in vacuolar degradation is the lysosome. Lysosomes are vesicles formed in the Golgi apparatus. They contain more than 50 enzymes that are capable of degrading not only proteins, but also lipids, carbohydrates, and nucleic acids (Lee and Marzella 1994). There appear to be 15 to 20 proteolytic enzymes in the lysosomes. The lysosomal

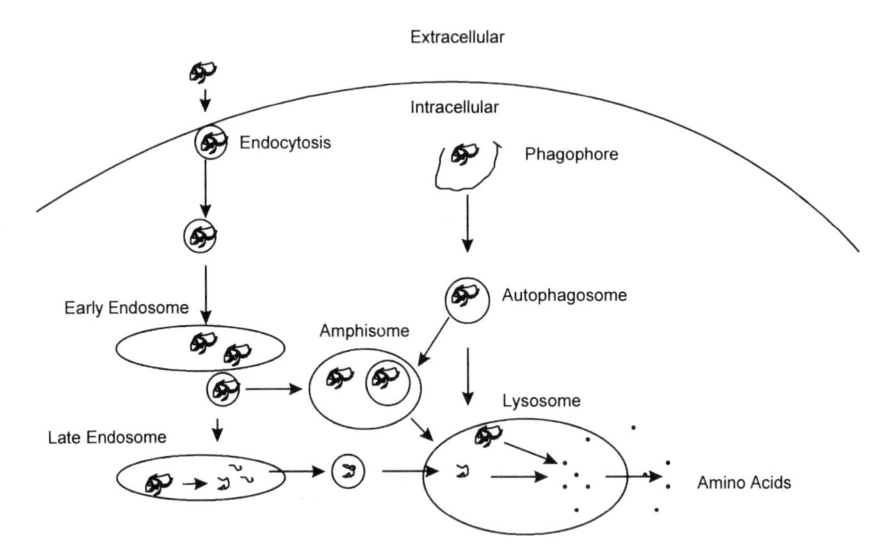

FIGURE 2.1. Vacuolar proteolysis. Some extracellular or membrane proteins may be returned to cell membrane intact after early endosomal phase (not shown). Partial degradation can occur in late endosomes, but complete degradation occurs only in lysosomes. Adapted from Seglen and Bohley (1992).

proteases require an acidic environment for optimal activity, so that a pH gradient must be maintained across the lysosomal membrane. This gradient is achieved with a proton pump driven by an ATPase (Lee and Marzella 1994). The combined action of these enzymes is sufficient to degrade proteins to individual amino acids or very small peptides. The amino acids diffuse into the cytosol where they can be used in various pathways, including reutilization for protein synthesis. Dipeptides also can cross the lysosomal membrane, and are degraded to amino acids by cytosolic peptidases.

Autophagic processes have been categorized as macroautophagic, microautophagic, or crinophagic (Lee and Marzella 1994). In macroautophagy, autophagosomes engulf entire organelles and their surrounding cytoplasm, then fuse with existing lysosomes or with amphisomes (Seglen and Bohley 1992), which contain both autophagosomes and endocytosed material. Amphisomes fuse with lysosomes for complete degradation of the contents. Macroautophagy accounts for most of the protein digested by lysosomes and is a nonselective pathway of protein degradation. In microautophagy, invaginations of the lysosomal membrane are pinched off to form small intralysosomal vesicles, which degrade cytosolic proteins taken up in this manner, as well as other cytosolic components such as glycogen and ribosomes. There may be mechanisms that cause certain proteins to be selectively degraded by microautophagy, such as binding to the lysosomal membrane (Benyon and Bond 1986). In crinophagy, vesicles containing secretory proteins fuse with lysosomes and the secretory proteins are degraded. Proteins sequestered within lysosomes are rapidly degraded, with a half-life of about 8 minutes (Bohley and Seglen 1992).

The first step in heterophagic protein degradation is transfer of the protein from the extracellular to the intracellular compartment. This step is accomplished by endocytosis, a process in which invaginations of the plasma membrane are pinched off to form small vesicles containing the extracellular material, which transport the material to early endosomes (Figure 2.1). In some cases, the endosomes contain ligands attached to plasma membrane receptors. The receptors can be sorted from the ligands and recycled back to the plasma membrane (Berg, Gjoen, and Bakke 1995). Material destined for degradation is transported in vesicles from the early endosomes to late endosomes or amphisomes. The late endosomes contain lysosomal enzymes and perform some degradation. The endosomes either mature into lysosomes, or, more likely, parts of the endosomes bud off and fuse with lysosomes for complete degradation of the endocytosed material. There may be partial degradation of extracellular proteins by extracellular proteases before endocytosis. For example, collagenases partially degrade collagen in the extracellular matrix.

Nonlysosomal degradation of proteins to small peptides occurs in proteasomes (Peters 1994; Goldberg 1995; Jennissen 1995; Hilt and Wolf 1996; Coux, Tanaka, and Goldberg 1996). Proteasomes are cylindrical particles with a hollow core constructed from many protein subunits (Figure 2.2). They are found in the cytoplasm and nucleus. In the older literature, proteasomes were called multicatalytic

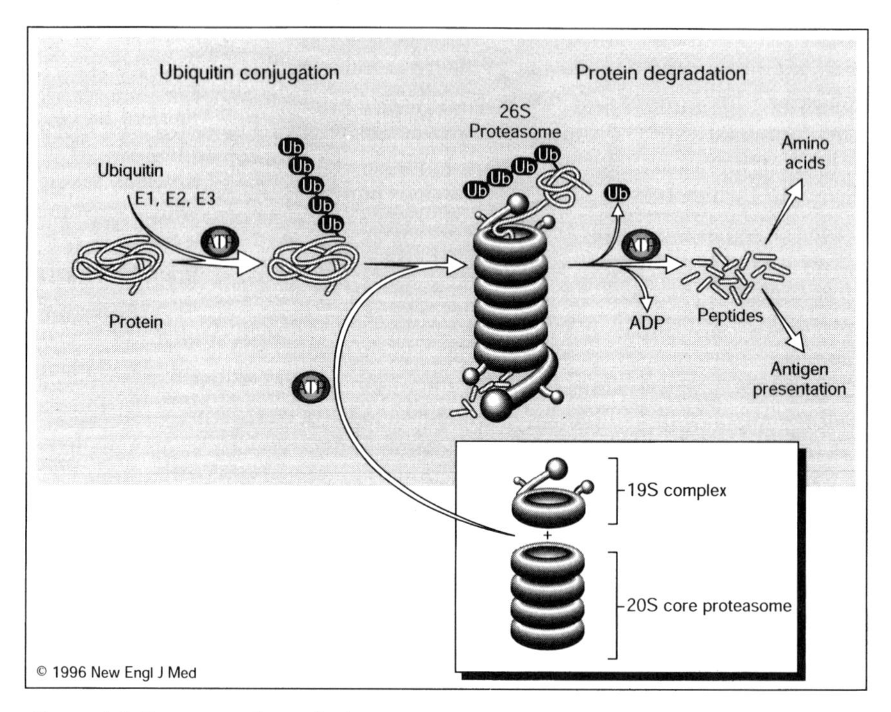

FIGURE 2.2. Proteasomal protein degradation. Polyubiquitination is ATP-dependent process catalyzed by conjugating enzymes (E1-E3) and is required for entry of the target protein into the core of the proteasome. Hydrolysis of proteins in the proteasome is incomplete. The peptide fragments are hydrolyzed to individual amino acids by cytosolic peptidases. From Mitch and Goldberg (1996). "Mechanisms of Disease: Mechanisms of Muscle Wasting—The Role of Ubiquitin-Proteasome Pathway, "The New England Journal of Medicine, Vol. 335, pp. 697–1905. Copyright © 1996 Massachusetts Medical Society. All rights reserved.

proteinase complexes. The proteolytic unit of the proteasome is a 700 kDa complex with a sedimentation coefficient of 20S. Several types of proteolytic activity are present within the proteasome, including cleavage after basic, acidic, and hydrophobic amino acid residues. Degradation by the 20S proteasome is not dependent on ATP, but the 20S proteasome itself appears to degrade only denatured and oxidized proteins. Most proteins are degraded in an ATP-dependent pathway in 26S proteasomes, of which the 20S proteasome is a component. The small peptides generated by the proteasomes can be completely hydrolyzed by cytosolic peptidases.

The 20S proteasome, a component of the 26S proteasome, also has been found as a component of other larger complexes whose function is unknown (Hilt and Wolf 1996; Peters 1994). The 26S proteasome is a 2000 kDa complex composed of the 20S proteasome and a 19S particle (Goldberg 1995; Coux, Tanaka, and Goldberg 1996). The 19S particle contains five different

ATPases, and a binding site for ubiquitin chains, which apparently mark at least some proteins for degradation. It engulfs the proteins destined for degradation and feeds them to the 20S proteasome, and also stimulates the peptidase activity of the 20S proteasome.

Conjugation with ubiquitin is required for some, if not most, proteins to be degraded by the 26S proteasomes. Ubiquitin is an 8.5 kDa protein that is conjugated to the ε-amino group of lysine residues of proteins by an ATP-dependent system involving separate activating, conjugating, and ligating enzymes. Substrates can be conjugated to a single ubiquitin molecule (monoubiquitination), to multiple ubiquitin molecules at multiple lysine residues (pluriubiquitinated), or to chains of ubiquitin molecules attached to a lysine residue (polyubiquitination) (Jennissen 1995). Ubiquitin also can be conjugated to the α-amino end of proteins with a free amino group, but most eukaryotic proteins are modified at the N terminal amino acid, thus preventing such conjugation. Although ubiquitin is synthesized as a polyubiquitin protein with peptide bonds connecting the adjacent ubiquitin units (which then is cleaved into individual ubiquitin molecules), polyubiquitinated substrates are formed by the stepwise addition of individual ubiquitin molecules by isopeptide bonds at a lysine residue in the ubiquitin molecule. Polyubiquitination appears to be the most common way of tagging proteins for ubiquitin-dependent proteolysis. The 19S complex of the 26S proteasome has a binding site for polyubiquitin chains. Some reviews state that the ubiquitin-proteasome pathway is primarily involved in degradation of abnormal or short-lived proteins. However, there is substantial evidence that this pathway is involved in degradation of the relatively long-lived myofibrillar proteins in skeletal muscle, at least under certain conditions (Wing, Haas, and Goldberg 1995; Argiles and Lopez-Soriano 1996) and may be responsible for as much as 90% of total proteolysis in some cultured mammalian cells (Coux, Tanaka, and Goldberg 1996). Ubiquitination of plasma membrane proteins may serve as a signal for endocytosis and vacuolar degradation of these proteins (Hicke 1997). Ubiquitination of a protein does not necessarily lead to its destruction. There are many deubiquitinating enzymes that can remove ubiquitin and ubiquitin-like peptides from proteins (Wilkinson 1997). In addition to disassembling the polyubiquitin degradation signal after proteasomal degradation of the marked protein, some of these enzymes may be part of a reversible ubiquitination process that serves to regulate protein activity or transport.

Intracellular calcium ion concentrations may be an important regulator of proteolysis. Calcium-dependent neutral proteases (calpains) are ubiquitous enzymes, but their primary role appears to be one of limited cleavage of selected proteins leading to altered function rather than complete degradation (Saido, Sorimachi, and Suzuki 1994). However, even limited cleavage by calpains may render proteins more susceptible to degradation by other enzymes, so that calpain action may be an important determinant of the complete degradation of some proteins. At least in muscle, elevated intracellular calcium increases bulk proteolysis, and calcium also may be necessary for lysosomal proteolysis (Zeman et al. 1985; Furuno and Goldberg 1986).

Caspases are proteases that are important in regulating apoptosis (programmed cell death) (Barinaga 1998). Caspases appear to be activated by the mitochondrial protein cytochrome c, although other triggers also may be involved. Activation of caspases leads to a chain of events resulting in the total destruction of the cell and complete degradation of its proteins by scavenger cells.

Oxidative damage to proteins can increase their susceptibility to proteolytic degradation (Stadtman 1995). Under physiological conditions, the rate of oxidative damage to proteins is very slow, so that oxidative damage is probably a minor determinant of overall rates of proteolysis. However, enhanced removal of damaged proteins helps to ensure that protein quality is high, particularly quality of slowly turning over proteins that have a higher probability of undergoing oxidative or other chemical modifications. Newly synthesized polypeptide chains appear to be much more susceptible to proteolysis than proteins that have undergone their posttranslational modifications and have found their proper subcellular locations (Wheatley, Grisolia, and Hernandez-Yago 1982). Thus, proteins are most vulnerable to degradation both immediately after synthesis and after they have existed long enough to accumulate enough oxidative damage to mark them for degradation.

Molecular chaperones are proteins that serve to promote the proper refolding of proteins that have been denatured by heat or other stressors, promote the proper folding of newly synthesized proteins, stimulate assembly or disassembly of multimeric proteins, and facilitate translocation of proteins across membranes. Chaperones also appear to be involved in protein degradation. There is evidence for their involvement in protein degradation by intramitochondrial proteases, in degradation of some abnormal proteins by the ubiquitin-proteasome system, and in transporting certain proteins into lysosomes (Hayes and Dice 1996).

The relative importance of lysosomal and nonlysosomal pathways to overall intracellular protein degradation is hard to decipher from the various reviews of this topic. Lysosomal degradation is most active under conditions of nutrient deprivation. As pointed out by Seglen and Bohley (1992), much of the cell culture research that elucidates basic mechanisms is done under conditions that tend to minimize lyosomal degradation. They concluded that nonlysosomal degradation was fairly constant, from 1% to 1.5%/h in most cell types, whereas lysosomal degradation varied from zero to 4%/h depending on nutritional and hormonal conditions. Although subsequent chapters will make it clear that human protein turnover is much slower than 1%/h, this conclusion makes the point that both lysosomal and nonlysosomal pathways can be important. Their relative contribution to overall proteolysis may vary according to type of cell, nutritional status, and presence or absence of various regulators of the protein degradative pathways.

Disposal of Amino Acids Generated by Proteolysis

Amino acids derived from proteolysis can be reused for protein synthesis. In adult humans, after an overnight fast, about 80% of the amino acids derived from protein degradation are used for protein synthesis. This 80% figure is a whole-body average and may not reflect amino acid reutilization for protein synthesis in all cell types under all conditions. Amino acids not used for protein synthesis are deaminated, and their carbons can be oxidized for ATP production, incorporated into glucose (some amino acids) or fat (probably a very minor pathway under most conditions), or used in other synthetic pathways. Most of the nitrogen from oxidized amino acids is excreted in urine in urea, although some can be excreted as ammonia or in other nitrogenous compounds. Excretion of amino acids in the urine is minimal. The pathways of degradation of the individual amino acids are described in any good biochemistry textbook (Stryer 1995; Coomes 1997) and will not be described here. Figure 2.3 summarizes the disposal of the carbon skeletons of deaminated amino acids.

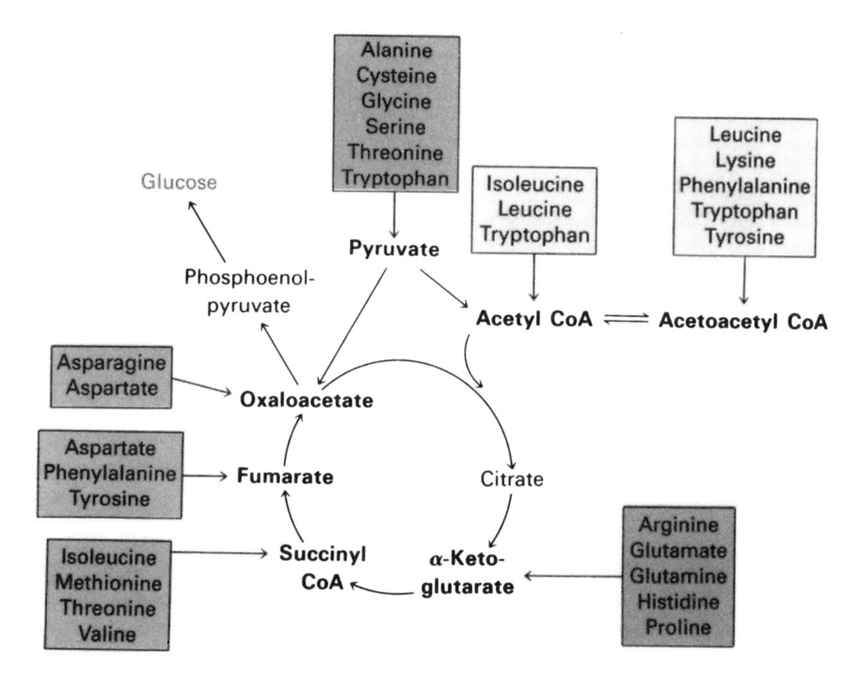

FIGURE 2.3. Fates of the carbon skeletons of amino acids. From Biochemistry 4th edition by Stryer © 1995 by Lubert Stryer. Used with permission of W.H. Freeman and Company.

Protein Synthesis

In contrast to the multiple pathways of protein degradation, there is only one known general pathway for protein synthesis, at least up to the point of translating mRNAs into polypeptide chains. Of course, there is as much diversity in protein synthesis rates as there is in protein degradation rates, but this phenomenon is more easily explained by what is known about protein synthesis mechanisms. Much of the material in this section was derived from textbooks that will not be cited repeatedly (Freifelder and Malacinski 1993; Stryer 1995; Glitz 1997).

The human genome contains about 10^5 genes that encode proteins, but any particular cell may express only about 10^4 proteins, the other genes being permanently suppressed in terminally differentiated cells. The genes are located in the 23 pairs of chromosomes present in each diploid human cell. Each member of a pair of chromosomes contains the same set of genes (except for the XY pair in males), although there are polymorphisms and mutations so that the DNA strands of each member of the pair are not identical. The DNA comprising these chromosomes is packaged in the nucleus of the cell, along with various proteins. Individual chromosomes are visible by microscopy only during certain stages of cell division. Otherwise the DNA-protein complex appears as a heterogeneous mass known as chromatin. Although the nucleus looks as though it is a bag of disorganized chromatin, there actually are important structural arrangements within the nucleus that ensure proper gene expression (Lamond and Earnshaw 1998). Mitochondria also contain a small amount of DNA that encodes a few of the mitochondrial proteins.

The process of protein synthesis begins with the transcription of the genes that encode the messenger RNAs (mRNAs), the ribosomal RNAs (rRNAs), and the transfer RNAs (tRNAs). Each mRNA encodes a specific protein, whereas the rRNAs and tRNAs are required for synthesis of all proteins. All RNAs are chains of ribonucelotides (one of four bases attached to a ribose monophosphate) attached by phosphodiester bonds (Figure 2.4). The bases are the purines adenine (A) and guanine (G) and the pyrimidines cytosine (C) and uracil (U). A, G, and C are also bases in the DNA of the genes, but DNA contains thymine (T) rather than U. DNA serves as the template[1] for RNA synthesis as shown in Table 2.1. There are three major classes of RNA polymerase that transcribe DNA in eukaryotic cells. RNA polymerase I produces rRNAs, RNA polymerase II produces mRNAs, and RNA polymerase III produces tRNAs and 5S RNA (a small rRNA). The transcribed genes contain both introns and exons. Although only the exons encode the amino acid sequences of proteins, both the exons and the interspersed introns are transcribed by the RNA polymerase. Thus the primary gene transcript must be

[1] By convention, gene sequences are published as the sequence of the DNA strand complementary to the one actually used as the template for RNA synthesis, so that the published sequence is the same as that of the RNA (except that the T in DNA is replaced by U in RNA).

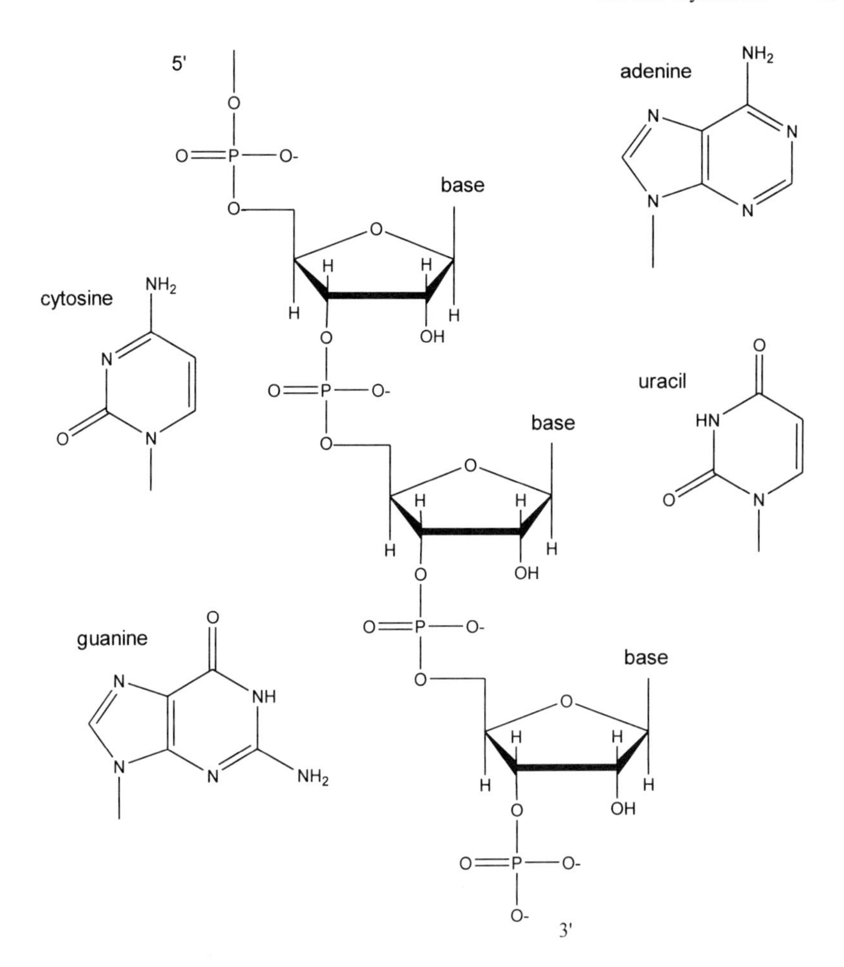

FIGURE 2.4. Structure of RNA. The bases are shown as radicals to indicate site of attachment to ribose units. Free bases have hydrogen at this position. Messenger RNA is translated in 5' to 3' direction.

TABLE 2.1. Transcription code.

DNA base	Codes for addition of following base to RNA strand during transcription:
Adenine (A)	Uracil (U)
Guanine (G)	Cytosine (C)
Cytosine (C)	Guanine (G)
Thymine (T)	Adenine (A)

processed to remove the introns to form the mature mRNA, a process termed splicing. In some cases, there is alternative splicing so that the same primary transcript can yield different mature mRNAs. Before splicing, the mRNA is capped at its 5' end (the end at which transcription starts) with a methylated guanosine derivative, 7-methylguanosine, and most mRNAs are polyadenylated. Polyadenylation refers to the addition of about 20 to 250 adenosine monophosphates to the 3' end of the RNA (the end at which transcription ends) after cleavage of part of the 3' end of the primary transcript. The mature mRNA contains a 5' untranslated region, the translated region that encodes the amino acid sequence of a protein, and a 3' untranslated region that usually ends with a polyA tail. Processing of the RNA to form the mature mRNA occurs in the nucleus, after which the mRNA is exported to the cytoplasm. Each molecule of mRNA can be translated many times in the cytoplasm.

In addition to protein-coding regions, the genes have regulatory regions that determine the transcription rate. These regulatory regions are promoters that are the site of RNA polymerase binding for transcription initiation and enhancers that increase the rate of transcription initiation. Transcription factors, proteins that bind to the promoters and enhancers, are required for such effects. Some genes are expressed constitutively, being transcribed at a fairly constant rate under a wide range of conditions. The transcription of other genes is regulated by hormones or other stimuli and can vary several-fold. One major reason that different proteins have different synthesis rates is that the abundances of the mRNAs can range from very few molecules per cell for some transcripts to thousands of molecules per cell for other transcripts. The total rate of protein synthesis by a cell probably is not determined primarily by its total mRNA abundance, but the different mRNAs are translated by the same protein synthetic machinery, and therefore the relative synthesis rates of the various proteins is determined to a large extent by the relative concentrations of their mRNAs. The abundance of an mRNA is determined not only by its transcription and nuclear processing, but also by its stability in the cytoplasm. As is the case for proteins, different mRNAs have different half-lives.

Translation of the mRNAs to produce proteins is a complex process that requires more than a hundred different macromolecules, including rRNAs, ribosomal proteins, tRNAs, and numerous enzymes. The process involves initiation of peptide chain synthesis, elongation of the peptide chain, and termination of translation. The translated regions of the mRNAs determine the sequences of the proteins because each of the codons in the mRNAs (3 consecutive bases) directs the addition of a particular amino acid to the growing polypeptide chain. Table 2.2 shows the codons that direct the addition of each of the amino acids used in protein synthesis.

The largest structure involved in translation is the ribosome, a complex of rRNAs and proteins. The entire ribosomal complex has a sedimentation coefficient of 80S and dissociates into subunits of 60S and 40S. The 60S subunit contains three rRNAs (5S, 5.8S, and 28S) and about 50 proteins. The 40S subunit contains an 18S rRNA and about 30 proteins. Ribosomes are found in

TABLE 2.2. Translation code.

5' base	Second base U	C	A	G	3' base
U	Phe	Ser	Tyr	Cys	U
	Phe	Ser	Tyr	Cys	C
	Leu	Ser	Stop	Stop	A
	Leu	Ser	Stop	Trp	G
C	Leu	Pro	His	Arg	U
	Leu	Pro	His	Arg	C
	Leu	Pro	Gln	Arg	A
	Leu	Pro	Gln	Arg	G
A	Ile	Thr	Asn	Ser	U
	Ile	Thr	Asn	Ser	C
	Ile	Thr	Lys	Arg	A
	Met	Thr	Lys	Arg	G
G	Val	Ala	Asp	Gly	U
	Val	Ala	Asp	Gly	C
	Val	Ala	Glu	Gly	A
	Val	Ala	Glu	Gly	G

the cytosol and attached to the cytosolic side of the endoplasmic reticulum. Initiation of protein synthesis occurs when an mRNA molecule binds to an initiation complex, which is a 40S ribosomal subunit attached to an initiator methionyl-tRNA[2] and several enzymes (Figure 2.5). The initiation complex binds to the 5' end of the mRNA and then scans the mRNA until it reaches the initial AUG codon, which encodes methionine. An 80S initiation complex is then formed by addition of the 60S ribosomal subunit. At this point, there is a ribosome attached to the initial AUG codon of the mRNA, with methionyl-tRNA attached at its P site (peptidyl site, where the growing peptide chain is attached to the ribosome) and an empty A site (aminoacyl site, where aminoacyl-tRNAs are attached prior to peptide bond formation). Several eukaryotic initiation factors (eIFs) are involved in the initiation steps (Table 2.3).

Elongation (Figure 2.6) is the stepwise addition of amino acids to the peptide chain. A peptide bond is formed between the carboxyl group of the last amino acid on the chain (or the initial methionine) and the amino group of the next amino acid being added to the chain, with the loss of an oxygen atom from the carboxyl group and two hydrogen atoms from the amino group

[2] Each tRNA can be attached to an amino acid, and the resulting molecule is named according to the particular amino acid attached. The suffix -yl replaces -ine in the name of the amino acid. For example, leucine attached to its tRNA is leucyl-tRNA, and phenylalanine attached to its tRNA is phenylalanyl-tRNA. The initiator methionyl-tRNA is designated as methionyl-$tRNA_i$ and is not the same as the methionyl-tRNA used for adding methionine during elongation.

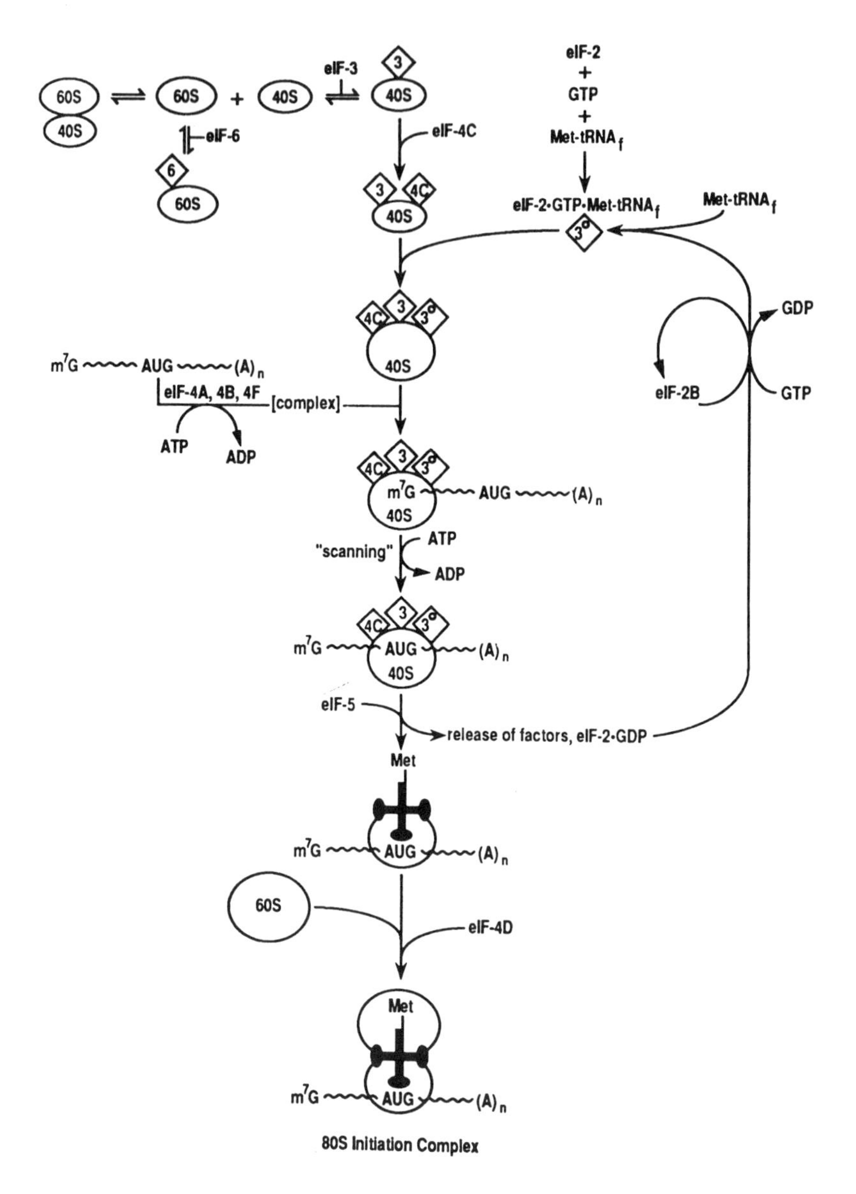

FIGURE 2.5. Formation of initiation complex. 40S=40S ribosomal subunit; 60S=60S ribosomal subunit; m^{7}G=7-methylguanosine cap on mRNA; $(A)_n$=polyadenylated tail on mRNA; AUG=initiation codon of mRNA; see Table 2.3 for identity of other initiation factors. From Merrick "Mechanism and regulation of eukaryotic protein synthesis," Microbiological Revews. Vol. 56 (2), pp. 291–315, 1992, Permission granted by American Society for Microbiology Journals Department.

TABLE 2.3. Eukaryotic translation factors.

Factor	Function
eIF-1	Pleiotropic; only weak stimulation of several steps
eIF-1A (eIF-4C)	Stabilizes 40S preinitiation complex; promotes dissociation of 80S ribosomes
eIF-2	Methionyl-tRNA binding to 40S ribosomal subunit; activity GTP-dependent; 3 subunits (α,β,γ); activity regulated by phosphorylation of α subunit – phosphorylation inhibits initiation; p67, which is regulated by glycosylation, protects it from phosphorylation
eIF-2A	AUG-dependent methionyl-$tRNA_i$ binding to 40S ribosomal subunit; in vivo role unclear; only functional in AUG-dependent assays
eIF-2B (GEF, guanine exchange factor)	GTP/GDP exchange on EF-2-GDP complex; phosphorylation may increase activity; 5 subunits (α,β,γ,δ,ε); phosphorylated eIF-2-GDP binds more tightly to eIF-2B, but does not exchange GDP for GTP
eIF-2C (co-eIF-2A)	Stabilizes ternary complex of GTP-eIF2-methionyl-$tRNA_i$ at low concentrations of eIF-2 and methionyl-$tRNA_i$; prevents disruption of ternary complex by naked mRNA
eIF-3	Promotes dissociation of 40S and 60S subunits by binding to 40S subunit; stimulates formation of ternary complex of GTP-eIF2-methionyl-$tRNA_i$ at low concentrations of eIF-2 and methionyl-$tRNA_i$; 8 subunits; largest of initiation factors
eIF-3A (eIF-6)	Promotes dissociation of 40S and 60S subunits; binds to 60S subunit
eIF-4 (eIF-4F)	Binding of 40S ribosomal subunit to 5' cap of mRNA; 3 subunits (eIF-4E,eIF-4A,eIF-4G)
eIF-4A	Promotes binding of 40S ribosomal subunit to mRNA; RNA helicase – unwinds secondary structure of mRNA; subunit of eIF-4F; also acts independently of eIF-4 complex; ATPase
eIF-4E (eIF-4α)	Cap binding; subunit of eIF-4; inactive when bound to PHAS-I; phosphorylation of PHAS-I by MAP kinase releases eIF-4E to allow activity; least abundant initiation factor; phosphorylation increases activity
eIF-4G (eIF-4γ, p220)	Subunit of eIF-4; binding 40S ribosomal subunit to mRNA - bridge between eIF-4E and 40S ribosomal subunit; binds Pab1p
eIF-4B	Release of eIF-4E from cap for reutilization; stimulation of RNA-dependent ATPase and helicase activities of eIF-4A
eIF-5	Ribosome-dependent GTPase activity - hydrolyzes GTP bound to eIF-2, promoting joining of 40S and 60S ribosomal subunits to form 80S initiation complex
eIF-5A (eIF-4D)	Stimulates synthesis of first peptide bond; spermine-mediated conversion of a lysine residue to hypusine required for activity
eEF-1α	Binding of aminoacyl-tRNA to ribosome; GTPase; very abundant protein; has activities unrelated to translation elongation; isoform called S1 or EF-1α2 expressed in postmitotic tissues
eEF-1βγ	Replaces GDP with GTP on EF-1α
eEF-2	Translocation of peptidyl-tRNA from A site to P site of ribosome, of deacylated tRNA to E site, and of ribosome by one codon on mRNA; GTPase; phosphorylation inhibits activity
eRF	Release of polypeptide from ribosome when stop codon encountered; GTPase – ribosome can dissociate into subunits upon GTP hydrolysis

Source: Compiled from various sources (Hershey 1991; Merrick 1992; Redpath and Proud 1994; O'Brien 1994; Hentze 1997; Glitz 1997).

Abbreviations: eIF=eukaryotic initiation factor; eEF=eukaryotic elongation factor; eRF=eukaryotic releasing factor.

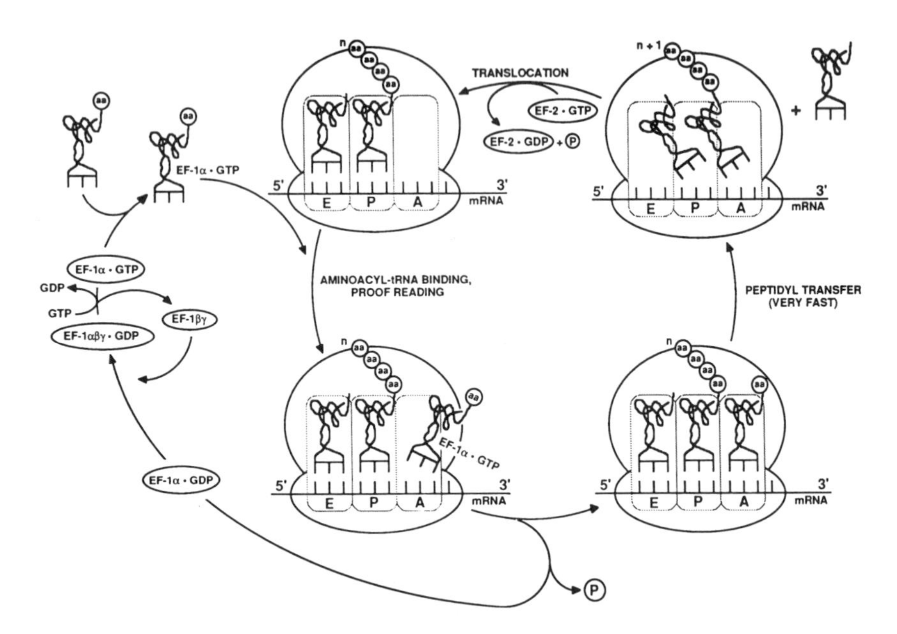

FIGURE 2.6. Elongation phase of protein synthesis. A=aminoacyl-tRNA binding site on ribosome; P=peptidyl-tRNA binding site on ribosome; E=exit site for deacylated tRNA; aa=amino acid; see Table 2.3 for identity of other factors. From Merrick "Mechanism and regulation of eukaryotic protein synthesis,"Microbiological Revews. Vol. 56 (2), pp. 291–315, 1992, permission granted by American Society for Microbiology Journals Department.

to form H_2O. Thus the initial methionine has a free amino group and the final amino acid of the polypeptide chain has a free carboxyl group. Peptide bond formation is catalyzed by a peptidyl transferase enzyme complex (part of the ribosome) when an aminoacyl-tRNA, charged with an amino acid corresponding to the codon at the A site, binds to the ribosome. The bond between the growing peptide chain and the tRNA at the P site is cleaved when the peptide bond is formed, leaving an uncharged tRNA at the P site and a peptide chain attached to the tRNA at the A site. Translocation then occurs, a process in which the uncharged tRNA at the P site moves to the E site, the peptidyl-tRNA moves from the A site to the P site, and the ribosome moves to the next codon. An aminoacyl-tRNA then occupies the A site, and the process is repeated. The eukaryotic elongation factors (eEFs) that catalyze the elongation steps are summarized in Table 2.3. Other enzymes essential for translation are the synthetases that catalyze the formation of aminoacyl-tRNAs from amino acids and uncharged tRNAs. There is a specific synthetase for each amino acid. Because of the degeneracy of the genetic code (Table 2.2), there are more than 50 different tRNAs in most cells, and a synthetase can aminoacylate a small family of tRNAs.

Membrane proteins and proteins destined for secretion usually are synthesized by ribosomes attached to the endoplasmic reticulum. Attachment of the ribosome to the endoplasmic reticulum occurs only after protein synthesis has started, with the initial 15 to 30 amino acid residues determining whether the ribosome attached to the peptide chain will attach to a signal recognition particle (SRP). The SRP then binds to a docking protein on the cytosolic side of the endoplasmic reticulum and halts protein synthesis until docking is completed. The elongation continues after the ribosome attaches to a translocon in the endoplasmic reticulum membrane, and the peptide chain is pushed through the pore of the translocon into the lumen of the endoplasmic reticulum as it grows. Proteins that are misfolded or otherwise denatured in the endoplasmic reticulum are exported to the cytoplasm for degradation, probably through a process that involves ubiquitination during export from the endoplasmic reticulum (Reizman 1997; Sommer and Wolf 1997).

The process of elongation is quite rapid. A ribosome can add six amino acids to the peptide chain each second. A typical adult synthesizes over a trillion trillion (10^{24}) peptide bonds each day. Most of the energy used for protein turnover is devoted to elongation of peptide chains. Two high-energy phosphate bonds are required to aminoacylate a tRNA molecule, another to position the aminoacyl-tRNA on the ribosome, and another for translocation. In someone whose cells are producing about 3 moles of peptide bonds daily (about 330 g protein), the energy requirement is about 240 kcal (3 moles × 4 high energy phosphate bonds × 20 kcal required to generate a mole of high energy phosphate bonds). In a typical young subject, this amount of energy represents about 14% of the resting metabolic rate, or most of the 20% of resting metabolic rate that appears to be required for protein turnover in healthy young subjects (Welle and Nair 1990b).

A single mRNA strand can be attached to many ribosomes simultaneously. Once the ribosome has moved far enough from the initiating codon, another initiation complex can bind and start another round of peptide chain synthesis while the other ribosomes are completing the synthesis of other protein molecules. Ribosomes usually are spaced along the mRNA molecule at 80 to 100 nucelotide intervals, although they can bind as closely as every 30 nucleotides (Hershey 1991). The complex of multiple ribosomes bound to a mRNA molecule is called a polyribosome, or sometimes a polysome.

When a ribosome reaches a termination codon (Table 2.2), a releasing enzyme (eRF, Table 2.3) removes the completed peptide. The ribosome dissociates into the 40S and 60S subunits for further rounds of peptide synthesis.

Posttranslational Modifications

After the complete peptide chain encoded by the mRNA has been synthesized, further steps are necessary to produce a functional protein. These steps are referred to as posttranslational, although sometimes they are cotranslational, meaning that they occur while the polypeptide still is attached to a ribsome for elongation. Such modifications include cleavage of peptide bonds, modifications of the terminal amino and carboxyl groups, and modifications of amino acid side chains. The oxidative and other damaging modifications discussed in Chapter 1 are slow and nonspecific, but most posttranslational modifications are created by rapid and highly specific enzymatic processes. Whereas the mRNAs encode the inclusion of 20 amino acids in the polypeptide, there appear to be hundreds of modified amino acid residues in mature proteins. The functional significance of these modifications is not always clear, but the fact that specific enzymes have evolved to produce them indicates that they must be useful for determining protein activity, stability, or subcellular localization. A few such modifications will be mentioned. Readers interested in more detail should consult other sources (Glitz 1997; Krishna and Wold 1993; Han and Martinage 1993; Rucker and McGee 1993).

Although the initiator codon always encodes methionine, most proteins do not have a methionine residue at their N terminal. One or more amino acids from the N terminal are cleaved from most peptide chains. The N terminal often is acetylated, sometimes is acylated with formate, myristate, palmitate, or other molecules. Arginyl residues can be added at the N terminal in the absence of ribosomes. Carboxyl terminal modifications include amidation, ADP-ribosylation, acylations, prenylation, and addition of tyrosyl residues.

Numerous modifications of amino acid side chains have been described. For example, certain histidine residues of actin and myosin are methylated to 3-methylhistidine, and proline residues of collagen are converted to hydroxyproline. There are many other amino acid side chain modifications, including acetylation, glycosylation, acylation, carboxylation, uridylation, and sulfation. As mentioned in the section on proteolysis, many proteins are ubiquitinated, which can increase their rate of degradation. However, ubiquitination may have functions other than marking proteins for degradation. For example, ubiquitination of histones may be involved in regulating chromatin structure.

Some posttranslational modifications are readily reversible and are used to regulate protein activity. Phosphorylation at hydroxyl groups on amino acid side chains is common, and specific enzymes phosphorylate and dephosphorylate proteins to regulate their activity. Reversible O-linked glycosylation also may regulate the activity of proteins (Hart 1997).

Many proteins synthesized by ribosomes attached to the endoplasmic reticulum are glycoproteins, meaning that carbohydrate structures are attached after translation. The carbohydrates are added in the endoplasmic reticulum and are further processed in the Golgi apparatus before being transported to their destinations.

Proteins must be folded into the proper three-dimensional configuration to be functional. Although the final protein structure depends on the primary structure of the polypeptide chain, in the complex and crowded cellular environment misfolding can occur. Certain proteins, known as chaperones, help to prevent misfolding. Some proteins are made up of two or more separate polypeptide chains, usually referred to as subunits. Each subunit can also be considered as a separate protein, but some proteins were given a single name for the entire multisubunit structure before it was even known that there were multiple subunits. Often the subunits are attached covalently by disulfide bonds, which are formed by the oxidation of two cysteine residues. Disulfide bonds also can form between cysteine residues within the same polypeptide chain. Other amino acid residues also can participate in the formation of crosslinks within or between proteins.

Some proteins must bind to metals or other cofactors for activity. For example, hemoglobin contains heme, an organic molecule that contains iron. The proteolytic enzyme carboxypeptidase A requires binding to zinc for activity.

Some functional proteins are fragments of a larger protein. For example, the prohormone proinsulin is cleaved into insulin and C-peptide. Proopiomelanocortin is cleaved to produce adrenocorticotropin (ACTH), β-endorphin, β-lipotropin, α-lipotropin, melanocyte stimulating hormone, and other proteins. In lower organisms, a few proteins have been identified that are spliced by proteolytic cleavage followed by reformation of a peptide bond between the segments (Davis and Jenner 1995).

Mitochondrial Protein Synthesis

Most mitochondrial proteins are encoded by chromosomal DNA, synthesized in the cytosol as described above, and imported into mitochondria. However, a few proteins are encoded by mitochondrial DNA and are synthesized within the mitochondria (Glitz 1997). Human mitochondrial DNA is a small circular molecule with approximately 17,000 base pairs. It encodes 13 proteins, 22 tRNAs, and two mitochondrial rRNAs. The general scheme of protein synthesis is the same as in the cytosol, but mitochondria have distinct ribosomes, tRNAs, tRNA synthetases, and other components. The codon usage is slightly different in mitochondria. N terminal sequences of polypeptides synthesized in the cytosol designate whether they are to be transported into the mitochondria, and their location within the mitochondria. Proteolytic processing of precursor peptides to functional proteins occurs within the mitochondria.

Potential Sites of Regulation of Protein Turnover

Hormonal and metabolic control of protein turnover will be discussed in later chapters, but it is worth noting the most likely control points at the molecular level. For protein degradation, initiation of proteolysis appears to be the rate limiting step. Once the degradation of a protein molecule begins, its complete degradation occurs rapidly. Ubiquitination appears to be the most common way of marking proteins for degradation by proteasomes. The abundance of lysosomes also appears to be regulated, but the nature of the molecular signal is unclear. Most proteins have a monoexponential decay pattern, suggesting that selection of individual molecules for degradation is random. For most proteins, the rate of proteolysis does not occur at a fixed rate, but at a fixed fraction of the protein concentration. However, the rate constant for degradation can be regulated. Although the turnover rate of most proteins depends on the global proteolytic activity of the cell, specific mechanisms can regulate the rate of degradation of specific proteins, so that they can be removed rapidly even when overall proteolysis is not accelerated.

Protein synthesis can be regulated both globally and at the level of individual proteins. At the global level, total abundance of mRNA is rarely a rate-limiting factor, although the relative abundances of the specific mRNAs determine the pattern of proteins that are synthesized. The availability of ribosomes for translating the messages may be rate limiting under some conditions, and changes in ribosome concentrations often correlate with changes in the global rate of protein synthesis. However, overall protein synthesis often is regulated by the rate of initiation of translation, even when ribosomal concentrations are constant. The most likely rate-limiting initiation factors are those occurring in low concentrations or those whose activity is regulated by phosphorylation, such as eIF-2 and eIF-4E. Elongation also can be rate limiting, although initiation is a more likely site of regulation. The activity of the elongation factor eEF-2 is inhibited by phosphorylation and is a potential site of regulating global rates of protein synthesis.

At the level of specific proteins, alterations of the abundance of a specific mRNA relative to other mRNAs by increased or decreased transcription is an obvious control point. This mechanism is frequently involved in hormonal regulation. The control of transcription of specific genes involved directly in protein synthesis or of enzymes that regulate the activity of the proteins involved in protein synthesis is a way that hormones can regulate global protein synthesis by affecting specific genes. There is a growing appreciation of the importance of posttranscriptional events in the control of synthesis of specific proteins. There are many examples of highly specific mechanisms for regulating the stability and translatability of particular mRNAs. The 3'-untranslated regions of mRNAs probably are very important for this type of regulation. The importance of translational control is evident from a report that protein synthesis encoded by about 12% of the mRNA species in T cells appears to be primarily regulated by translation rather than mRNA abundance (Garcia-Sanz et al. 1998).

3
Methods for Studying Protein Metabolism in Humans

Measuring Changes in Protein Mass of the Body

Nitrogen Balance

The earliest studies of human protein metabolism involved determination of nitrogen balance (N balance). By determining the amount of N ingested, and subtracting N lost in urine, feces, sweat, and minor miscellaneous losses, one can calculate the N balance. Because almost all of the N in the human body is contained in proteins, a positive N balance generally indicates that the protein mass of the body has increased, and a negative N balance that it has decreased. However, the nonprotein N content of the body can vary enough to invalidate short-term N balance as a measure of protein balance. Each gram of N represents about 6.25 g of protein (Forbes 1987a). Because, in theory, N balance could be measured with a precision of a few tenths of a gram each day, and a 70 kg man contains about 1800 g of N (Forbes 1987a), this method could detect changes in total protein mass on the order of 0.1%.

In practice, however, there are usually errors in this method, which tend to produce values for N balance that are on the positive side. One problem is ensuring that all excreta are completely collected. Urine must be collected into acid to retain ammonia. The ingestion of N may be difficult to determine precisely. Subjects must eat every bit of food provided to them, and only the food provided by the investigator. Another problem is uncertainty about whether the N content of the aliquot of food provided to the subject is identical to that of the food actually eaten. There is some loss of N from the skin, usually estimated to be 5 to 8 mg N p/day p/kg body weight. N also is lost in sweat and in the menstrual flow. Obviously, precise N balance requires a metabolic ward setting and extremely cooperative subjects. In general, control experiments should be done to show that N balance is approximately zero when normal subjects are weight stable and not doing anything that might alter their body composition. If N balance is not close to zero under these conditions, then systematic errors in the method are likely.

Changes in Body Composition

In long-term studies, changes in whole-body protein mass can be estimated with various body composition methods. Total body N determination, which is done by neutron activation, should be the most precise method to estimate changes in protein mass. However, the equipment for such measurements is not generally available and requires exposure to ionizing radiation. Total protein mass can be estimated from other body composition methods (underwater weighing, dual energy X-ray absorptiometry, total body potassium counting, and others) if it is assumed that the protein content of the fat-free body mass is constant. Changes in protein mass of the body need to be on the order of ~3% to 5% or more to make meaningful conclusions based on such methods, because of random errors in these measures. Changes in protein mass of individual organs can be estimated by calculating organ volume from imaging methods such as computerized tomography or magnetic resonance imaging, but again the assumption must be made that protein concentration of the organ is constant.

A drawback of all of these methods is that nothing is learned about the dynamics of protein metabolism. There could be a change in the rate of protein synthesis that is matched by a change in protein breakdown, and the N balance or body composition methods will not provide any evidence that protein metabolism has changed. If you are interested only in protein mass, this limitation is not a problem. Another drawback of these methods is that no information can be gained about acute changes in protein metabolism, those occurring within a few hours after an intervention. The kinetic methods described below are necessary to overcome these limitations.

Whole-Body Protein Turnover

Most investigators studying protein metabolism in humans use heavy isotopes of amino acids, either radioactive or stable, as tracers. The basic principle of using an isotope to trace protein degradation is illustrated in Figure 3.1. Essential amino acids, those that cannot be synthesized *de novo* in human cells, are derived only from degradation of the body's proteins, or from dietary sources (proteins or free amino acids). For now, assume that all of the free amino acids coming from protein breakdown mix instantaneously with the amino acids in the blood. Thus, there are only two pools of amino acids, those in protein and those in the free amino acid pool, and the circulating amino acids reflect the entire free pool of amino acids in the body. Other assumptions of tracer methods for whole-body protein turnover are:

1. The tracer behaves the same as the traced amino acid in all biochemical pathways. Because the labeled amino acid is chemically identical to the unlabeled amino acid, this assumption is quite reasonable. However, there are examples of isotope discrimination in biological systems.

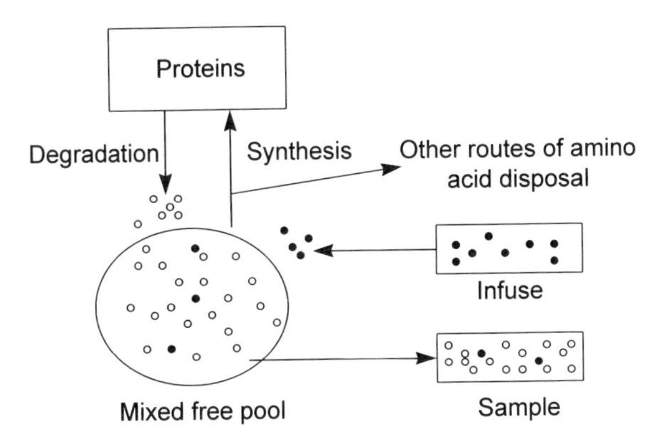

FIGURE 3.1. Principle of tracer dilution method of measuring whole-body protein turnover, according to simplified model with a single pool of free amino acids in which intracellular and extracellular free amino acids mix freely. Open circles represent unlabeled amino acids. Filled circles represent labeled amino acids. Amino acids can come from breakdown of endogenous proteins, or hydrolysis of food proteins in the gut. The rate of infusion of the tracer is known, so that determination of the ratio between labeled and unlabeled molecules in the sample indicates the rate of appearance of unlabeled molecules from protein degradation.

2. There is negligible recycling of the tracer from newly synthesized proteins to the free pool of amino acids during the course of the tracer infusion, so that essentially all of the amino acids coming from protein breakdown are unlabeled. There are some proteins with a very short half-life from which there will be extensive recycling even within a few hours of starting a tracer infusion. However, the mass of these proteins is small enough that the calculation of whole-body protein turnover is not affected very much during tracer infusions of a few hours. With prolonged tracer infusion, recycling can have a measurable effect on the calculated whole-body protein turnover.

Steady State Protein Turnover (Protein Metabolism Not Changing During the Study)

Protein Breakdown

First consider the simple case of a postabsorptive person—one who has not eaten for several hours so that there is no dietary source of free amino acids. If an essential amino acid labeled with a stable isotope, e.g.[^{13}C]leucine is infused into this person's blood, the ratio of [^{13}C]leucine to unlabeled leucine in the blood is determined by the amount of unlabeled leucine entering the blood from degradation of proteins in that person's body (Figure 3.1). If there is no protein degradation, all of the leucine entering the blood carries the ^{13}C label. If there is a very high rate of protein degradation, the ratio of [^{13}C]leucine to total leucine (the *isotopic enrichment* of leucine) is very small. Thus during a constant infusion of [^{13}C]leucine, the rate of protein degradation in the whole body is inversely proportional to the ^{13}C enrichment of the circulating leucine. The rate at which the amino acid enters the circulation and dilutes the tracer is usually called the rate of appearance (R_a) of the amino acid, and is calculated as:

$$R_a = (I/E) - I$$

where I is the infusion rate of the tracer and E is the isotopic enrichment of the tracer in the circulation.[1] This calculation assumes that the R_a is constant during the experiment. If protein breakdown changes while the tracer infusion is constant, then obviously E changes and no single value for R_a can be calculated. Usually, this type of experiment is done under "steady state" conditions, meaning that experimental conditions are such that the rate of protein turnover is nearly constant during the study. If multiple measurements of E show no change over time during a constant tracer infusion, then the R_a must be constant. In reality, there are always some fluctuations in E over time, but if they are small enough, they usually are ignored and the results from the multiple time points are averaged to derive R_a.

Even under steady state metabolic conditions, the value for E does not reach a steady value immediately after the tracer infusion is started. When the tracer infusion begins, there already are free unlabeled amino acids in the body. Therefore, the initial E will be lower than the value that is eventually reached after a long period of tracer infusion. As the initial free amino acid pool is depleted by amino acid catabolism or reincorporation into protein and is replenished by amino acids coming from protein breakdown and from the tracer infusion, then the enrichment of the tracer will gradually rise toward its final value. As shown in Figure 3.2, when E nears the final value the rate of change is very slow. At this point, measurement errors are such that no systematic change in E can be detected. The term "plateau" is frequently used to describe the section of the E versus time curve over which no significant increase or decrease in E can be detected. How long does it take to reach a steady level of isotope enrichment? The answer to this question depends on the half-life of the free amino acid that is being traced. Each half-life brings E

---

[1] If a radioactive isotope is used, the ratio of radioactive to nonradioactive amino acid is termed the *specific activity* of the amino acid. This formula is accurate for both stable and radioactive isotopes. However, when radioactive isotopes are used their mass is negligible, so that the infusion rate of the tracer typically is not subtracted from first term. The specific activity is measured as disintegrations per minute (dpm) of the tracer divided by concentration of the amino acid, and the infusion rate is expressed as dpm infused per minute rather than as the mass of tracer infused per minute. Equations will be given for stable isotope tracers, because they are applicable to both stable and radioactive tracers. Investigators using radioactive tracers can disregard terms in the equation that are designed to remove the influence of tracer mass.

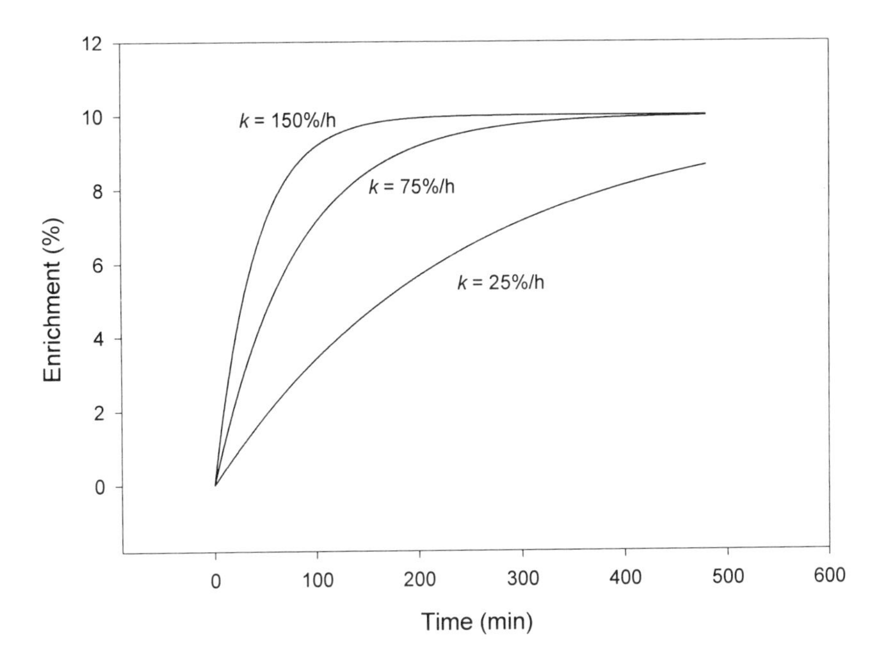

FIGURE 3.2. Time course of isotopic enrichments of a circulating amino acid during an unprimed, constant tracer infusion, as a function of the fractional disappearance rate (k, the fraction of the circulating pool of traced amino acid that is removed from the circulation per unit of time). Based on constant tracer infusion equal to 10% of the rate of removal of the traced amino acid from the circulation.

50% closer to the final value. Thus, it takes two half-lives to reach 75% of the final E, and five to reach 97% of the final E.

The use of a priming dose of tracer is a common way to approach the steady state E more rapidly. A priming dose is a rapid injection of enough tracer to bring the E close to the expected plateau. However, unless the priming dose is exactly correct, it does not reduce the time it takes to reach the final plateau E. It only minimizes the error caused by not reaching the steady state (Figure 3.3).

When the goal of the experiment is to trace protein metabolism rather than the metabolism of a specific amino acid, it is necessary to use an essential amino acid as the tracer. Otherwise, the appearance of unlabeled amino acid could reflect *de novo* synthesis of the amino acid rather than protein breakdown. For whole-body studies, labeled leucine is often used as a tracer because it facilitates measurement not only of protein breakdown, but also protein synthesis. Other amino acids used commonly as tracers of protein breakdown are phenylalanine and lysine. The rate of protein breakdown is estimated from the R_a of the amino acid by estimating the contribution of that amino acid to the total protein mass being degraded. The amino acid composition of the thousands of different proteins in the body obviously varies substantially, but reasonable estimates of total protein breakdown can be made

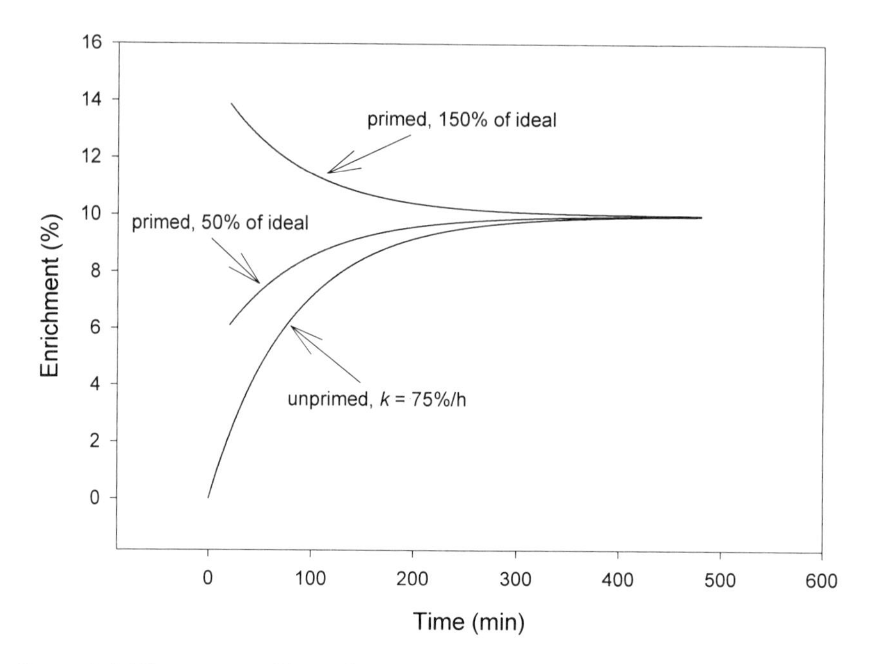

FIGURE 3.3. Time course of isotopic enrichments of a circulating amino acid during a primed, constant tracer infusion, as a function of the initial priming dose (given as a bolus at the start of the constant infusion). Based on constant tracer infusion equal to 10% of the rate of removal of the traced amino acid from the circulation. The "ideal" priming dose is one that immediately raises the isotopic enrichment to the final value. The shape of the curves in the first few minutes after priming (not shown) depends on how rapidly the bolus of tracer mixes completely with circulating amino acid.

based on the amino acid composition of proteins from various animal organs (Munro and Fleck 1969; FAO 1970; Fuller et al. 1989). Table 3.1 shows the approximate amino acid composition of whole-body proteins, which can be used to estimate protein breakdown from the R_a of an essential amino acid:

$$\text{Protein breakdown rate} = R_a/\text{concentration of amino acid in proteins}$$

It is important to remember that the appearance of amino acids from the proteins being degraded may not correspond exactly to the amino acid composition of the body, because individual proteins have very different amino acid compositions and very different turnover rates.

Plasma R_a of Essential Amino Acids Underestimates True R_a

The discussion above is based on the assumption that the infused tracer and the amino acids coming from protein breakdown freely mix together in a homoge-

TABLE 3.1. Approximate amino acid composition of whole-body proteins.

Amino acid	μmol/g protein	% protein mass
Leucine	663	7.5
Isoleucine	388	4.4
Valine	484	4.8
Lysine	585	7.5
Methionine	163	2.1
Phenylalanine	274	4.0
Tyrosine	203	3.3
Threonine	380	3.8
Alanine	689	4.9
Glycine	768	4.4
Arginine	363	5.7
Histidine	182	2.5
Proline	441	4.3
Serine	416	3.6

Average values for beef (FAO 1970) and pig (Munro and Fleck 1969). For beef, values of flesh and offals (viscera) were averaged. For pigs, liver and intestine were averaged to obtain a value for viscera, which then was averaged with values for muscle to obtain whole-body estimate. Original sources expressed data as g/16 g nitrogen. Each g is assumed to represent the mass of the free amino acid, as analyzed in the hydrolyzate, rather than the mass of the amino acid residue in the intact proteins (mass of free amino acid minus mass of water). To calculate the contribution of the amino acid to the protein mass. The molecular weight of the amino acid residue in proteins was used. It was also assumed that 16 g nitrogen is equivalent to 100 g protein.

neous free pool, and that the fluid being sampled for tracer enrichments (almost always blood plasma) reflects the enrichment of this pool. In reality, the situation is more complex. The tracer is infused into the blood, then must cross the capillary, extracellular matrix, and plasma membrane to enter the cells where proteins are being degraded to amino acids. Although small molecules such as amino acids should freely diffuse between capillary cells and through the extracellular matrix, amino acids are taken into cells by specific transporters. Moreover, even within the cell it is possible that the tracer might not mix rapidly enough with amino acids from proteolysis to accurately reflect the rate of proteolysis. For example, if tracer molecules entering the cells from the extracellular compartment are more likely to flow back into the circulation than the amino acid molecules coming from protein degradation, then the enrichment in plasma would be higher than it would be if mixing were instantaneous. In this case, protein breakdown would be underestimated. Several studies have shown that the degree of tracer dilution within tissues is greater than the tracer dilution in plasma. Hence the R_a of an essential amino acid into plasma underestimates the protein degradation rate. Under most conditions, this problem does not necessarily invalidate using the appearance rate of an essential amino acid as an *index* of whole-body proteolysis, as long as the investigator realizes that absolute protein degradation rate is underestimated to some extent.

To minimize this problem of underestimating the protein degradation rate during infusion of labeled leucine, Matthews et al. (1982) developed the method of measuring the isotopic enrichment of leucine's ketoacid, α-ketoisocaproate (usually abbreviated as KIC). The KIC can only be formed intracellularly from deamination of leucine, a reversible reaction that proceeds rapidly in cells that express leucine aminotransferase, which includes most of the cells in the human body. The KIC formed in the cells can be sampled in the plasma, permitting a noninvasive determination of intracellular tracer dilution, at least in theory. Usually, the isotopic enrichment of the plasma KIC is about 80% of the plasma leucine enrichment. However, plasma KIC isotope enrichment is higher than the leucine isotope enrichment in tissue, which suggests that even use of plasma KIC specific activity does not completely prevent the problem of underestimating protein breakdown. Another potential problem with using KIC is that leucine aminotransferase activity varies among tissues, and tissues with the highest activity may contribute a disproportionate amount to the circulating KIC. In spite of these problems, the rate of whole-body protein breakdown calculated from plasma KIC isotope enrichment is closer to the actual value than the rate calculated from plasma leucine enrichment. Use of plasma KIC enrichment during labeled leucine tracer infusion has been termed the "reciprocal pool model."

Determination of Protein Breakdown in Fed Subjects

So far the discussion has dealt only with determination of whole-body protein breakdown in a postabsorptive person. The situation is more complicated when amino acids are entering the body from food. In this case the tracer infused into the systemic circulation will be diluted not only by amino acids from degradation of body proteins, but also by amino acids entering the circulation from absorption of digested food proteins or absorption of dietary free amino acids. The appearance of amino acid from breakdown of the endogenous proteins of the body ($R_{a\text{-}end}$) is then calculated as:

$$R_{a\text{-}end} = R_{a\text{-}total} - R_{a\text{-}dietary}$$

If there is parenteral amino acid administration, the infusion rate of the amino acid being traced is substituted for $R_{a\text{-}dietary}$. There are different ways to calculate $R_{a\text{-}dietary}$. One way is to assume that all of the protein consumed is hydrolyzed to its constituent amino acids and that all of the amino acids enter the systemic circulation. When the protein is given as a single meal, the rate of digestion and absorption can vary, so that this approach could produce large errors. However, if the food is given as small meals at regular intervals, then eventually the rate of dietary amino acid entry into the systemic circulation should stabilize at a rate approximating the rate of amino acid ingestion. This approach will result in an overestimate of the $R_{a\text{-}dietary}$, because not all protein will be absorbed from the gut as its constituent amino acids (a small amount will not be absorbed at all, a small amount will be absorbed as small peptides) and not all amino acids absorbed from the gut will enter the systemic circulation. Some of the amino acids entering the portal vein from the gut will be extracted and metabolized by the liver, and will never enter the systemic circulation.

A better way to determine the rate of appearance of amino acids from dietary sources is to use a separate tracer, one that can be distinguished from the tracer used to determine total systemic R_a. If the ratio of dietary tracer to unlabeled amino acid in the diet is known, then the systemic appearance of amino acid from the diet can be calculated:

$$R_{a\text{-}diet} = R_{a\text{-}total} \times E_p/E_d$$

where E_d is the isotopic enrichment of the dietary tracer, and E_p is the enrichment of the dietary amino acid tracer in the plasma from the systemic circulation. If the meals contain only free amino acids rather than protein as their amino acid source, then the investigator can be certain that the tracer reflects total entry of the traced amino acid from the gut into the systemic circulation. However, if the tracer is a free amino acid and the traced amino acid is fed as a protein, the tracer and the unlabeled amino acid might not behave identically. The free amino acid tracer would be absorbed more rapidly and more completely, leading to an overestimate of the contribution of the diet to the total systemic appearance of the amino acid. If the unlabeled amino acids are fed as protein, the tracer also should be fed in the form of protein (Boirie et al. 1996).

The first-pass splanchnic extraction of the dietary tracer can be calculated by examining the ratio of whole-body R_a calculated from the dietary tracer with R_a calculated from the systemic tracer. The decrease in R_a with the dietary tracer reflects the splanchnic extraction.

Protein Synthesis

The term turnover refers not only to the degradation of proteins to amino acids, but also the reutilization of these amino acids for protein synthesis. Over relatively long periods of time, the rate of degradation of most proteins is equal to the rate of synthesis, so that protein mass does not change. However, over short periods of time the degradation and synthesis rates can be unequal, with protein mass increasing or decreasing. In postabsorptive humans, whole body protein breakdown is always more rapid than whole-body protein synthesis. Amino acids not used for protein synthesis can be oxidized, used in other metabolic pathways, or can accumulate in the cells or extracellular fluids. For now, only steady state conditions are being considered, so that amino acid levels would be constant and the accumulation (or depletion) of the free amino acid pool would not be an issue. When the free amino acid pool is constant, the R_a of an amino acid from protein breakdown is equal to the rate of disappearance (R_d) for protein synthesis or other pathways of amino acid metabolism. The incorporation of the amino acid into all of the body's proteins cannot be determined directly because all proteins cannot be sampled for tracer incorporation. However, with some tracers the protein synthesis rate can be derived by subtracting the disposal in other pathways from the total R_d. The use of L-[1-^{13}C]leucine is the most common application of this approach for determining whole-body protein synthesis. The first step in the pathway of leucine oxidation is deamination to KIC. This step is reversible and does not result in loss of the ^{13}C label, but when the carboxyl group of the KIC is removed as CO_2 to form isovaleryl-coenzyme A, the ^{13}C label is irreversibly lost from leucine. The only fates of the ^{13}C label with this tracer are incorporation into proteins or loss as CO_2 (oxidation). Thus, leucine incorporation into proteins (R_s) can be calculated if leucine oxidation rate (R_{ox}) and R_a are measured:

$$R_a = R_d \text{ (steady state conditions)}$$

$$R_s = R_d - R_{ox}$$

Leucine oxidation rate is determined as:

$$R_{ox} = \text{Total } CO_2 \text{ production rate} \times (E_{CO2}/E_{KIC})/rf$$

where E_{CO2} is the enrichment of ^{13}C in expired CO_2, E_{KIC} is the plasma KIC enrichment, and *rf* is a recovery factor that accounts for the fact that some of the $^{13}CO_2$ is sequestered in a bicarbonate pool with a slow turnover. The KIC enrichment rather than the leucine enrichment is used for this calculation, because the ^{13}C label is removed from KIC and not leucine. It is possible that the plasma KIC enrichment does not precisely reflect the enrichment of all of the KIC that produced the $^{13}CO_2$, but it should generally be a good approximation. The value for the recovery factor is often given as 0.8 for studies lasting a few hours. However, the actual *rf* varies according to metabolic conditions. It is about 0.75 in postabsorptive subjects, 0.8 to 0.85 in fed subjects, and approaches 1 in exercising subjects. With the uncertainties in estimating the true R_a, as discussed earlier, and the uncertainties in measuring R_{ox}, it follows that this method of determining whole-body protein synthesis gives an index rather than a precise value.

In fed subjects the interpretation of whole-body protein synthesis is more complicated. Under steady state conditions, R_d is equal to R_a, as it is in postabsorptive subjects. In one respect, the calculation of protein synthesis is easier than the calculation of protein breakdown in fed subjects, because it is not important whether R_a derives from protein breakdown or amino acids coming from the digestive tract. However, the problem of first-pass hepatic extraction of dietary leucine coming from the gut does lead to an underestimate of true whole-body R_a, and therefore of R_d. The calculation of the oxidation of the dietary leucine is not affected by this problem. If first-pass hepatic extraction of the dietary leucine is determined with a second tracer, as described above, then a better measure of whole-body protein synthesis is obtained.

Labeled phenylalanine also can be used to estimate whole-body protein synthesis (Thompson et al. 1989b). The first step in phenylalanine catabolism in humans is hydroxylation to tyrosine. Thus the incorporation of phenylalanine into proteins (R_s) is the R_d (equal to R_a under steady state conditions) minus the conversion to tyrosine ($R_{phe\text{->}tyr}$):

$$R_s = R_d - R_{phe\text{->}tyr}$$

$R_{phe\text{->}tyr}$ can be determined by multiplying the total R_a of tyrosine (determined by standard tracer dilution methods) by the fraction of tyrosine coming from phenylalanine hydroxylation. If ring-$^2H_5$-phenylalanine is used as the tracer for phenylalanine R_a, it will be converted to ring-$^2H_4$-tyrosine by hydroxylation. The fraction of tyrosine coming from phenylalanine hydroxylation can be calculated as E_{tyr}/E_{phe}, where E_{tyr} is the enrichment of $^2H_4$-tyrosine and E_{phe} is the enrichment of $^2H_5$-phenylalanine. The tracer to be used for determining the total R_a for tyrosine must be distinguished from the one derived from phenylalanine hydroxylation. When $^2H_5$-phenylalanine is infused to trace phenylalanine R_a and conversion to tyrosine, $^2H_2$-tyrosine can be used to measure total tyrosine R_a.

The problems inherent in using labeled phenylalanine as a tracer are similar to those inherent in using leucine as a tracer. The enrichment of the tracer in plasma generally is less than the enrichment in the cells, so that total R_a (and therefore R_d) is underestimated. Phenylalanine does not have a metabolite comparable to KIC that can be used to estimate intracellular enrichment of the tracer. Although it is converted to tyrosine in the liver and perhaps to a small extent in some other tissues, it is not converted to tyrosine in many tissues. Thus tyrosine cannot be used as an index of whole-body protein turnover during labeled phenylalanine infusion in the same way that KIC can be used during labeled leucine infusion. The same problem of first-pass hepatic extraction of phenylalanine also occurs in fed subjects. When the goal of a study is to measure only whole-body protein turnover, there does not appear to be any advantage in using phenylalanine over leucine as a tracer (except that an isotope ratio mass spectrometer is not needed for measuring breath $^{13}CO_2$ enrichment). However, when muscle protein metabolism studies are being done simultaneously, use of a phenylalanine tracer allows simultaneous determination of muscle and whole-body turnover. This application is described later.

Bolus Tracer Injection Rather Than Continuous Infusion

Proper use of the equations listed above requires a constant infusion of tracer and requires that tracer enrichment has reached a reasonable plateau. It also is possible to calculate the R_a of an amino acid under steady state metabolic conditions after a bolus injection of a tracer. Assuming a single pool of free amino acid as before, the R_a is calculated as:

$$R_a = \left(I / \int E \right) - I$$

where I is the dose of tracer and $\int E$ is the area under the curve of the enrichment from the time the tracer is injected until E returns to the baseline (pretracer value). E does not return to the true pretracer baseline during a study, but the area under the curve can be approximated after only a few hours because the decay curve of the tracer enrichment is fairly well defined. Intuitively, it should be obvious that if R_a is very rapid, the injected tracer will be rapidly diluted, yielding a small area under the enrichment curve and a small denominator in the above equation. In practical terms, the simplicity of a bolus tracer injection is offset by the difficulty of the increased number of samples and the more complicated calculations required to define the area under the curve. Thus the bolus injection method usually is reserved for studies using multicompartmental models, which will be discussed later.

Nonsteady State Conditions

A true metabolic steady state probably is never really achieved. After an overnight fast there are gradual changes in protein, carbohydrate, and fat metabolism as the postabsorptive period becomes longer. After meals there are rapid changes in hormones and substrates that can lead to rapid changes in protein metabolism. In postabsorptive studies, the metabolic changes generally are gradual enough that the assumption of steady state conditions does not cause significant errors. In fed subjects, studies often are designed to give small meals at frequent intervals rather than one large meal, to establish a quasisteady state. Steady state equations are adequate for such studies. However, having to rely on steady state conditions can impose significant limitations on the study design. Rapid changes in protein metabolism, those occurring before steady state is reached, cannot be evaluated with steady state equations. Thus, sometimes it is desirable to estimate protein breakdown and synthesis under nonsteady state conditions. Under such conditions, both the mass of the traced amino acid in the free pool and the tracer enrichment of the amino acid vary over time. Intuitively, you might think that simply using the steady state equations for each individual time point would give the R_a and R_d for that particular time point. This approach would be accurate only if changes in protein metabolism were instantly reflected by a jump to new amino acid enrichments and if concentrations were not changing. In reality, there is some delay between a change in protein breakdown and its ultimate effect on plasma tracer enrichment. Moreover, if the free pool of amino acids is changing, as reflected by changing plasma amino acid concentrations, then $R_a \neq R_d$, so that estimates of the change in the mass of the amino acid in the free pool must be made to calculate R_d. Steele developed equations to measure nonsteady state glucose metabolism, and these have been adapted to estimate protein turnover (Miles et al. 1983):

$$R_a = \frac{I - \left(\frac{P_1 + P_2}{2}\right)\left(\frac{E_2 - E_1}{t_2 - t_1}\right)}{\left(\frac{E_1 + E_2}{2}\right)} - I$$

where R_a is the rate of appearance of the traced amino acid at a point in time midway between t_1 and t_2, I is the tracer infusion rate, P_1 and P_2 are the masses of the free pool of the amino acid being traced at times t_1 and t_2, and E_1 and E_2 are the plasma enrichments of the tracer at times t_1 and t_2. It is difficult to precisely determine the values for P, which is estimated from the plasma concentration times the volume of distribution of the tracer. The volume of distribution can be determined experimentally by injecting a known amount of tracer and determining its initial concentration. A volume of distribution also can be estimated based on the volume of some physiological compartment such as extracellular water or total body water. Because the plasma amino acid concentration might be different than the intracellular amino acid concentration, the apparent volume of distribution as determined experimentally might not be the same as the volume of any identifiable physiological compartment. Generally, the same qualitative results will be obtained with different assumed volumes of distribution, although the results may differ quantitatively. This equation should be applied only if the changes in P and E between consecutive time points are fairly linear, which is generally true if blood sampling is done frequently enough, such as every 10 to 30 minutes. However, if there is a long delay between consecutive samples and there is a significant acceleration or deceleration of R_a during this interval, then the usefulness of the equation is diminished.

R_d at a point in time midway between t_1 and t_2 can be estimated during nonsteady state conditions, after calculating R_a as described above, with the following equation:

$$R_d = R_a - [(P_2 - P_1)/(t_2 - t_1)]$$

Venous Versus Arterial (or Arterialized) Blood Sampling

The compartment being sampled in human studies of whole-body protein turnover is blood plasma. The tracer is infused into the venous circulation. Amino acids derived from proteolysis flow into the venous circulation and are well mixed with the tracer by the time they reach the left ventricle for delivery to the tissues in the arterial circulation (Figure 3.4). As these circulating amino acids pass through a tissue, the tracer is again diluted with unlabeled amino acids coming from protein breakdown in that tissue. Thus the tracer enrichment is higher in the arterial than in the venous plasma (except for the venous pathway by which the newly infused tracer flows to the

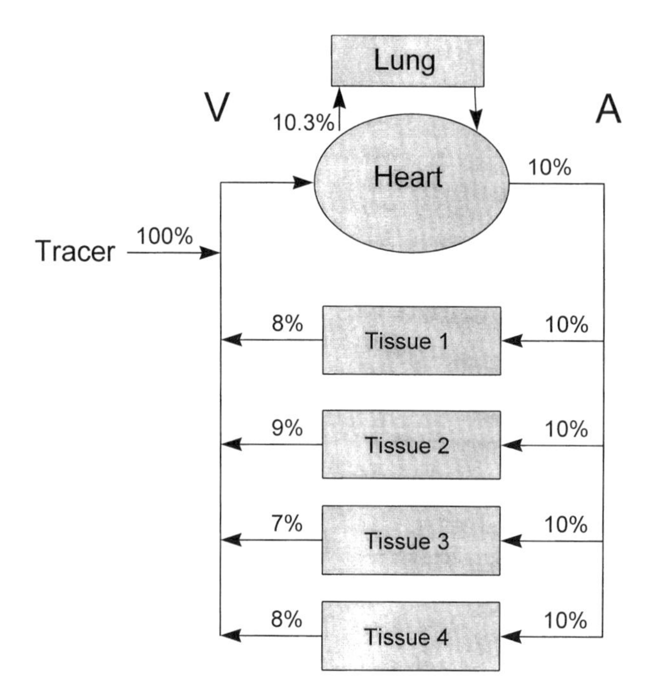

FIGURE 3.4. Schematic representation of isotopic enrichments in arterial (A) and venous (V) circulation during constant intravenous infusion of a tracer amino acid. The systemic arterial circulation has the same enrichment throughout, assuming that there is no dilution of tracer in transit to the tissues by the arterial vessels. The venous effluents from different tissues have different enrichments, according to the extent to which the tissue dilutes the tracer with unlabeled amino acid from proteolysis. This difference is used to calculate protein metabolism in specific organs, but complicates the calculation of whole-body protein metabolism. The continuing infusion of tracer into the venous circulation results in a higher arterial enrichment than what would be obtained if the venous effluents of the various tissues were pooled. Thus, whole-body protein turnover is underestimated to some extent when the arterial enrichment is used as the index of [illegible]

heart). The difference in enrichments between the artery and vein depends on blood flow and tissue proteolysis and can be used to calculate protein turnover in an organ, as discussed later. For the purpose of calculating whole-body protein turnover, the arterial enrichment is usually used because it avoids the problem of local dilution of the tracer in a tissue, which varies from one tissue to another. In practice, arterial sampling is a much more difficult and invasive procedure than venous sampling, so that most investigators use arterialized blood sampling. This procedure involves heating the hand so that blood flows through so rapidly that there is negligible fractional extraction or dilution of the tracer, making the venous blood of the hand very close to the arterial blood in terms of tracer enrichment. When a vein is sampled, it usually is a superficial one that drains the skin and subcutaneous adipose tissue, and the tracer enrichment is slightly less than that of the artery or an arterialized hand vein. It is unlikely that a different qualitative result would be obtained using superficial venous rather than arterialized plasma in a whole-body protein turnover study, although absolute turnover values will be lower with arterialized plasma.

Arterial (or arterialized) E gives a value for R_a that is too low because newly infused tracer, in its first pass through the arterial circulation, has not yet been diluted with unlabeled amino acid from proteolysis. Use of plasma KIC enrichment during labeled leucine infusion is a way around this problem. The ideal method would be to infuse the tracer into the pulmonary artery and to sample for tracer enrichment in the right atrium. Obviously, this is not a realistic approach for human studies. Alternatively, a correction can be made based on how much the newly infused tracer raises arterial E (Katz and Wolfe 1988), which can be estimated from cardiac output. The greater the ratio of cardiac output to tracer infusion rate, the smaller the correction. Most investigators do not bother with this correction, because R_a is only an index of protein breakdown and most interventions do not alter cardiac output enough to have a major impact on this index. However, conditions that greatly raise the cardiac output, such as exercise, could result in a lower arterial E (apparent increase in R_a) even if proteolysis is unchanged.

^{15}N Tracers

Thus far the chapter has dealt with the use of carbon-labeled essential amino acids to trace whole-body protein turnover. There is a large literature on the use of ^{15}N labeled amino acids, particularly [^{15}N]glycine, for this purpose. The methods described above for carbon-labeled or hydrogen-labeled essential amino acids are more straightforward in their interpretation than the ^{15}N methods and are recommended whenever their use is practical. However, when repeated blood sampling is impractical, the ^{15}N methods must be used because urine sampling can be used instead of blood sampling. Of course, one could use a ^{15}N-labeled essential amino acid in the same way as a ^{13}C-labeled tracer, with constant infusion and blood sampling when enrichment has reached a plateau. However, the problem with ^{15}N tracers is that the label can be removed from the tracer via transamination, so that enrichment can be diminished by increased transamination rather than by increased protein turnover. The extent of the problem varies according the extent to which the particular tracer is transaminated. Transamination of lysine is limited, whereas that of leucine is extensive. Leucine N flux is about twice the C flux (Matthews et al. 1981). Because of this problem, an alternative model based on the excretion of the ^{15}N in urea or NH_3 is used to measure whole-body protein turnover when ^{15}N amino acids are used as tracers. These are referred to as "end product methods." A more detailed description of this approach can be found in the book by Waterlow, Garlick, and Millward (1978).

The ^{15}N end product methods require the same assumptions that were described earlier for carbon labeled tracers. An additional assumption is that N in the free amino acid pool is in isotopic equilibrium with N excreted as urea or NH_3 (Figure 3.5). There is no requirement that the tracer be an essential amino acid, and [^{15}N]glycine is the tracer used most often. The ideal tracer is one whose ^{15}N behaves the most like that of the "average" amino acid. If lysine, which tends to retain its N more than other amino acids, is used with this method, an erroneously high rate of turnover is calculated because protein synthesis sequesters its N more readily than the N of the average amino acid (Garlick and Fern 1985). Some ^{15}N labeled amino acids give lower values for protein turnover than [^{15}N]glycine when this method is used, and some

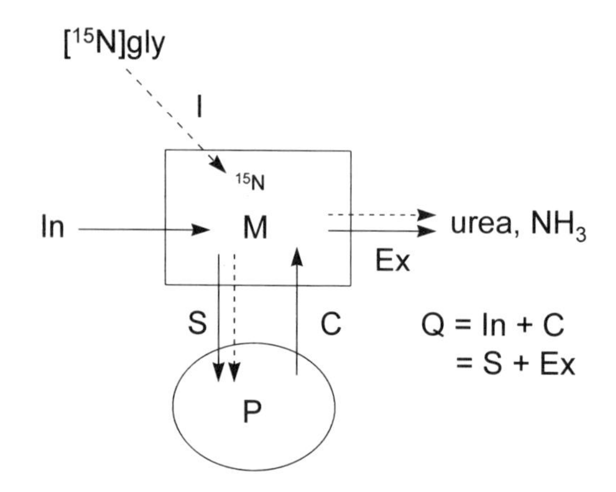

FIGURE 3.5. Model for estimating whole-body nitrogen flux (Q), protein breakdown (C), and protein synthesis (S) with [^{15}N]glycine as the tracer. The ^{15}N (dotted lines) is diluted by unlabeled N in the metabolic pool (M) of free amino acids and NH_3, by amino acids and NH_3 coming from protein breakdown and food intake (In). The ratio of ^{15}N to unlabeled N in the excreted end products represents tracer dilution by C and In. If M and P are constant, then S + Ex = In + C. Note that not all of the ^{15}N stays with glycine and can be incorporated into protein in other amino acids, as well as being excreted as urea or NH_3 or used for synthesis of other nitrogenous compounds.

give higher values (Figure 3.6). [^{15}N]Glycine gives an intermediate value for protein turnover, one that agrees fairly well with values obtained when ^{15}N labeled wheat proteins are fed as a tracer (Garlick and Fern 1985).

During constant infusion of [^{15}N]glycine or some other ^{15}N-labeled amino acid, the whole-body N flux[2] (Q) is calculated as follows:

$$Q = I/E_u$$

where I is the tracer infusion rate and E_u is the isotopic enrichment of the urea or NH_3 in urine at isotopic plateau. Note that Q includes the contribution of the tracer itself when this formula is used. When no priming of the free N or urea pool is done, it takes about 2 days for an isotopic plateau to be reached. Thus the method is applied to fed subjects, because the assumption of a steady metabolic state cannot be met during periods of prolonged fasting.

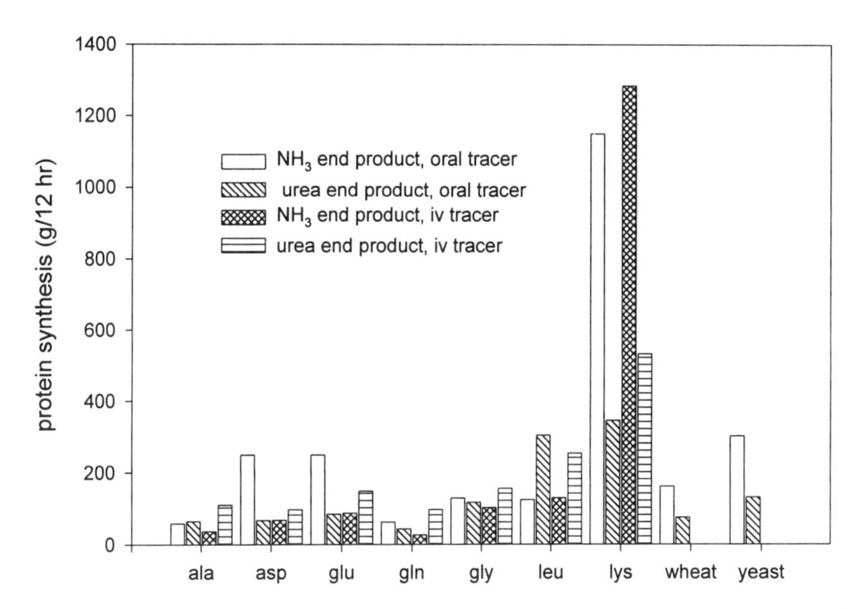

FIGURE 3.6. Apparent rates of whole-body protein synthesis in a single subject measured by different amino acids and proteins labeled with ^{15}N. Each amino acid was given orally (left two bars for each amino acid) and intravenously (right two bars for each amino acid). Labeled wheat or yeast proteins were given orally. The calculated protein synthesis rate is highly dependent on the end product used (NH_3 or urea). Data of Fern, Garlick, and Waterlow (1985).

---

[2] The literature on the use of the ^{15}N end product method usually employs the term flux, or N flux, rather than R_a, because the R_a of the amino acid is not being measured, only the dilution of its N in the free N pool.

Often the tracer is given as frequent oral doses rather than intravenously, because of the long period required and because the method is often applied when a more noninvasive procedure is desired. The total flux of N in fed subjects is derived from N intake (*In*) and endogenous protein breakdown (*C*, for catabolism). When the there is a constant pool of free N, the total flux coming from *In* and *C* must equal the removal of N from the free pool, for protein synthesis (*S*) or excretion (*Ex*). Thus:

$$In + C = S + Ex = Q$$

$$C = Q - In$$

$$S = Q - Ex$$

The rate of N coming from whole-body protein breakdown and the rate of N entry into proteins can therefore be derived from measurement of N intake, N excretion, and isotopic enrichment of N in a urinary end product. N mass is usually converted to protein mass by multiplying N by 6.25.

There are a few problems with this method. A long period of time is required for isotopic plateau to be reached. This problem has been addressed by variations on the basic method, as discussed below. Another problem is that different results are obtained with urea and NH_3 as the end products (Figure 3.6). The ^{15}N enrichment of NH_3 is generally higher than that of urea when [^{15}N]glycine is the tracer, leading to lower values for *Q* (Fern et al. 1981; Wolfe 1992) when NH_3 is used as the end product. Most of the NH_3 excreted in urine is produced in the kidneys from glutamine. Because glutamine is produced primarily in muscles, the higher ^{15}N enrichment in NH_3 than in urea could reflect the fact that muscles have a slower fractional protein turnover than liver (Wolfe 1992). The ratio of NH_3 enrichment to urea enrichment is not always the same. For example, in fed subjects the ^{15}N enrichment of NH_3 was higher than that of urea whether [^{15}N]glycine was given intravenously or orally (Fern et al. 1981). In postabsorptive subjects, the same situation occurred when the tracer was given intravenously, but not when given orally. Fern et al. (1981) suggested averaging the results from the two end products to obtain the best index of whole-body protein synthesis, but it is unclear whether or not this approach is valid. Another problem is that, at least in theory, the higher fractional turnover of protein in liver compared with the whole-body average (see Chapter 4) could lead to overestimates of whole body protein turnover because urea is produced in the liver. Although comparable whole-body protein turnover rates have been obtained with the [^{15}N]glycine method and the [^{13}C]leucine method described earlier (Garlick and Fern 1985), in some cases high hepatic protein turnover could invalidate the [^{15}N]glycine method (Wolfe 1992). Another potential problem is that long periods of tracer administration can increase recycling of ^{15}N from proteins that were synthesized during the study, which violates the assumption that protein breakdown yields only unlabeled N.

In rats, protein synthesis calculated with the [^{15}N]glycine method was compared with direct measurement of ^{15}N incorporation into whole-body protein homogenates (Stein et al. 1980). Calculations of whole-body protein synthesis based on direct incorporation of [^{15}N]glycine gave values that were twice as high as those obtained by the end product method. Thus, the end product methods might yield values for whole-body protein synthesis that are too low.

The long period required for isotopic plateau with the original method led to modifications to shorten the length of time needed to estimate protein turnover. The use of a bolus dose of tracer rather than a continuous infusion can be used in a fashion analogous to using a bolus of an essential amino acid labeled with ^{13}C or ^{14}C. The area under the enrichment curve between tracer administration and return of enrichment to baseline is comparable to plateau enrichment during constant tracer administration. Because the free NH_3 pool is a smaller, more rapidly turning over pool than the urea pool, use of NH_3 rather than urea as the end product results in a more rapid return to baseline enrichment. Waterlow, Golden, and Garlick (1978) proposed that a study could be completed within 12 h with the single tracer dose method and use of urinary NH_3 as the end product. Unfortunately, this method generally gives lower rates of protein turnover than the original method. A further modification reduced the procedure to 9 hours for the urea end product calculations, by correcting the measured urea excretion and enrichment for amounts of labeled and unlabeled urea remaining in the body at 9 hours after a single dose of [^{15}N]glycine (Fern et al. 1981). This modification did not eliminate the discrepancy between values obtained with urea and those obtained with NH_3. Priming of the urea pool with ^{15}N can shorten the period of time required for the continuous infusion method. For example, giving 30 times the hourly tracer infusion dose as a bolus at the beginning of a [^{15}N]glycine infusion can shorten the time to reach a urinary urea plateau enrichment to as little as 6 hours (Jeevanandam et al. 1985), although this dose substantially overprimes the NH_3 pool. There is real danger of falsely concluding that an early plateau has been reached when measuring isotope enrichments of slowly turning over pools, and this danger is exacerbated when a priming dose is used. If consecutive measurements show little or no difference in enrichment over a brief period, it is tempting to conclude that a plateau has been achieved. However, biological or analytical variations may obscure a trend toward increasing or decreasing enrichments that would be uncovered with more prolonged measurements.

Multicompartmental Models

The methods I have discussed to this point ignore the multiple compartments through which amino acids may travel in the course of protein synthesis and breakdown. Waterlow, Garlick, and Millward (1978) called these models "stochastic," a term that is sometimes used today. The term stochastic refers to random sampling from a probability distribution, so the term "single-pool" is preferred to stochastic. Single-pool models assume that amino acids derived from protein breakdown, and those that are used for protein synthesis enter and exit a single homogeneous pool of free amino acids. However, there is not a single, well-mixed, free amino acid pool in the body. Thus, there have been attempts to refine the understanding of protein metabolism with multicompartmental models. The mathematics of these models are beyond the scope of this book. Readers should consult original papers using such methods for more details (Irving et al. 1986; Cobelli et al. 1991; Carraro, Rosenblatt and Wolfe 1991).

The basic idea of multicompartmental modeling is that the enrichment versus time curve after a bolus of a tracer is determined by the rates of entry of unlabeled tracee and disposal of tracer and tracee into and out of multiple compartments that have unique turnover rates. If the tracer enters and exits from a single free amino acid compartment, there is a monoexponential enrichment decay curve after initial mixing of the tracer and tracee. However, if there are multiple compartments, the enrichment decay curve is more complex and depends on the sizes of the compartments and the transfer rates of tracer and tracee among the compartments. In reality, each kind of protein in each organ could be considered a separate compartment for influx and appearance of an amino acid, but the resolution of the analytical and mathematical methods is such that only a limited number of compartments can be resolved. A major limitation is that only a few compartments can be sampled directly. Many of the compartments in such models are significant only with respect to the kinetics of a particular amino acid rather than with protein metabolism in general. Thus the models would be useful for protein metabolism studies only to the extent that they increase the precision or validity of the determination of the whole-body R_a, R_d, or irreversible tracer loss for calculating whole-body protein turnover. However, it is difficult to determine whether the values for R_a, R_d, R_{ox}, or other components relevant to protein metabolism are any more valid with a multicompartment model than with the single-pool model described earlier. Although the models are generated by minimizing the difference between observed and predicted parameters, it is possible to get a good mathematical fit of the data to the model even if the model does not have any physiological basis. Because of the requirement of a large number of samples, extremely precise enrichment or specific activity data, and mathematical complexity of multicompartmental models, most studies of whole-body protein metabolism probably will continue to be done with the single-pool model. If future studies can validate a model that can noninvasively determine protein turnover in a compartment that has a definite physiological basis, then the model would be very useful. For example, if a multicompartmental model could determine muscle protein synthesis and breakdown without requiring a muscle biopsy or insertion of arterial and deep venous catheters, such a method would be a major advance.

Whole-Body Protein Oxidation

The most common way of estimating protein oxidation is to measure urinary N excretion, with the assumption that each g of excreted N reflects the oxidation of 6.25 g of protein. Although some N is excreted in urine from sources other than amino acids coming from protein, these sources usually are minor compared with the N from protein. This method determines the rate at which the N from amino acids is converted to urea and NH_3, and therefore does not directly reveal the rate at which the carbon skeletons of the amino acids are oxidized as a fuel source. Some of this carbon is converted to glucose via the gluconeogenic pathway, and therefore may be oxidized as glucose rather than entering the Krebs cycle directly. Some of the carbon can be incorporated into fat via de novo lipogenesis, then stored or oxidized as fat, but this pathway accounts for a negligible portion of overall amino acid carbon disposal. Some of the amino acid metabolites could be used in other synthetic pathways and retained or eliminated without being oxidized. These other fates of the carbon skeletons are probably minor compared with oxidation under most conditions.

Most of the N generated by amino acid deamination is excreted as urea. Thus, urea excretion rather than total N excretion sometimes is used to calculate protein oxidation. Approximately 70% to 90% of the excreted N is in the form of urea, but this fraction can be reduced under acidotic conditions that increase NH_3 excretion, such as prolonged fasting. There is some degradation of urea in the gut by bacteria, so urea excretion underestimates the total rate of urea production (Hamadeh and Hoffer 1998). Because the N generated by urea hydrolysis is recycled back to the liver or the rest of the body, urea hydrolysis does not invalidate the use of N or urea excretion as indices of net protein oxidation. Urea excretion may exceed urea production, or vice versa, especially over short time intervals, resulting in changes in the whole-body urea pool. This problem can be addressed by measuring the total urea pool at the beginning and end of an experimental period, by measuring blood urea N concentrations and multiplying by the measured or estimated total body water.

Urea production rate can be measured isotopically, by infusing ^{15}N, ^{13}C, or ^{18}O labeled urea (Wolfe 1992). ^{15}N labeled urea gives a lower value for urea production rate because of recycling of ^{15}N from urea degraded in the gut, so ^{13}C or ^{18}O labeled urea is better for determining total urea production. Urea production may be greater than net amino acid deamination, because some of the N from urea hydrolysis in the gut is used to reaminate ketoacids. Simultaneous use of [^{15}N]urea and another tracer can be used to determine the recycling from the gut. A problem with tracer determination of the urea production rate is the large pool of urea in the body. The tracer must be infused for many hours to achieve an isotopic plateau, unless an accurate priming dose is administered, in which case a few hours may be adequate. When urea production is altered during a continuous tracer infusion, a significant change in tracer enrichment can be detected within an hour (Wolfe 1992), but a new plateau is not achieved for many hours (Hamadeh and Hoffer 1998).

Some investigators have equated the oxidation of a particular amino acid with overall protein oxidation. As discussed earlier, the tracer L-[1-^{13}C]leucine often is used to measure whole-body proteolysis and protein synthesis, with the difference between total leucine disappearance and leucine oxidation reflecting protein synthesis. It often is erroneously assumed that leucine oxidation must therefore reflect protein oxidation. The problem with such an approach is that the oxidation of any particular amino acid does not necessarily reflect oxidation of all other amino acids coming from protein catabolism. Leucine could be preferentially oxidized or preferentially spared from oxidation, relative to other amino acids under various conditions. The problem is exaggerated if there is absorption of amino acids from protein meals that do not have the same amino acid composition as the amino acid composition of body proteins. For example, consumption of a protein meal can increase levels of essential amino acids much more than it increases levels of nonessential amino acids. Hence leucine oxidation after a protein containing meal may increase much more than oxidation of many other amino acids. Although oxidation rates of specific amino acids will generally correlate with overall protein oxidation, there is no single amino acid for which the oxidation rate always reflects total protein oxidation.

Protein Metabolism in Specific Organs or Limbs

In studies lasting from several days to weeks or months, changes in the protein mass of specific organs or limbs can be estimated using imaging methods such as magnetic resonance imaging or computerized axial tomography. If the protein concentration of an organ remains the same, then changes in the volume of the organ indicate the change in protein balance over the experimental period. One should be cautious in making conclusions about protein balance with such methods, for there is no guarantee that conditions affecting organ size do not also affect the protein concentration of the organ. No information about whether there are changes in protein turnover can be obtained in this manner. Moreover, acute changes in protein metabolism would be undetectable because of the long time period needed for changes in protein breakdown or synthesis to produce a measurable change in organ volume.

In theory, rates of incorporation of a tracer into proteins of specific organs or regions of the body and disappearance of the tracer could be done with nuclear magnetic resonance or positron emission tomography (PET). Many technical problems would have to be overcome. Paans et al. (1996) reviewed the potential of [^{11}C]tyrosine to trace protein synthesis by PET.

Amino Acid Balance

Limb-balance or organ-balance methods generally are used to evaluate protein metabolism in limbs and organs. The earliest applications of this method did not involve the use of isotopic tracers and simply examined the net balance of amino acids in a limb or organ. To use the method, one must be able to catheterize an artery (or obtain arterialized blood) and a vein that drains the organ of interest and to measure the rate of flow of blood into that vein. For example, to determine leg amino acid balance, you would catheterize the femoral vein. To determine renal amino acid balance, you would catheterize the renal vein. Splanchnic amino acid balance requires catherization of the hepatic vein, whereas hepatic amino acid balance requires catheterization of both the portal and hepatic veins. The mass of an amino acid entering an organ or limb is the arterial blood concentration times the blood flow rate into the organ, and the mass leaving the organ is the venous concentration times the flow rate:

$$Balance = F \times (C_a - C_v)$$

where F is blood flow rate through the limb or organ, C_a is the arterial blood concentration of the amino acid, and C_v is the venous concentration of the amino acid. All of these variables must be measured precisely for the method to be useful.

The correct amino acid balance will be obtained if blood flow is multiplied by the difference in whole-blood amino acid concentrations. However, often only the plasma amino acid concentrations, rather than the whole-blood amino acid concentrations, are measured in such studies. The question, then, is whether to multiply the difference in plasma concentrations by the blood flow or by the plasma flow. The answer depends on the degree to which erythrocytes participate in the arterial-venous difference in amino acid concentrations. Very little change might be expected in the concentration of amino acids within erythrocytes in the few seconds it takes for them to travel from the artery to the vein, even if plasma concentrations change significantly. In this case, it would be appropriate to multiply the arterial-venous plasma concentration difference by the plasma flow rather than by the blood flow. If erythrocyte concentrations were to instantly equilibrate with plasma concentrations, it would be appropriate to multiply the arterial-venous plasma concentration difference by the blood flow. In the case of partial equilibration of plasma and erythrocyte amino acid concentrations during transit of blood from the capillary to the venous sampling site, tissue amino acid balance cannot be calculated exactly on the basis of plasma concentration measurements. However, use of plasma flow in this case would still give a measure of plasma exchange, which reflects the net effect of the tissue and the erythrocytes on the amino acid concentrations. The relative importance of erythrocyte participation in amino acid transport to and from a tissue varies according to the particular amino acid, the tissue, and the amino acid concentration (Tessari et al. 1996e; Felig, Wahren, and Raf 1973; Aoki et al. 1976; Barrett et al. 1987). The safest strategy to ensure that amino acid balance reflects only the tissue metabolism is to use whole-blood amino acid concentrations.

The utility of amino acid balance to study protein metabolism depends on the particular amino acids and organs being studied. Under conditions in which the free amino acid pool of the organ is constant, net release (negative balance) of certain amino acids reflects negative protein balance, whereas net uptake (positive balance) of these amino acids reflects positive protein balance. However, this conclusion can be made only for certain amino acids. For example, the liver could remove alanine from the circulation and convert it to glucose. This process could lead to positive alanine balance in the liver, even though protein balance might be negative. Muscle might remove leucine from the circulation and oxidize it or release its ketoacid. This process would yield a positive leucine balance, even though protein balance could be negative. Conclusions about protein balance can be made only when studying amino acids that cannot be produced or metabolized in the organ of interest, except via protein breakdown or incorporation into proteins. Essential amino acids cannot be synthesized in any tissue and therefore are most useful. However, even the essential amino acids can be removed via pathways other than protein synthesis. For studies of forearm or leg, the phenylalanine balance reflects the balance between protein breakdown and protein synthesis, because phenylalanine is not metabolized by the limbs. However, the liver can metabolize phenylalanine after converting it to tyrosine, so that phenylalanine balance does not necessarily reflect protein balance in the liver. One must be very cautious about drawing conclusions about protein balance from amino acid balance studies, and must carefully consider the contribution of pathways of disposal other than incorporation into proteins.

The arterial-venous difference in concentrations of 3-methylhistidine can be used to determine the rate of breakdown of myofibrillar proteins. 3-Methylhistidine is formed in actin and myosin by posttranslational modification of histidine, as will be discussed later. It cannot be used for protein synthesis or converted back to histidine once it is released by proteolysis. Its only known route of elimination is entry into the circulation for renal excretion. Although there is some controversy about how much 3-methylhistidine in urine is derived from cells other than striated muscle fibers, there is no doubt that almost all of the arterial-venous 3-methylhistidine difference in a limb reflects proteolysis of actin and myosin in skeletal muscle. This method does not provide any information about protein synthesis.

Amino Acid Balance Combined with Tracer Infusion

The amino acid balance method cannot indicate the rates of protein breakdown and protein synthesis in the organ. To obtain such data, you must combine the organ balance method with a tracer infusion. Ring-[$^2H_5$]phenylalanine (D_5-phe) is often used to determine protein breakdown and synthesis in the limbs (Nair, Schwartz, and Welle 1992). [^3H]phenylalanine was originally used for this purpose (Barrett et al. 1987). Phenylalanine is the ideal amino acid to trace in limb balance studies, because it can appear only through proteolysis and disappear only through protein synthesis in these tissues (Barrett et al. 1987). Figure 3.7 illustrates the method.

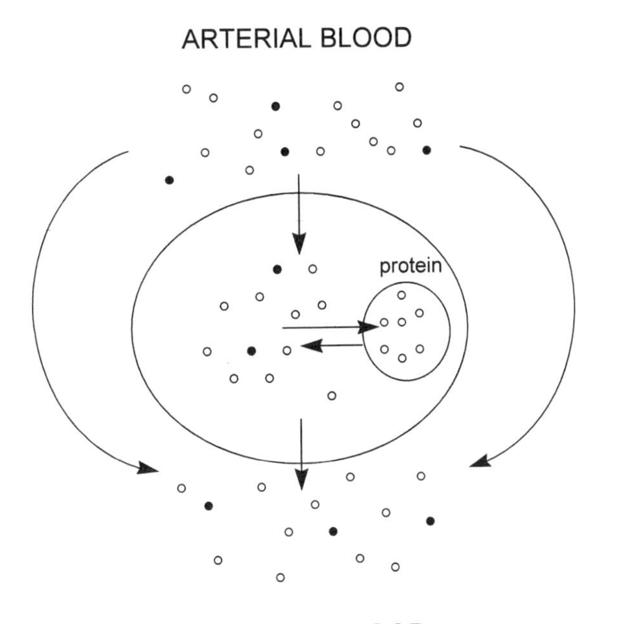

FIGURE 3.7. Schematic representation of D_5-phe dilution method of determining protein metabolism in a tissue (e.g., muscle) that does not metabolize phenylalanine except for incorporation into proteins. Open circles represent unlabeled phe. Filled circles represent D_5-phe. Enrichment with D_5-phe is highest in the arterial circulation during constant infusion of this tracer. Most of the phe in the arterial blood flows into the venous circulation without entering the cells. During the first few hours of tracer infusion, the amount of D_5-phe in proteins is negligible, so proteolysis reduces the D_5-phe enrichment in the cells relative to that of the artery. D_5-phe enrichment in the venous circulation is used to estimate the enrichment of the phe entering proteins. See text for calculations based on this model.

In general, the concept of the D_5-phe balance method is the same as that of the amino acid balance described above. The advantage of adding a tracer to the balance method is that in studies lasting only a few hours, a negligible fraction of the tracer that is incorporated into protein returns to the free amino acid pool via proteolysis (note that this assumption is true for the total protein mass, but is not necessarily true for individual proteins). Hence the extraction of the D_5-phe by the limb reflects protein synthesis. Because the net balance of phenylalanine is the difference between protein synthesis and protein degradation, measurement of net balance and tracer extraction allows calculation of protein breakdown. The method assumes that the intracellular free pool of phenylalanine is constant during the study. The primary problem with the method, as with other methods for protein synthesis described elsewhere in this chapter, is that there is some uncertainty about how much unlabeled phenylalanine enters the protein pool along with the D_5-phe. If all of the phenylalanine used for protein synthesis were derived from the circulation, then the isotope ratio (ratio of D_5-phe to unlabeled phe) in the arterial plasma would be an adequate indicator of phenylalanyl-tRNA enrichment. However, there is significant dilution of the D_5-phe within the limb because of proteolysis, as indicated by a lower isotope ratio in the venous circulation than in the arterial circulation. The venous isotope ratio is more likely to reflect the isotope ratio of the phenylalanine used for protein synthesis. Nevertheless, because some of the venous phenylalanine comes directly from the arterial circulation, without entering and leaving the cells within the limb, even the venous isotope ratio could be higher than the isotope ratio within the pool of free phenylalanine used for protein synthesis. Thus, the method will give a minimal estimate of protein turnover, but still is very useful for evaluating the extent to which changes in net phenylalanine balance are the result of changes in protein synthesis and protein degradation.

The equations for determining protein synthesis and breakdown with the D_5-phe method are as follows.

$$FE = ([D_5\text{-}phe_A] - [D_5\text{-}phe_V])/[D_5\text{-}phe_A]$$

where FE is the fraction of phenylalanine that is extracted by the limb and $[D_5\text{-}phe_{A\ or\ V}]$ is the arterial or venous concentration of D_5-phe. The concentration is determined by multiplying the total phenylalanine concentration by the isotopic enrichment (E). The total amount of tracer incorporated into protein is the amount entering the limb:

$$\text{Tracer} \rightarrow \text{limb} = FE \times F \times [D_5\text{-}phe_A]$$

The total amount of phenylalanine used for protein synthesis is the amount of tracer incorporated divided by the venous isotope enrichment (E_V), which is an index of the tracer enrichment of the phenylalanyl-tRNA. Phenylalanine disappearance (R_d) is equivalent to its incorporation into protein in the limbs.

$$R_d = (\text{Tracer} \rightarrow \text{limb})/E_V$$

The total net balance of phenylalanine is given by the equation

$$\text{Balance} = R_d - R_a$$

where R_a is the release of phenylalanine from proteolysis. R_a therefore can be calculated because balance is known.

$$\text{Balance} = F \times ([phe_A] - [phe_V])$$

$$R_a = R_d - \text{Balance}$$

These equations give the same results as equations described by Nair et al. (1992), but the format has been changed to facilitate understanding the concepts underlying the equations. Subtracting the amount of D_5-phe incorporated into proteins from R_d is not recommended, because the small increase in total phenylalanine concentrations resulting from the D_5-phe infusion is unlikely to stimulate protein synthesis.

Whether it is adequate to use plasma flow, plasma E, and plasma concentrations rather than whole blood values is unclear. In the dog hindlimb, use of plasma flow and [^{3}H]phenylalanine specific activity yielded values for phenylalanine R_d and R_a that were not significantly different from the results obtained with whole blood measurements (Barrett et al. 1987). However, there is evidence that in human studies the use of plasma measurements underestimates the phenylalanine balance and postprandial R_d (Tessari et al. 1996e). As discussed above for net balance, the only way to ensure that the calculated phenylalanine kinetics reflect only the tissue protein metabolism is to use whole blood concentrations and tracer enrichments. Under some conditions the plasma measurements may be adequate, but unless this assumption is tested under the particular conditions of the experiment there is always some uncertainty about the erythrocyte contribution.

In limb studies, it is possible to obtain a muscle biopsy to estimate the tracer enrichment of the free amino acid pool within the tissue. This enrichment is lower than that of the venous plasma, and probably gives a more accurate estimate of protein metabolism (Wolfe 1992). The obvious disadvantage is the need to obtain a muscle biopsy. The limb tracer balance method measures protein metabolism not only in the muscles, but also in bone, skin, and fat in the limb. Hence the tracer enrichment in a muscle biopsy does not necessarily give the true value for all tissues contributing to limb protein metabolism. Moreover, there is no guarantee that the enrichment of the free amino acid in the muscle tissue is the same as the aminoacyl-tRNA enrichment. Thus the method described here suffers from the same uncertainties about aminoacyl-tRNA enrichment as the direct incorporation studies of protein synthesis, which are described later.

The D_5-phe method is much more complicated in studies of splanchnic or hepatic protein metabolism, because the D_5-phe is removed not only by incorporation into proteins, but also by conversion to tyrosine. The tyrosine can then be catabolized, released into the circulation, or incorporated into proteins. The fraction of tyrosine that is derived from phenylalanine can be determined by measuring the ratio of D_4-tyr enrichment to D_5-phe enrichment in the hepatic vein. The hepatic tyrosine balance and irreversible loss of tyrosine in the liver by protein synthesis or tyrosine oxidation must be measured with a separate tracer, such as D_2-tyr, using equations analogous to those listed above for phenylalanine. The relevant equations and their derivation are presented by Nair et al. (1995).

Tracers other than phenylalanine certainly can be used for tracer balance studies. The key point is that all routes of appearance and disposal of the particular amino acid being traced, in the limb or organ being studied, must be accounted for. The critical assumption, and the basis for using a tracer rather than just measuring the amino acid balance, is that tracer incorporated into proteins or otherwise removed from the free amino acid pool cannot find its way back into the free pool of amino acids during the course of the tracer infusion. Leucine tracers are often used in limb and organ balance studies (Cheng et al. 1985; Nair et al. 1995). The R_a of leucine reflects proteolysis, just as it does for phenylalanine. However, the routes of disappearance of the leucine include not only use for protein synthesis, but also deamination to KIC. The KIC can be reaminated to form leucine or it can be oxidized. If the leucine is labeled with ^{15}N, then the label is lost during transamination. If the carbons are labeled, then oxidation by the organ or limb can be estimated by measuring labeled CO_2 efflux. The amount of leucine gained or lost by transamination can be estimated by labeling carbons or hydrogens and determining the concentrations and isotope enrichments of both leucine and KIC in arterial and venous blood. Equations for using this method are presented elsewhere (Cheng et al. 1985). The D_5-phe (or ^3H-phe) method is more straightforward and technically easier than the leucine tracer method when studying limbs or organs in which there is negligible conversion of phenylalanine to tyrosine.

To convert amino acid balance, R_a, and R_d to protein values, you must estimate the amino acid composition of the proteins in the tissue or organ being studied. Values for muscle, liver, and intestines of rats, pigs, and cattle have been published (Waterlow, Garlick, and Millward 1978; Munro and Fleck 1969; FAO 1970). Phenylalanine and leucine are the tracers used most often to estimate protein kinetics with the balance method. For these tracers, it appears that the values given in Table 3.1 would be adequate not only for the whole body, but also for both muscle and splanchnic proteins

Incorporation of Tracer into Tissue Proteins or Specific Proteins

In animal and cell culture studies of protein metabolism, the most common method is to determine the incorporation of a tracer amino acid, usually radioactive, into the mixture of proteins produced by the cells or tissue or into specific proteins. The same method can be used with human subjects. However, the need to obtain a tissue sample has limited its use for in vivo human studies. Most such studies in man have examined the synthesis of circulating proteins or muscle proteins, although a few studies have involved removal of other tissues for measuring tracer incorporation. The ease of collecting circulating proteins should be obvious. Muscle protein synthesis is of great interest because muscle contains about half of the body's total protein. Although obtaining muscle tissue is more invasive than obtaining blood, needle biopsies can be obtained with very little physical or psychological trauma to the subjects.

The concept of determining protein synthesis by tracer incorporation is very simple. If synthesis of a protein is rapid, then the rate of tracer enrichment of that protein is rapid, relative to the tracer enrichment of the pool of free amino acid available for protein synthesis. If protein synthesis slows down, the rate of tracer enrichment of the protein also slows down, if the enrichment of the precursor amino acid remains constant. Although the method is conceptually straightforward, a major problem with interpreting such studies in humans is uncertainty about the enrichment of the pool of free amino acid being used for protein synthesis. Unless you know how many molecules of unlabeled amino acid enter the protein for every molecule of tracer entering the protein, you cannot determine the protein synthesis rate. If you could measure the aminoacyl-tRNA enrichment, the problem would be solved. Unfortunately, the pool of aminoacyl-tRNA is very small, so generally there is not enough of it in the tissue sample to measure tracer enrichment. In the case of circulating proteins, many of which are synthesized in the liver, usually it is not possible to obtain any tissue at all to address the problem of aminoacyl-tRNA enrichment. Several approaches can be taken to estimate aminoacyl-tRNA enrichment in specific tissues during tracer infusions:

Enrichment

Plasma Tracer Enrichment

During tracer infusions lasting only a few hours, the mixture of endogenous proteins in a tissue has a very low enrichment relative to plasma tracer enrichment. Even though a few proteins with a rapid turnover become highly enriched, proteolysis produces a lower enrichment of the free amino acid pool within the cell than the enrichment of that amino acid in plasma (the compartment into which the tracer is infused). The extent to which the plasma tracer enrichment is higher than the aminoacyl-tRNA enrichment depends on the relative contributions of plasma-derived amino acids and proteolysis-derived amino acids to the aminoacyl-tRNA pool. The aminoacyl-tRNA tracer enrichment typically is significantly lower than that of the corresponding plasma amino acid. For example, the plasma leucine enrichment during infusion of L-[1-^{13}C]leucine is 1.28 to 1.78 times higher than the enrichment in leucyl-tRNA extracted from skeletal muscle (Table 3.2). If this ratio were constant, then plasma amino acid enrichment could be used with a correction factor. However, there is no guarantee that this ratio is the same under different nutritional or physiological conditions. Thus the plasma amino acid enrichment can only give a minimum boundary for the rate of protein synthesis.

When the total concentration of the amino acid being traced (labeled plus unlabeled) is very high, then the enrichments of the intracellular and extracellular compartments are very close. The flooding dose method, described below, is based on the similarity of the enrichments of all free amino acid pools after a large load of amino acid is administered.

Table 3.2. Plasma amino acid enrichment, plasma KIC enrichment, and tissue free amino acid enrichment as indices of aminoacyl-tRNA enrichment.

Species, tissue	Condition	Amino acid	$E_p/E_{aa\text{-}tRNA}$	$E_{pKIC}/E_{aa\text{-}tRNA}$	$E_{tissue}/E_{aa\text{-}tRNA}$
Human, erector spinae muscle[a]	Surgery	leu	1.28 v	1.12 v	0.92
Human, vastus lateralis muscle[b]	Overnight fast	leu	1.77 a	1.47 a	1.00
	Carbohydrate meals	leu	1.78 a	1.45 a	0.94
	Mixed meals	leu	1.59 a	1.32 a	1.08
Pig, muscle[c]	12 h fast, anesthetized	leu	1.33 a	1.16 a	1.26
			1.22 v	1.12 v	
		phe	1.57 a		1.10
			1.35 v		
Pig, heart[c]		leu	1.07 a	0.93 a	1.04
			0.98 v	0.90 v	
		phe	1.11 a		1.24
			0.95 v		
Pig, liver[c]		leu	1.57 a	1.37 a	1.06
			1.43 v	1.31 v	
		phe	1.24 a		0.99
			1.06 v		
Pig, muscle[d]	Fasting	leu	1.29 a	1.04 v	0.97
	Amino acid infusion	leu	1.03 a	1.04 v	1.02
Rat, muscle[a]	Anesthetized	leu	1.20 a	1.12 a	0.87
			1.18 v	1.09 v	
Rat, liver[a]		leu	1.75 a	1.63 a	0.59
			1.72 v	1.59 v	
Rat, muscle[e]	Overnight fast, anesthetized, after 90 min tracer infusion	phe	2.15 a		0.82
			1.42 v		
Rat, heart[e]			1.21 a		0.79

[a]Data of Watt et al. (1991).
[b]Data of Ljungqvist et al. (1997).
[c]Data of Baumann et al. (1994).
[d]Data of Watt et al. (1992).
[e]Data of Young et al. (1994).

Abbreviations: E_p - plasma enrichment; E_{pKIC} - plasma KIC enrichment; $E_{aa\text{-}tRNA}$ - enrichment of aminoacyl-tRNA in tissue; E_{tissue} - enrichment of free amino acid in tissue; a - arterial plasma; v - venous plasma.

Some cells may rely primarily on circulating amino acids for protein synthesis. For example, immunoglobulin-secreting cells produce much more protein each day than they contain, and it is likely that almost all of the immunoglobulin synthesized by these cells is comprised of amino acids taken up from the circulation (Thornton et al. 1996). Plasma tracer enrichment may be an adequate index of aminoacyl-tRNA enrichment in such cases.

Plasma KIC Enrichment

Because KIC is formed by intracellular deamination of leucine, in theory the KIC enrichment should reflect intracellular leucine enrichment, assuming that there is a homogenous pool of free leucine that is used both for tRNA charging and deamination. Of course, a problem is that the plasma KIC enrichment is a composite of the KIC enrichments of different tissues, each of which can have a different level of intracellular leucine and KIC enrichment. Because muscle appears to account for most of the circulating KIC in humans, plasma KIC enrichment has been particularly popular for studies of muscle protein synthesis. Two studies in humans have examined the ratio of plasma KIC enrichment to leucyl-tRNA enrichment of muscle during [^{13}C]leucine infusions (Table 3.2). Although the plasma KIC enrichment is closer than plasma leucine enrichment to the muscle leucyl-tRNA enrichment, it is somewhat higher. These results are similar to those obtained in pigs and rats (Table 3.2), although the discrepancy between plasma KIC tracer enrichment and leucyl-tRNA enrichment may be larger in humans than in pigs.

Not only is leucine deaminated to KIC, but KIC is reaminated to leucine. During infusion of labeled KIC, there is production of labeled leucine within muscle cells and other tissues. Therefore you can estimate intracellular leucine enrichment during infusion of labeled KIC by measuring the plasma leucine enrichment. It was reported that the plasma leucine enrichment was only 8% higher than the muscle tissue free leucine enrichment during labeled KIC infusion in postabsorptive subjects (Chinkes et al. 1996). In the same subjects, the plasma KIC enrichment was 32% higher than the muscle tissue free leucine enrichment during infusion of labeled leucine. Whether there is an advantage of using labeled KIC rather than labeled leucine to measure muscle protein synthesis in fed subjects has not been examined.

Because many plasma proteins are synthesized in the liver, there is much interest in whether plasma KIC enrichment reflects intrahepatic leucyl-tRNA enrichment. Under postabsorptive conditions, the enrichment of circulating KIC was one third higher than that of intrahepatic leucyl-tRNA in pigs (Table 3.2). In rats, the plasma KIC tracer enrichment was 63% greater than the intrahepatic leucyl-tRNA tracer enrichment. Comparable studies of hepatic tracer enrichment have not been performed in human subjects. However, some insight about the relation between plasma KIC tracer enrichment and leucyl-tRNA tracer enrichment can be obtained from the tracer enrichment of apolipoprotein B-100 in very-low density lipoprotein particles (VLDL). VLDL is synthesized in the liver and has a rapid turnover rate, so that after 8 to 10 hours of a constant infusion of labeled leucine the enrichment of VLDL apolipoprotein B-100 reaches a plateau. The plateau indicates that VLDL apolipoprotein B-100 enrichment has achieved the same enrichment as the leucyl-tRNA in the hepatocytes producing the protein. In postabsorptive subjects, plasma KIC enrichment was found to be within 2% of VLDL apolipoprotein B-100 enrichment during prolonged intravenous infusion of

[$^2H_3$]leucine (Reeds et al. 1992) However, when the subjects were fed, the plasma [$^2H_3$]KIC enrichment was ~30% greater than the enrichment in VLDL apolipoprotein B-100. In another study with intravenous [$^2H_3$]leucine infusions, plasma [$^2H_3$]KIC enrichment was ~25% higher than VLDL apolipoprotein B-100 enrichment when subjects were fed protein-free meals, and ~ 60% higher when they were fed meals containing protein (Cayol et al. 1997). When the same tracer was infused intragastrically, apolipoprotein B-100 enrichment was higher than that of KIC. Thus, plasma KIC enrichment may be an adequate indicator of intrahepatic leucyl-tRNA enrichment under postabsorptive conditions when the tracer is delivered intravenously. When subjects are fed, or when the tracer is delivered intragastrically, plasma KIC is not a reliable index of leucyl-tRNA in the liver.

Plasma KIC enrichment seems to be a good index of leucyl-tRNA enrichment in the pancreas of postabsorptive subjects (Bennet, O'Keefe and Haymond 1993).

Tissue Free Amino Acid Enrichment

The enrichment of the free amino acid in a tissue is a good index of its aminoacyl-tRNA enrichment. This fact has been demonstrated in human muscle and in various tissues in pigs (Table 3.2). Although most of the free amino acid in a tissue sample comes from intracellular fluid, some comes from extracellular fluid. The extracellular compartment generally has a higher tracer enrichment. There is some evidence for preferential use of amino acids taken up from the extracellular compartment for charging of tRNA, which could explain why enrichment of free leucine and phenylalanine in the tissue is so close to aminoacyl-tRNA enrichment even though the extracellular tissue compartment is not excluded.

Enrichment of a Rapidly Turning Over Protein

The turnover rate of the body's entire mass of proteins is only ~ 0.1%/h in an adult. If you infuse enough [^{13}C]leucine to label 10% of the leucyl-tRNA for 10 h, then only ~ 0.1% of leucine in the body proteins will be labeled at the end of the infusion. However, there is great variability among proteins in their turnover rates. These conditions will lead to labeling of 6.5% of the leucine in a protein with a turnover rate of 10%/h and labeling of 9.7% of the leucine in a protein with a turnover rate of 30%/h. The higher the turnover rate of the protein, the more rapidly the tracer enrichment of the protein will achieve a value that approximates the enrichment of the aminoacyl-tRNA. As the protein enrichment approaches the aminoacyl-tRNA enrichment during a constant tracer infusion, the tracer enrichment of the protein reaches a plateau. Thus the tracer enrichment of a protein at this plateau represents the aminoacyl-tRNA enrichment of the cells synthesizing the protein. This value then can be used to calculate the synthesis rate not only of the rapidly turning over protein, but also of other proteins synthesized by the same cells.

There are certain limitations to this method. The rapidly turning over protein must be abundant enough to purify in large enough quantities to determine its tracer enrichment. The protein must be very pure, or else aminoacyl-tRNA enrichment could be underestimated. For example, if a rapidly turning over protein is 10% enriched, but is contaminated with an equal mass of proteins with an average enrichment of only 0.1%, the apparent plateau enrichment will be only ~ 5% (the enrichment of the contaminating proteins is increasing so slowly that it appears a plateau has been reached).

Hippuric Acid Enrichment as an Index of Hepatic Glycyl-tRNA Enrichment

Hippuric acid (benzoylglycine) is formed in the liver from glycine and benzoic acid. It is not necessary to administer benzoic acid or a precursor of benzoic acid, because there is enough endogenous benzoate production to supply enough hippuric acid for measurement of its isotopic enrichment in plasma or urine. Because both hippuric acid and glycyl-tRNA are formed from glycine in hepatocytes, their isotopic enrichment should be similar when labeled glycine is infused. There is some experimental support for the validity of this method, but also evidence against. When [^{15}N]glycine was infused for 5 days, the plateau enrichment of glycine in fibronectin (a protein synthesized in the liver, although not exclusively in the liver) was the same as that of urinary hippuric acid (Carraro, Rosenblatt, and Wolfe 1991). However, there is a delay of several hours before the hippuric acid enrichment plateaus, which limits the method to very long tracer infusions, although plasma enrichment may plateau sooner than urinary enrichment (Arends et al. 1995; Carraro, Rosenblatt, and Wolfe 1991). In another study, the enrichment of plasma hippuric acid significantly exceeded that of glycine in VLDL apolipoprotein B after 8 hours of [^{15}N]glycine infusion, even though both enrichments appeared to have achieved an isotopic plateau. These data suggest that the glycyl-tRNA enrichment can be lower than that of hippuric acid. However, it is possible that the VLDL apolipoprotein B glycine had not yet achieved its final plateau value, that some other protein(s) with a lower enrichment contaminated the lipoprotein or that some apolipoprotein B from denser lipoprotein particles contaminated the lipoprotein preparation. In the absence of further information about the mismatch between the isotopic enrichment of hippuric acid and that of glycine in plasma VLDL apolipoprotein B, the validity of hippuric acid enrichment as an index of hepatic glycyl-tRNA enrichment is questionable.

Urea Enrichment as an Index of Hepatic Arginyl-tRNA Enrichment

Arginine is an intermediate in the formation of urea. The guanidine carbon comes from CO_2, so that administration of labeled CO_2 (as bicarbonate) yields production of labeled arginine and then carbon-labeled urea in the liver, the only site of urea production. Administration of [^{15}N]glycine or some other source of ^{15}N that can transfer N to urea via the urea cycle will lead to ^{15}N labeling of hepatic arginine. Arginine of hepatic origin cannot be distinguished from arginine from other tissues by blood sampling, so urea enrichment must be used as an index of hepatic arginine enrichment. Thus, urea enrichment should reflect hepatic arginyl-tRNA enrichment, but only if the pool of free arginine used to charge tRNA is identical to the pool of arginine used for urea synthesis. Both processes take place in the cytosol, so the assumption is reasonable. A practical problem is that it takes a long time to obtain steady enrichment of the urea pool, even though production of labeled urea from labeled CO_2 or N begins as soon as tracer is infused. Thus, plasma or urinary urea enrichment will underestimate intrahepatic urea enrichment for several hours after the start of labeled CO_2 or ^{15}N administration, unless the urea pool is primed appropriately.

Flooding Dose Method

If a very large dose of an amino acid is infused into the circulation, then a very large proportion of the free amino acid pool in the cells is derived from the circulation and relatively little is derived from proteolysis. If the infused dose has some tracer in it, then the enrichment of the amino acid in all intracellular and extracellular pools is close to the plasma enrichment. Under such conditions, the plasma enrichment is a good index of the aminoacyl-tRNA enrichment in all tissues. Another advantage of this method is that enough tracer can be infused so that E is much higher than it is with low-dose tracer administration, shortening the time required to achieve adequate enrichment of the proteins. The main problem with this approach is uncertainty about whether or not infusion of a large amount of an amino acid alters protein synthesis. There is considerable controversy about this point. Arguments both for and against the notion that a flooding dose alters protein synthesis have been presented (Garlick et al. 1994; Rennie, Smith, and Watt 1994; Chinkes, Rosenblatt, and Wolfe 1993; Smith et al. 1998a). Flooding dose methods have produced values for muscle protein synthesis that are higher than those obtained with low-dose tracer infusions. Most, but possibly not all, of the discrepancy in calculated protein synthesis rates between studies using low-dose tracer infusion and studies using the flooding dose method seems to be explained by overestimation of aminoacyl-tRNA enrichment during low-dose tracer infusions. For example, the synthesis rate of human vastus lateralis muscle in postabsorptive subjects usually is reported to be ~1.2%/d when the low-dose leucine tracer method is used with plasma KIC enrichment as the index of enrichment of intramuscular leucyl-tRNA. True leucyl-tRNA enrichment appears to be only about two thirds of the plasma KIC enrichment under these conditions (Ljungqvist et al. 1997), so that the true synthesis rate is probably ~1.8%/d. The latter value is only slightly less than what has been reported for muscle protein synthesis calculated with the flooding dose procedure (1.86%/d using plasma leucine enrichment, 1.95%/d using KIC enrichment) (Garlick et al. 1989). On the other hand, it is difficult to explain why flooding with leucine, valine, phenylalanine, or threonine increases incorporation into muscle proteins of tracers (amino acids not used for flooding) being infused at a low dose, unless high levels of these amino acids somehow alter the rate of protein synthesis (Smith et al. 1998a). Unfortunately, we do not have measures of protein synthesis comparing flooding dose and low-dose tracer methods using directly measured aminoacyl-tRNA enrichments to settle the question.

Whether a flooding dose of an amino acid increases the rate of protein synthesis probably is not too important for studies investigating the effect of some intervention on protein synthesis, unless the intervention has the same effect as a flooding dose. Whether or not the low-dose tracer infusion method causes underestimation of protein synthesis also is not too important as long as experimental treatments do not affect the degree of underestimation. However, the ratio of plasma tracer enrichment (or KIC enrichment) to true intracellular aminoacyl-tRNA enrichment is more likely to be variable with the low-dose tracer. At least in theory, the ratio of plasma to aminoacyl-tRNA enrichment could be altered by factors that influence proteolysis or amino acid uptake. When the body has been flooded with a large dose of tracer, such effects are less likely to be quantitatively significant.

It has been argued that protein synthesis might be increased after a flooding dose by a factor that is proportional to the increase in the intracellular concentration of the amino acid being used to flood the body (Chinkes, Rosenblatt, and Wolfe 1993). However, there is no reason why the rate of protein synthesis should increase in proportion to the increase in the intracellular concentration of a single amino acid. The only exception might be a situation in which the preflooding-dose intracellular concentration is extremely low and rate-limiting for protein synthesis.

With prolonged tracer infusions, proteins that turn over rapidly reach a high level of tracer enrichment early, but have a diminished rate of increase in tracer enrichment as the aminoacyl-tRNA enrichment is approached. Consequently, brief tracer infusions can give higher values for the rate of tissue protein synthesis than prolonged tracer infusions. Because flooding dose studies usually have a shorter period of tracer incorporation, the flooding dose method could result in higher (and more correct) values for tissue protein synthesis than low-dose tracer methods requiring many hours of tracer infusion. When synthesis of individual proteins is being studied, rather than mixtures of different tissue proteins, this issue is not relevant.

Calculations

Fractional Rate of Synthesis

The fractional rate refers to the synthesis rate in relation to the amount of protein already present in the tissue. Implicit in this method of expressing data is the assumption that the amount of the protein is fairly constant during the tracer infusion. Expression of synthesis rates as fractional values is common because very often the mass of the protein is unknown and the absolute synthesis rate cannot be calculated. Expressing data in this manner does not imply that synthesis is regulated to be a constant fraction of the protein mass. When the enrichment of the protein at the end of the tracer infusion is much less than the enrichment of the aminoacyl-tRNA during the tracer infusion, the loss of tracer from the protein due to proteolysis of newly synthesized protein can be ignored. In this case the fractional rate of synthesis (*FRS*) is calculated as follows:

$$FRS = \left(EP_{t2} - EP_{T1}\right) / \int E_{aa-tRNA}$$

where $EP_{t2} - EP_{t1}$ is the change in the tracer enrichment in the protein between time 1 and time 2 and $\int E_{aa\text{-}tRNA}$ is the area under the curve of the aminoacyl-tRNA enrichment[3] (or the index of this enrichment as described above) between time 1 and time 2. If multiple determinations of tracer enrichment in the protein are made between time 1 and time 2, then the slope of the regression line relating *EP* to time can be used to determine the change in *EP* between time points.

If the *FRS* is constant between times 1 and 2, the $E_{aa\text{-}tRNA}$ does not have to be constant, as long as enough time points are measured to define the area under its enrichment curve during this interval. If plasma KIC enrichment is used as an index of leucyl-tRNA enrichment in muscle, for example, it can be measured frequently enough to define the shape of the $E_{aa\text{-}tRNA}$ versus time curve. However, the shape of this curve often cannot be determined precisely because the index of $E_{aa\text{-}tRNA}$ is not easily obtained. For example, if the muscle tissue free amino acid tracer enrichment is used to estimate the $E_{aa\text{-}tRNA}$, usually only two muscle samples will be obtained because of the need to minimize muscle sampling in human subjects. If the initial muscle sample were taken shortly after administration of a priming dose of tracer that either overprimed or underprimed the free amino acid pool in muscle, then the mean value for tissue free amino acid enrichment might not accurately reflect the true area under the curve. The best strategy for designing such experiments is to obtain the tissue or plasma protein samples between time points when the $E_{aa\text{-}tRNA}$ is fairly constant, so that a mean value will accurately reflect the area under the curve.

[3] Some investigators advocate using the tracer to tracee ratio rather than the enrichment (i.e., ratio of tracer to tracee + tracer) of the protein and aminoacyl-tRNA index when using a stable isotopic tracer. However, this method assumes that there is an increase in protein synthesis that is proportional to the increase in the mass of the free amino acid pool caused by the stable tracer. In my opinion, there is not adequate theoretical or experimental evidence to support such an assumption, and I prefer to use the enrichment values. The two methods will deviate significantly only when the tracer enrichment is high, such as during flooding dose experiments.

When the difference in *EP* between two time points is used, the average *FRS* will be correctly calculated even if *FRS* is not constant, as long as $E_{aa\text{-}tRNA}$ is constant. However, if the $E_{aa\text{-}tRNA}$ is changing while the *FRS* is changing, the calculated *FRS* will be incorrect. Specifically, if *FRS* is slowest while $E_{aa\text{-}tRNA}$ is at its highest level, then mean *FRS* will be underestimated. If *FRS* is fastest while $E_{aa\text{-}tRNA}$ is at its highest level, then mean *FRS* will be overestimated.

If only two samples of protein are obtained to measure *FRS*, it is important to determine that isotope incorporation into the protein is not delayed beyond the time at which the initial protein sample is obtained (i.e., tracer incorporation is linear between the two time points). This problem is illustrated in Figure 3.8. With secreted proteins there often is a significant delay between the start of the tracer infusion and the initial appearance of labeled protein in the circulation. This delay is mostly the result of the time required for posttranslational processing before secretion.

In some cases, the turnover rate of the protein is rapid enough that a significant amount of the newly synthesized protein is degraded during the tracer infusion. If a constant $E_{aa\text{-}tRNA}$ is maintained and if protein mass and *FRS*

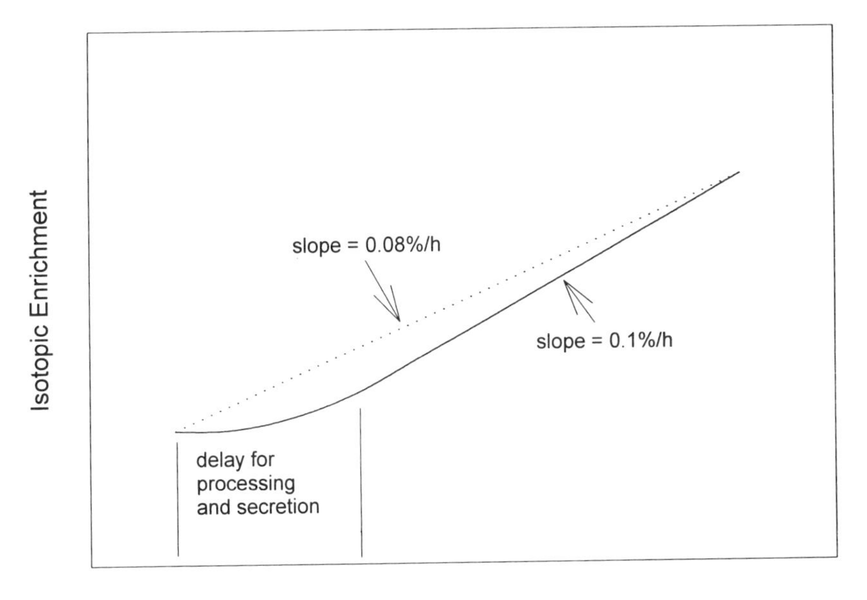

FIGURE 3.8. Illustration of why it is important to define the linear portion of the enrichment versus time curve when determining plasma protein synthesis. After the start of tracer infusion, there can be a significant delay before newly synthesized proteins begin to be exported into the circulation. Thus, isotope incorporation versus time is linear only after the initial delay.

are constant, then EP should approach $E_{aa\text{-}tRNA}$ in a monoexponential fashion. In this case, FRS can be calculated by solving for it in the equation[4]:

$$EP_t = E_{aa\text{-}tRNA} \times (1 - e^{-FRS \times t})$$

where EP_t is protein enrichment at time t. This equation should be used whenever EP is more than ~10% of $E_{aa\text{-}tRNA}$, or otherwise FRS will be underestimated by more than 5% because of loss of tracer from protein degraded during the experiment. This equation generally is used when studying proteins with a rapid turnover. A curve fitting procedure is used to find the value for FRS that produces the closest match between the observed EP_t values and the monoexponential curve produced by using that particular FRS value in the above equation.

When the FRS or $E_{aa\text{-}tRNA}$ are not constant during the experiment, then more complex data sets are needed. In this case, the investigator can sample EP and the $E_{aa\text{-}tRNA}$ more frequently and apply the simple equations to short time intervals. If these intervals are short enough, a close approximation of the FRS during a particular time interval can be determined by using the mean of the $E_{aa\text{-}tRNA}$ values at the start and end of the interval. Readers interested in more details should consult Toffolo et al. (1993).

Absolute Rate of Synthesis

If information is available about the mass of protein present, then the absolute synthesis rate can be calculated as FRS times the protein mass. If the protein mass changes significantly during the tracer infusion, then the term "fractional rate of synthesis" loses its meaning because the denominator is changing. In this case, the calculation of absolute synthesis rate would require a more complex mathematical model and knowledge of the fractional degradation rate (Toffolo, Foster, and Cobelli 1993). For most proteins, or mixtures of proteins, that are examined in studies of human protein synthesis, the total protein mass is large in relation to the amount synthesized or degraded over the course of a few hours of tracer infusion. Hence significant changes in protein mass during the experiment usually do not occur and absolute synthesis can be estimated fairly accurately from fractional synthesis and protein mass.

---

[4] The solution for FRS in this equation is $FRS = (-1/t)\ln[(E_{aa\text{-}tRNA} - EP_t)/E_{aa\text{-}tRNA}]$. This solution has limited usefulness because it applies to a situation in which there is only a single sample of enriched protein. Ideally, if nonlinear tracer incorporation is expected, multiple samples should be taken over time to ensure a monoexponential increase in tracer enrichment. If this equation is used to estimate FRS between two time points at which the protein is enriched with tracer relative to the pre-tracer baseline, t refers to the time difference between these points, EP_t refers to the difference in enrichment, and $E_{aa\text{-}tRNA}$ refers to the difference between $E_{aa\text{-}tRNA}$ and EP at the first time point.

Protein Degradation Rates

If protein mass is constant, then the rate of protein degradation is the same as the rate of protein synthesis. If protein mass is changing, then protein degradation could theoretically be calculated as the synthesis rate minus the change in protein mass. However, estimating degradation in short experiments, using this reasoning, can be inaccurate. In most cases, small errors in the measurement of protein mass lead to large errors in the estimation of protein degradation. For example, suppose that you measure the rate of muscle protein synthesis over a 6 h period and find that the *FRS* is 0.05%/h, or 0.3% over 6 h. Suppose that you know that muscle protein mass is 5 kg, so that the absolute synthesis rate of muscle proteins over 6 h is 15 g. You are unable to detect any change in muscle mass over that 6 h, so you conclude that protein degradation rate also is 15 g/6 h. However, it is unlikely that your estimate of a change in muscle protein mass is any more accurate than about 2% to 3% of the initial value, so that the actual degradation rate could range from zero to more than 100 g/h (although the latter value would indicate an extremely unlikely catabolic state!). You might argue that, over a long period of time, muscle protein mass is fairly constant so that degradation must be approximately equal to synthesis. However, this argument would be valid only to the extent that your measurement of protein synthesis over 6 h is representative of the average protein synthesis over a long period. It is likely that there are periods of imbalance between protein synthesis and degradation over periods of several hours, even though protein mass is constant over periods of weeks and months.

Choice of Tracer for Incorporation Studies

When incorporation of tracer into a pure protein is being studied, the same *FRS* should be obtained regardless of which amino acid is being used as a tracer. In this case, the primary consideration may be how easy it is to estimate $E_{aa\text{-}tRNA}$ with a particular tracer, as discussed above. Another consideration may be the abundance of the amino acid in the proteins of interest. For example, proline is much more abundant in collagen than in most other proteins. Thus, tissue protein synthesis based on a proline tracer would give values weighted more toward the collagen synthesis rate than other tracers would. Because each tracer has a different abundance in the various tissues proteins, and these proteins have different synthesis rates, it follows that different tracers will yield different values for *FRS*. This finding is expected, and obtaining different values with different tracers does not necessarily mean that one of the values is erroneous.

Polyribosome Abundance as a Measure of Protein Synthesis

Polyribosomes are aggregates of ribosomes and their associated proteins bound to mRNA molecules (Chapter 2). Ribosomes or their subunits that are not bound to mRNA are not involved in protein synthesis, whereas those bound to mRNA are synthesizing proteins. Polyribosomes can be separated from free ribosomes and ribosomal subunits by density gradient centrifugation and quantified by absorption of ultraviolet light. Studies applying this method as an index of protein synthesis in human skeletal muscle have been described by Wernerman et al. (1985, 1986). Its main advantage over tracer methods is that no assumptions need to be made about aminoacyl-tRNA tracer enrichment. The primary problem with the method is that it is not quantitative. It would be quite possible to have an increase in protein synthesis without an increase in polyribosome abundance, for example, if both initiation and elongation were increased so that residence time of the ribosome on the mRNA were reduced. A shift in the size distribution of the polyribosomes can be informative about the status of protein synthesis, when combined with tracer data. If initiation rate is increased without an increase in elongation, then polyribosomes should become heavier as more ribosomes accumulate on each mRNA. However, a reduction in the elongation rate without a change in the initiation rate could have the same effect. To date, this method has been applied to humans only for studies of muscle protein synthesis. Unfortunately, it is more difficult to extract polyribosomes from muscle than from most other tissues, and it is possible that monoribosomes and ribosomal subunits are extracted more efficiently than polyribosomes (Waterlow, Garlick, and Millward 1978). Thus, using the ratio of polyribosomes to monoribosomes as an index of protein synthesis potentially problematic, although Wernerman's group has obtained reproducible values for this ratio (Wernerman, von der Decken, and Vinnars 1985, 1986).

3-Methylhistidine Excretion as a Measure of Myofibrillar Proteolysis

Myofibrils are threads of proteins that shorten to produce muscle contractions. Because myofibrils account for about one third of the total protein mass of the body, estimation of the rate of myofibrillar proteolysis is of great interest in studies of protein metabolism. Myofibrils consist of multiple proteins, but most of their mass is comprised of actin and myosin. There is a posttranslational conversion of specific histidine residues in actin and myosin to 3-methylhistidine (Young and Munro 1978). There is one 3-methylhistidine residue per molecule of myosin heavy chain in fast-twitch muscle fibers, but slow-twitch fibers and cardiac myocytes do not methylate histidine in myosin. There is one 3-methylhistidine residue per actin molecule in all types of muscle fibers. Each gram of adult human muscle protein contains ~ 4 μmol of 3-methylhistidine (Ballard and Tomas 1983; Bilmazes et al. 1978b), and each gram of muscle protein in infants contains ~ 3.2 μmol of 3-methylhistidine (Bilmazes et al. 1978b). There is some 3-methylhistidine in tissues other than muscle, because cytoskeletal actins in all types of cells contain 3-methylhistidine. The free 3-methylhistidine produced by proteolysis of myofibrils cannot be used for protein synthesis, and its only route of elimination appears to be excretion in urine. Thus, excretion of 3-methylhistidine is an index of the rate of actin and myosin degradation. Because of the lack of 3-methylhistidine in myosin of slow-twitch fibers, because there are about 3 times more molecules of actin than of myosin heavy chain in muscle (Yates and Greaser 1983), and because actin is present in tissues other than muscle, most of the 3-methylhistidine is in actin rather than myosin.

There are several problems inherent in the use of this method. The major concern is that actin is present in all cells, so that 3-methylhistidine is not derived exclusively from muscle or from myofibrils. Even though myofibrils account for most of the 3-methylhistidine in the body's proteins, proteolysis of nonmuscle actins may account for a disproportionate share of the 3-methylhistidine excretion because myofibrillar protein turnover is slower than the protein turnover of most tissues (Rennie and Millward 1983). Even so, in normal subjects skeletal muscle myofibrillar turnover probably accounts for at least 75% of the total 3-methylhistidine excretion (Ballard and Tomas 1983). When muscle wasting is present, a larger proportion of the 3-methyhistidine can be derived from nonmuscle sources. A practical problem is that 3-methylhistidine is absorbed when meat is digested, so that subjects must be placed on a meat-free diet when 3-methylhistidine is to be used as an index of myofibrillar proteolysis. Clearance of 3-methylhistidine is slow, so 2 to 3 days on a meat-free diet are needed before 3-methylhistidine excretion becomes a meaningful index of myofibrillar degradation.

Markers of Collagen Metabolism

Collagen is the most abundant protein in the body. It accounts for about 20% to 25% of the total protein mass, much of it in the skin, bones, and cartilage (Waterlow, Garlick, and Millward 1978). Collagen comprises, by weight, 4% of the liver, 10% of the lung, 12% to 24% of the aorta, 50% of cartilage, 64% of the cornea, 23% of cortical bone, and 74% of skin (Schultz and Liebman 1997). There are various types of collagen that differ according to tissue distribution, subunit composition, carbohydrate content, and posttranslational modifications (Schultz and Liebman 1997). Collagen contains many 4-hydroxyproline residues, because of a cotranslational enzymatic modification of proline residues and also contains 5-hydroxylysine to which carbohydrates are linked. Collagen is a fibrous protein that contains 3 chains in a triple helix, with covalent crosslinks that impart great stability. Preprocollagen polypeptides are modified in the endoplasmic reticulum and Golgi apparatus of fibroblasts, osteoblasts, or chondroblasts to form procollagen, which is released into the extracellular matrix by secretory vesicles (Glitz 1997). The procollagen is converted to collagen extracellularly, but the cross linkages made by newly synthesized collagen may not be fully established for days or weeks (Waterlow, Garlick, and Millward 1978). Collagen initially is degraded by specific collagenases, then the fragments are further degraded by other proteolytic enzymes. Some fragments are resistant to complete proteolysis and are excreted as peptides in the urine.

Unfortunately, there is no unique product of collagen metabolism that is quantitatively excreted in the urine to provide a measure of collagen turnover that is analogous to using 3-methylhistidine for myofibrillar turnover. Waterlow, Garlick, and Millward (1978) estimated collagen turnover by assuming that 10% of the hydroxyproline resulting from collagen degradation was excreted in the urine (Chapter 4). This estimate is only an educated guess. Hydroxyproline also is derived from collagen present in meat and could represent more than 10% of the total hydroxyproline excretion in persons getting most of their protein from meat (Itoh and Suyama 1996). A number of other serum and urinary markers of collagen turnover have been used (Risteli and Risteli 1997; Zanze et al. 1997), but can be used only as indices of relative differences in collagen turnover rather than as quantitative measures of collagen degradation or synthesis. Markers of type I collagen synthesis and degradation are most often used in studies of bone turnover, but it is important to remember that type I collagen is not restricted to bone.

Radioiodinated Protein Administration to Measure Protein Half-Life

Radioiodine (^{125}I or ^{131}I) can be covalently attached to proteins purified from plasma or produced in vitro. The labeled protein can then be injected into the circulation for measurement of its half-life, volume of distribution, or multicompartmental kinetics. The method is limited to studying metabolism of circulating proteins. Other limitations of the method include the exposure of the subject to radiation and uncertainty about whether or not the iodinated molecules are metabolized exactly the same as the unlabeled protein molecules. In theory, any covalent modification to a protein could be used to trace its metabolism, as long as the modification does not alter the metabolism of the protein and does not occur naturally. A requirement of any such tracers is that the labeled residue cannot be reused for protein synthesis, so that there is no recycling of the label into newly synthesized proteins. This requirement is the reason that amino acids labeled with radioactive or stable isotopes cannot be used in a manner analogous to the radioiodination method.

Analytical Methods

A review of the analytical methods for studying protein metabolism is beyond the scope of this book. The book by Wolfe (1992) provides most of the information needed for using radioactive and stable isotope tracers. Because the term "enrichment" has been mentioned so often in this chapter, it is worth noting the instrumentation required to measure enrichment. With radioactive tracers, enrichment is called "specific activity" and is determined by separate determinations of tracer (scintillation counting) and tracee (various types of assays). With stable isotope tracers, the amounts of tracer and tracee are measured simultaneously by a mass spectrometer. Any instrumental errors in mass spectrometry generally affect the detection of both tracer and tracee to the same extent, so that the enrichment can be determined very precisely.

Traditionally, two types of mass spectrometry have been used—gas isotope ratio mass spectrometry (IRMS) and gas chromatography-mass spectrometry (GC-MS). IRMS can detect very small differences in isotope ratios, on the order of 0.001%. It requires more sample than GC-MS (nanomoles to micromoles) and analyzes pure gases at room temperature. Thus, you must purify the compound of interest, convert it into CO_2 or N_2, and purify the gas before it is introduced into the mass spectrometer. GC-MS can analyze isotope enrichments with subnanomolar quantities, and requires only that the compound can be converted to a volatile, thermostable derivative to be compatible with gas chromatography. The disadvantage is that enrichments of less than 0.1% cannot be reliably detected, and enrichments of less than 0.5% often are not very reproducible. A major problem with measuring small changes in isotope ratios with GC-MS analysis of large molecules is that the background abundance of the heavy isotope can be very high.[5] Use of tracers with multiple heavy atoms can improve the precision of GC-MS determinations of isotopic enrichment, because background enrichment is very low. Isotope enrichments of ~ 0.01% have been measured by GC-MS with such tracers (Patterson et al. 1997b). IRMS has been used for the ^{15}N end product method, and for measuring the isotope enrichment of tissue proteins that have low enrichments. GC-MS has been used to measure whole-body protein turnover with amino acid tracers, and to measure isotope enrichment of rapidly turning over proteins.

Recent developments have blurred the traditional line between IRMS and GC-MS. There are machines that interface a GC with an IRMS, automatically combusting the compounds to CO_2 (or generating N_2 from N in amino acids) on-line. They require much less sample than older IRMS instruments. A disadvantage is that derivatization of the compounds for GC separation adds more unlabeled carbons that lower the ^{13}C enrichment. This problem can be avoided by using ^{15}N tracers, since N is not introduced by derivatization. Moreover, using tracers with multiple ^{13}C atoms reduces the problem of tracer dilution by derivatization.

---

[5]The natural $^{13}C/^{12}C$ ratio is 1.1%, so a compound with 15 carbons has a 15% chance of having one or more ^{13}C atoms. Moreover, many reagents used for derivatizing compounds for GC contain silicon, and the $^{29}Si/^{28}Si$ ratio is about 5%. It is not unusual for the background "enrichment" to be 20% or more, so that detecting a difference of only 0.5% or less is difficult. With IRMS, the background enrichment is only 1.1% for CO_2 and 0.75% for N_2.

4
Normative Data from Infancy to Old Age

Whole-Body N Balance

In infants and children there is a positive N balance associated with growth. A typical infant gains about 7 kg of fat free mass during the first two years after birth (Forbes 1987a), which is equivalent to about 1400 g protein and 230 g N, for an average N balance of about 0.3 g/d. Between the ages of 2 and 10 years, children gain an average of about 15 kg of fat free mass, or about 3 kg protein and 500 g N. The mean N balance is 0.17 g/d with such a gain. From the age of 10 until linear growth has ended, a typical girl will gain about 16 kg of fat free mass, whereas a boy will gain about twice this amount. The average N balance during this period is about 0.3 g/d, with growth ending in females sooner than it does in males. Growth occurs in spurts, so that these average daily N balances do not necessarily reflect what would be observed in an individual child at any particular point in time.

A weight-stable, healthy adult generally gains or loses only a trivial amount of N over periods of several days. There are periods of positive and negative N balance within each day, and there are some days with a positive N balance and some days with a negative N balance. However, if a healthy adult shows a positive or negative N balance that reflects a protein mass change of more than 2% to 3% when body weight is stable, it is more likely that the N balance is in error than that the protein mass of the body has changed significantly. There could be some exceptions to this rule. For example, a low energy diet combined with a weightlifting program might increase muscle protein mass while reducing fat mass, with no significant weight change. Some hormone treatments may increase protein mass while reducing fat mass. But if there are no experimental conditions that might produce such effects, it is safe to assume that N balance is approximately zero as long as body weight is fairly stable.

There is a gradual loss of fat free mass with postmaturational aging, even if total body weight remains constant. Cross-sectional data suggest that between the ages of 25 and 75 there is a 10% to 15% loss of fat free mass (and therefore N), and these data are supported by some longitudinal observations

(Welle et al. 1996b; Forbes 1987a; Aloia et al. 1996). However, the rate of N loss of about 5 g/yr, or less than 0.02 g/d, is too slow to detect by N balance methods.

Protein Oxidation

When N balance is zero, the oxidation of proteins is equal to protein intake. Protein oxidation in this context refers to oxidation not only of the amino acids derived from proteolysis of endogenous proteins, but also the oxidation of amino acids derived from ingested proteins. In a "typical" American this represents a protein oxidation rate of about 1 to 1.5 g protein$\times$kg$^{-1}\times$d^{-1}, but can be half that value or more than twice that value if protein intake is unusually high or low. Protein oxidation can fluctuate significantly over short periods of time, depending on the total energy expenditure and the dietary protein intake. After an overnight fast, the healthy adult is in negative N balance and oxidizes amino acids derived from breakdown of endogenous proteins. A typical rate of protein oxidation under these conditions is about 45 mg$\times$kg$^{-1}\times$h^{-1} in normal-weight subjects, based on total N excretion (Tappy, Owen, and Boden 1988). Not all of the excreted N comes from urea (Tappy, Owen, and Boden 1988), but total N excretion is often used to calculate protein oxidation because some of the N derived from amino acid deamination is excreted as ammonia or is used to form other compounds. Urea production after an overnight fast in normal young adults is about 225 mmol$\times$kg$^{-1}\times$h^{-1} according to dilution of ^{13}C and ^{18}O labeled urea (Wolfe 1992), which translates into a protein oxidation rate of about 40 mg$\times$kg$^{-1}\times$h^{-1}. Protein oxidation increases significantly after a protein-containing meal is consumed, but this effect reflects oxidation of mostly dietary proteins rather than endogenous proteins. In fact, oxidation of endogenous proteins usually decreases after a meal. The role of diet as a determinant of protein oxidation is discussed in Chapter 5.

Infants and children are gaining protein, so you might expect a greater efficiency of dietary protein utilization for synthesis rather than for oxidation. However, infants and children also consume more energy per kg body weight than adults. If the protein concentration of their diet is similar to that of adults, then reduced oxidation of amino acids would not necessarily be required to support growth. Some insight into this issue can be gained from examining the data on leucine oxidation. Although in short term studies there is no guarantee that leucine oxidation reflects total protein oxidation (see Chapter 3), in the long run it would not be possible to have a significant change in leucine oxidation without an accompanying change in the disposal of other amino acids in pathways other than protein synthesis. In infants, the leucine oxidation rate per kg of body weight is similar to that of adults, but about half that of adults when expressed as a fraction of the leucine R_a (Poindexter et al. 1997). Among children and adolescents 7 to 17 years old, the pubertal growth spurt is associated with a reduction, by about one third, in the leucine oxidation rate per kg fat free mass[1] (Beckett, Jahoor, and Copeland 1997).

Healthy older subjects (60–75 years old) oxidize leucine at the same rate as young adults when the smaller fat free mass of older subjects is taken into account (Welle et al. 1993; Welle et al. 1994b). Moreover, older subjects increase their rate of leucine oxidation after meals to the same extent as young adults (Boirie, Gachon, and Beaufrere 1997; Millward et al. 1997).

Whole-Body Protein Turnover

In weight-stable adults, whole-body protein degradation is approximately equal to whole-body protein synthesis over a typical 24 h period. Most investigators who have measured protein turnover over periods of 24 h, or longer, have relied on the [^{15}N]glycine method (Chapter 3). Estimated protein turnover from various studies in adult subjects who were fed a weight-maintaining diet has ranged from about 1.9 $g \times kg^{-1} \times d^{-1}$ to about 5.3 $g \times kg^{-1} \times d^{-1}$. This wide variation reflects small numbers of subjects in most studies, variations in subject characteristics between studies, analytical variations between different laboratories, and in some cases an inadequate amount of time to reach a true plateau of tracer enrichment in the end product. In healthy young adults, with continuous or frequent administration of tracer over 48 h or longer, the average protein turnover is about 3 $g \times kg^{-1} \times d^{-1}$, or about 3.75 $g \times$ kg fat free mass$^{-1} \times d^{-1}$ (Morais et al. 1997; Jeevanandam et al. 1985; Uauy et al. 1978; Young et al. 1975; Winterer et al. 1976). Assuming 200 g protein per kg fat free mass, the fractional whole-body turnover is ~ 1.9%/d according to the [^{15}N]glycine (urea end product) method. There is no difference between men and women when results are adjusted for the lower fat free mass in women. With administration of tracer over many hours, rapidly turning over proteins are highly labeled and their breakdown does not reduce the ^{15}N enrichment of the end product. Thus the calculated protein turnover is somewhat slower than the actual value.

Most studies of protein kinetics based on the R_a of essential amino acids have been done over periods of only a few hours. There are significant variations during the day in the rate of protein synthesis and degradation. Degradation of endogenous proteins is fastest when no food is being absorbed from the gut, whereas protein synthesis is slowest at these times. Table 4.1 shows

[1] Because there is no protein in the triglyceride stores of adipose tissue, which comprise about 85% of adipose tissue, it is preferable to express data per kg of fat free mass (often abbreviated FFM) rather than per kg of total body weight. This is important in comparing women to men, young to old, and overweight to normal-weight, because differences in body weight between these groups do not correspond exactly to differences in fat free mass. However, the protein concentration of the fat free mass is similar in all of these groups. The term lean body mass also is used frequently to denote fat free mass, and these terms are used interchangeably in this book.

TABLE 4.1. Normal postabsorptive amino acid R_a in healthy, weight-stable, adults.

Amino acid	R_a (μmol×kg^{-1}×h^{-1})	g protein×kg^{-1}×d^{-1}
Leucine (plasma KIC enrichment)	115	4.2
Leucine (plasma leucine enrichment)	90	3.3
Valine	60	3.0
Lysine	95	3.9
Phenylalanine	37	3.2

typical values for the plasma R_a of several essential amino acids in postabsorptive subjects, and the estimated protein degradation rate based on the R_a and the amino acid composition of the body (Table 3.1). These R_a values are averages obtained from numerous published studies. The R_a of a particular amino acid is proportional to the concentration of that amino acid in body proteins (Figure 4.1). When proteolysis has been studied with different tracers in the same subjects, the different tracers generally yield similar results, at least within the experimental errors of estimating the R_a and the amino acid composition of the proteins being degraded (Reeds et al. 1992; Tessari et al. 1996b; Hoffer et al. 1997). As discussed in Chapter 3, plasma R_a underestimates the rate of proteolysis in tissues and using plasma KIC enrichment to

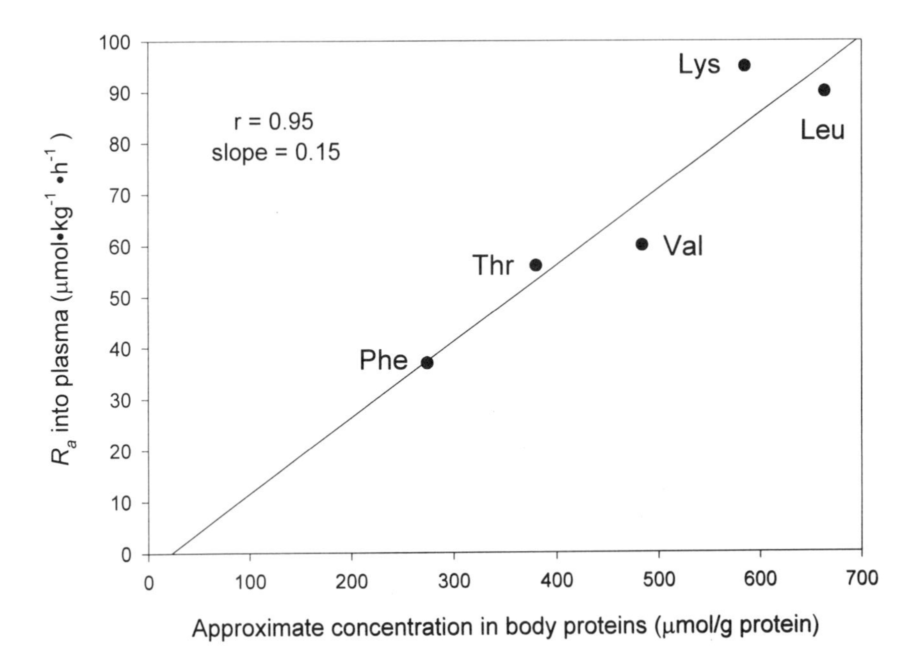

FIGURE 4.1. Relation between concentration of essential amino acid residues in body proteins (Table 3.1) and R_a of these amino acids into the circulation after overnight fasting (Table 4.2).

estimate the rate of leucine production from proteolysis gives a better indication of protein metabolism. This method indicates that postabsorptive proteolysis in normal-weight adults is about 4.2 g×kg^{-1}×d^{-1} (Table 4.1), or about 5.2 g×kg fat free mass^{-1}×d^{-1}. This rate of proteolysis represents a fractional rate of about 2.6%/d. It is not surprising that these values are higher than the value reported above for typical 24 h protein turnover using the [^{15}N]glycine method in fed subjects, because food intake suppresses the rate of proteolysis (see Chapter 5). Even the KIC enrichment method provides only a minimum estimate of proteolysis, because the enrichment of the free leucine in tissues is lower than the enrichment of plasma KIC (Table 3.2). In postabsorptive subjects, the rate of protein synthesis is about 80% of the rate of protein breakdown, based on the difference between leucine appearance from proteolysis and leucine oxidation. However, in fed subjects the rate of protein synthesis is greater than the rate of protein breakdown (Chapter 5).

Whole-body protein turnover, per kg body weight, is faster in infants and children than in adults. According to [^{15}N]glycine studies, protein synthesis is ~3 to 5 times faster in preterm infants than in adults (Young et al. 1975; Van Goudoever et al. 1995). In their second year, children's protein synthesis per kg is twice that of an adult (Young et al. 1975). Consistent with the data obtained from the [^{15}N]glycine method, the R_a of amino acids is faster in infants than in adults. Leucine and phenylalanine R_a both are about twice as fast in 3 d and 3 wk old normal infants than the values given for adults in Table 4.1 (Poindexter et al. 1997). Prepubertal children were reported (Arslanian and Kalhan 1996) to have higher rates of postabsorptive leucine R_a (240 mmol×kg fat free mass^{-1}×h^{-1}) than those reported above for adults (~ 145 mmol×kg fat free mass^{-1}×h^{-1}). Leucine R_a is less (per kg fat free mass) in pubertal adolescents than in prepubertal children (Arslanian and Kalhan 1996; Beckett, Jahoor, and Copeland 1997).

Older adults have a slower protein turnover per kg of body weight than young adults, but this difference disappears or diminishes when the data are adjusted for fat free mass (Morais et al. 1997; Boirie, Gachon. and Beaufrere 1997; Welle et al. 1993; Millward et al. 1997). The slower protein turnover per kg of body weight in the older subjects reflects the fact that older people tend to have more body fat and less active cell mass. Although the stromal component of adipose tissue is involved in protein turnover, most of the mass of adipose tissue is triglyceride that increases body mass without increasing protein turnover. One study did indicate that protein turnover per kg fat free mass was slighltly slower in middle-aged and older subjects than in young adults (Balagopal et al. 1997). Very old, frail women have reduced rates of protein turnover even after adjusting for their reduced fat free mass (Kohrt et al. 1997).

Protein Turnover in Organs and Tissues

Table 4.2 shows the rate of protein synthesis and degradation in various tissues and organs. Because of the more invasive nature of the methods needed

TABLE 4.2. Tissue protein turnover in postabsorptive humans.

Tissue	Method	Synthesis %/d	Breakdown %/d	Reference
Splanchnic	Leu balance/tracer	13[a]	12[a]	Tessari et al. (1996c)
		8[b]	8[b]	Nair et al. (1995)
		15[c]	12[c]	
			26[m]	Gelfand et al. (1988)
Liver	[^{15}N]amino acid incorporation	15[d]		Stein et al. (1978)
	[^{13}C]leu flooding dose	24[k]		Garlick et al. (1991)
	[$^{2}H_{5}$]Phe flooding dose	25[o]		Barle et al. (1997)
Stomach	[^{15}N]amino acid incorporation	25[d]		Stein et al. (1978)
Kidney	Leu balance/tracer	42[a]	61[a]	Tessari et al. (1996c)
Adipose	Phe balance/tracer	5[e]	8[e]	Coppack, Persson and Miles (1996)
Lymphocytes	[$^{2}H_{5}$]Phe flooding dose	7		McNurlan et al. (1996)
Colon	[^{15}N]amino acid incorporation	9[i]		Stein et al. (1978)
Rectal mucosa	[^{13}C]leu incorporation	31[j]		Hartl et al. (1997)
	[^{13}C]leu flooding dose	9		Heys et al. (1992)
Gut mucosa	[^{13}C]leu, [^{13}C]val incorporation	60		Nakshabendi et al. (1995)
	[^{13}C]leu flooding dose	10		Garlick et al. (1991)
	[^{13}C]leu incorporation	36[j]		Gore et al. (1997)
	[^{13}C]leu incorporation	40[l]		Bouteloup-Demange et al. (1998)
	[^{2}H]phe incorporation	88[l]		
Cardiac muscle	Phe balance/tracer	1[g]	6[g]	Young et al. (1991)
		5[h]	5[h]	
Limbs (forearm or leg)	Limb balance/ tracer methods	1.1	1.4	see text
Skeletal muscle	Tracer incorporation	1.2[n]		see text
		1.8[n]		

[a]Subjects with hypertension or cardiac valvular disease.
[b]Insulin-treated type I diabetic patients. My calculations using same assumptions used by Tessari et al. (1996c).
[c]Insulin-withdrawn type I diabetic patients.
[d]Normal tissue in patients undergoing surgery for GI tumors.
[e]My estimate from phe release and uptake, assuming 30g protein/kg adipose tissue and 274 mmol phe/100 g protein (Table 3.1).
[g]Anterior wall of left ventricle in patients with coronary artery disease. Authors did not show assumptions needed to calculate fractional rate from phe release and disposal.
[h]During infusion of branched chain amino acids.
[i]Normal tissue in patients undergoing surgery for GI tumors, mucosa and serosa not separated for analysis.
[j]Normal tissue in patients undergoing surgery for tumors.
[k]Healthy subjects and subjects with colon cancer.
[l]Tissue free amino acid enrichment used to estimate aminoacyl-tRNA enrichment for both tracers.
[m]My calculations using same assumptions used by Tessari et al. (1996c).
[n]Lower value based on plasma KIC enrichment as index of leucyl-tRNA in muscle; higher value based on assumption that plasma KIC enrichment is 68% of leucyl-tRNA enrichment in postabsorptive subjects (Ljungqvist et al. 1997).
[o]During cholecystectomy surgery.

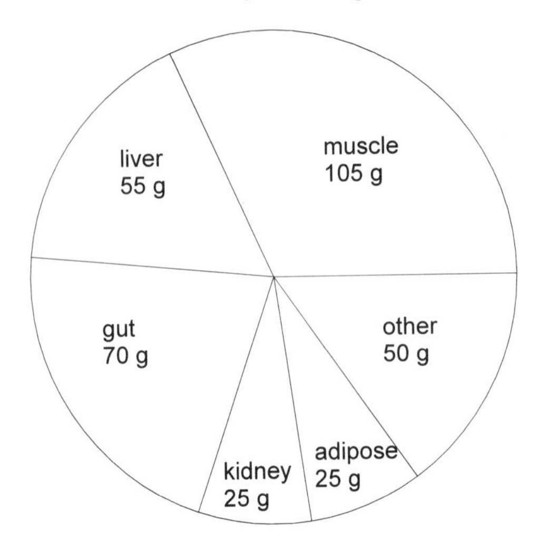

FIGURE 4.2. Estimated contribution of various tissues to total protein synthesis in a normal adult. These values are based on postabsorptive synthesis in a 75 kg man with 60 kg fat free mass. Total synthesis of 330 g/d based on assumption that synthesis is one third higher than the value calculated from plasma KIC enrichment during [^{13}C]leucine infusion (see Table 3.2). For muscle assume 5800 g protein and FRS of 1.8%/d. For liver assume 360 g protein and FRS of 15%/d. For gut, assume 700 g protein and FRS of 10%/d. For kidneys, assume 60 g protein and FRS of 42%/d. For adipose tissue, assume 500g protein and FRS of 5%/d.

for such studies, data are lacking for infants and children. Even in adults, much of the information is based on studies of diseased patients, although most of the data refers to healthy tissues that are not specifically influenced by the disease. However, protein metabolism of cancer patients might be affected by nutritional or hormonal changes associated with the disease, so some caution is needed in generalizing these results to the completely healthy population. Figure 4.2 shows the approximate contribution of selected tissues to the whole-body protein synthesis rate. This figure should be considered as an educated guess rather than a precise summary. The values for the liver and gut do not include secreted proteins.

Visceral and Adipose Tissue Protein Turnover

It is obvious from Table 4.2 that visceral tissues have a higher protein turnover rate than muscle. The high rate of gut mucosal protein turnover reflects not only intrinsic protein turnover, but also the very high rate of cell renewal in the intestinal epithelium (Waterlow, Garlick, and Millward 1978). The mass of intestinal epithelial cells replaced each day accounts for 50 to 60 g of protein synthesis. If intestinal protein mass is 700 g in an average adult male, a turnover rate at the lower end of the estimates for gut mucosa (10%/d) yields a daily protein synthesis of 70 g/d (Figure 4.2). A turnover rate at the high end of the estimates for gut mucosa (88%/d) seems impossibly high. The variable results for gut mucosa may reflect variability in protein turnover in different segments. If gut protein turnover were much higher than the value shown in Figure 4.2, then whole-body protein synthesis would have to be more rapid than values obtained with current tracer models.

Renal and adipose tissue protein turnover each have been measured only in a single study. Renal blood flow is very high, so that arterial-renal venous tracer enrichment differences are small and subject to some error. Thus, the renal value must be interpreted with some caution. Although adipose tissue has a slower fractional protein turnover than visceral organs, its large mass makes it an important contributor to whole-body protein turnover (Figure 4.2). Adipose tissue protein metabolism was measured only in subcutaneous tissue, and it is unclear whether or not visceral fat has a different rate of protein turnover.

Limb Protein Turnover

The value for limb protein turnover in Table 4.2 is a mean value derived from many sources (Gelfand and Barrett 1987; Thompson et al. 1989b; Fryburg et al. 1990; Louard, Barrett and Gelfand 1990; Bennet et al. 1991; Louard et al. 1992; Walker et al. 1993; Moller-Loswick et al. 1994; Newman et al. 1994; Barrett et al. 1995; Biolo, Fleming and Wolfe 1995; Biolo et al. 1995b; Fryburg et al. 1995a; Louard, Barrett and Gelfand 1995; Coppack, Persson and Miles 1996; Tessari et al. 1996e; Biolo et al. 1997). According to the phenylalanine tracer balance method in forearms or legs, the average rate of limb phenylalanine release (an index of proteolysis) is ~50 $nmol \times min^{-1} \times 100$ g $tissue^{-1}$ in healthy, postabsorptive, adult subjects. Phenylalanine disappearance into the limb tissue (an index of protein synthesis) averaged ~75% of the appearance value. There was no significant difference between forearm and leg tissue when data were expressed per volume of tissue. Fractional rates of turnover were calculated by assuming that the bone mineral and fat content of both arms and legs is 35% of the mass (Heymsfield et al. 1990), that protein comprises 20% of the remaining mass, and that limb proteins have the same concentration of phenylalanine as beef flesh (FAO 1970). The limb R_a of leucine (Cheng et al. 1987; Gelfand and Barrett 1987; Gelfand et al. 1988; Fryburg et al. 1990; Louard, Barrett, and Gelfand 1990; Heslin et al. 1992a; Louard, Barrett, and Gelfand 1995; Tessari et al. 1996e; Biolo et al. 1997) is ~2.25 times that of phenylalanine, which is exactly the ratio to be expected from the amino acid composition of muscle proteins (FAO 1970). The values based on these methods should be considered minimal estimates, because intracellular tracer enrichment is lower than tracer enrichment in the venous plasma (see Chapter 3 for discussion of methodological issues).

Although most of the protein synthesis and breakdown in the limbs occurs in muscle, the contribution of skin, fat, and bone should not be ignored. Protein accounts for only ~3% of adipose tissue mass, but these proteins appear to turn over several times faster than muscle proteins (Table 4.2). In someone with a large amount of fat in the limbs, adipose tissue could make a significant contribution to limb protein turnover. In dogs, skin accounts for 10% to 15% of total hindlimb protein turnover (Biolo et al. 1994). In rats, soluble proteins in bone account for about 15% of nonvisceral whole-body protein synthesis, skin proteins account for about 35%, and muscle proteins account for about 50% (Waterlow 1984).

Muscle Protein Turnover

Muscle accounts for only about 30% of the whole-body protein turnover rate, even though it accounts for half or more of the protein mass (Figure 4.2). Tracer incorporation into skeletal muscle proteins has been determined in many studies. The most common method has been to infuse [^{13}C]leucine and to use tracer enrichment of plasma KIC to estimate the enrichment of leucyl-tRNA. When this method is used, the fractional rate of synthesis of mixed muscle proteins in quadriceps muscle (vastus lateralis) of healthy, postabsorptive adults, appears to be ~1.25%/d, based on the average value from several studies (Gibson et al. 1987; Nair, Halliday and Griggs 1988; Halliday et al. 1988; Griggs et al. 1989; Welle, Jozefowicz, and Statt 1990; Carraro et al. 1990b; Essen et al. 1992b; Balagopal, Ljungqvist, and Nair 1997). This estimate is very similar to the estimate based on limb tracer balance studies. However, it has been shown that leucyl-tRNA enrichment is only 68% of the plasma KIC enrichment under these conditions (Ljungqvist et al. 1997), so that a better estimate (shown in Table 4.2) is 1.84%/d (1.25%/d ÷ 0.68). The latter value is similar to the average obtained from five studies that used a flooding dose of [^{13}C]leucine under similar experimental conditions (1.9%/d) (Garlick et al. 1989; McNurlan et al. 1991; Essen et al. 1992a; Smith et al. 1992; McNurlan et al. 1993). A flooding dose of [^{13}C]phenyalanine gives the same result as a flooding dose of [^{13}C]leucine (McNurlan et al. 1991). It has been suggested that a flooding dose of leucine or phenylalanine stimulates muscle protein synthesis (Smith et al. 1992; Smith et al. 1998a), but the analysis above suggests that much of the discrepancy between methods is explained by overestimation of the aminoacyl-tRNA enrichment when a tracer dose is used.

There do not appear to be major differences in the fractional rates of synthesis among different muscle groups. Biceps brachii (Chesley et al. 1992), tibialis anterior (Bennet et al. 1989; Smith et al. 1992), rectus abdominis (McNurlan et al. 1994b), and erector spinae muscles (Watt et al. 1991) all have fractional rates of synthesis within the range of values that have been reported for vastus lateralis.

The very low value for human cardiac muscle protein synthesis (Table 4.2), when subjects were not infused with branched chain amino acids, must be viewed with some skepticism. Because of the very high blood flow in the heart, the arterial-venous differences in amino acid concentrations and tracer enrichments are very small. Small analytical errors can lead to very large errors in calculated protein turnover when blood flow is high. The 95% confidence interval for the mean did not exclude a value as high as 5%/d. Moreover, animal studies all show that cardiac protein turnover is more rapid than skeletal muscle protein turnover (Table 4.3). Thus, it seems more likely that cardiac protein synthesis in postabsorptive humans is closer to 5%/d than 1%/d.

Fractional rates of synthesis of subfractions of skeletal muscle proteins have been determined. We have reported that a crude myofibrillar protein fraction (mostly actin and myosin) has a synthesis rate in postabsorptive young subjects of 1.3 %/d, when plasma KIC enrichment was used as the estimate of leucyl-tRNA enrichment during infusion of [^{13}C]leucine (Welle et al. 1993; Welle, Thornton, and Statt 1995). The water-soluble fraction (mostly sarcoplasmic proteins, but also any tissue protein soluble in water) was only 33% more enriched with the tracer (range 14-76%, n=5) than the myofibrillar fraction. In the first study of human muscle protein synthesis, Halliday and McKeran (1975) infused [^{15}N]lysine and used plasma lysine enrichment to estimate lysyl-tRNA enrichment in muscle. They reported a fractional rate of synthesis of 1.46%/d in myofibrillar proteins and 3.8%/d in sarcoplasmic proteins. Their subjects were fed intravenously during the study. The larger discrepancy between myofibrillar and sarcoplasmic proteins in their study than in ours could be related to different methods of obtaining the fractions, use of a different tracer, greater stimulation of sarcoplasmic protein synthesis than myofibrillar synthesis by intravenous feeding, or some combination of these factors. Some animal studies show very little difference between myofibrillar synthesis rates and sarcoplasmic protein synthesis rates, whereas others show wider discrepancies (Harris 1981).

Mitochondrial proteins in muscle are synthesized at a rate of 1.9%/d in young, postabsorptive subjects, when plasma KIC enrichment is used to estimate leucyl-tRNA enrichment during infusion of [^{13}C]leucine (2.6%/d when enrichment of free leucine in the tissue is used to estimate leucyl-tRNA enrichment) (Rooyackers et al. 1996). The fractional rate of mitochondrial protein synthesis is 95% faster than the fractional rate of synthesis of total muscle proteins. Under the same conditions, the fractional rate of synthesis of myosin heavy chain is 72% of that of total muscle proteins (Balagopal, Ljungqvist, and Nair 1997).

Synthesis of muscle proteins slows with aging. Fractional synthesis in erector spinae muscle was 217% and 73% faster in two teenagers than in the average adult (37–63 years old) (Watt et al. 1991). Men and women in their 60s and 70s have fractional rates of synthesis of total proteins and myofibrillar proteins in vastus lateralis muscle that are ~30% slower than those of

TABLE 4.3. Fractional rates of protein synthesis in vivo in various tissues of mammals.

	Rat (McNurlan, Pain and Garlick 1980)	Rat (Preedy, McNurlan and Garlick 1983)	Rat (Cherel et al. 1991)	Rat (Lewis Kelly and Goldspink 1984)	Rat (Mays, McAnulty and Laurent 1991)	Lamb (Attaix et al. 1988)	Lamb (Davis, Barry and Hughson 1981)	Pig (Baumann et al. 1994)	Pig (Garlick Burk and Swick 1976)	Dog (Biolo et al. 1994)
Liver	105	86	48			115	54	18	23	
Kidney	48								20	
Skeletal muscle	13	17	7	12[c]	16[c]	22	5	1.4	4	8[a] 7[b]
Heart	17		12	12[c]	27[c]		9	7	7	
Brain	17								8	
Jejunal serosa	69									
Jejunal mucosa	143									
Diaphragm			9							
Skin		64[d]	9		16[c]	24	35			13[a] 15[b]
Spleen	76								31	
Stomach	74									
Lung	33				27[c]				18	
Esophageal smooth muscle				25[c]						
Bone marrow		90								

All values expressed as %/d.

[a]Lysine tracer.

[b]Phenylalanine tracer.

[c]Values are given for 2-month-old rats.

[d]Only proteins soluble in dilute NaOH (excludes collagen, elastin, keratin).

subjects in their 20s (Welle et al. 1993; Yarasheski, Zachwieja, and Bier 1993; Welle et al. 1994b; Welle, Thornton, and Statt 1995). Mitochondrial protein synthesis and myosin heavy chain synthesis are ~40% slower in muscles of middle-aged (average age 54 yr) and older subjects (average age 73 yr) than in young adults (average age 24 yr) (Rooyackers et al. 1996; Balagopal et al. 1997). Actin synthesis may not decline with old age as much as myosin heavy chain synthesis does (Yarasheski et al. 1998).

Myofibrillar proteins account for about two thirds of the muscle protein mass. Myofibrillar degradation can be estimated from 3-methylhistidine excretion. The mean 24 h urinary excretion of 3-methylhistidine has ranged in various studies of normal adult subjects from about 110 to 250 μmol/g creatinine, with an average value across studies of about 160 μmol/g creatinine (Tomas, Ballard and Pope 1979; Dohm et al. 1985; Hickson and Hinkelmann 1985; Long et al. 1988; Pivarnik, Hickson, and Wolinsky 1989; Rathmacher, Flakoll, and Nissen 1995; Welle, Thornton, and Statt 1995). Because subjects in these studies were taking a meat free diet, each g of creatinine represents about 20 kg of muscle, or about 2.5 kg of myofibrillar proteins and 4 kg of total muscle proteins. Thus fractional turnover of actin + myosin heavy chain would appear to be about 1%/d, based on a 3-methylhistidine content of 4 μmol/g muscle protein (Chapter 3). Because some of the 3-methylhistidine comes from tissues other than skeletal muscle, the skeletal muscle turnover of actin + myosin heavy chain is probably slightly less than 1%/d. The ratio of 3-methylhistidine excretion to creatinine excretion is highest in childhood, and is fairly stable throughout adulthood (Tomas, Ballard, and Pope 1979). The fractional myofibrillar turnover of an infant is about twice that of an adult, and that of a 10 year old is about 50% faster. Although total 3-methylhistidine excretion is less in older individuals than in young adults, there is no difference per unit of creatinine excretion (Welle, Thornton, and Statt 1995; Uauy et al. 1978), suggesting that fractional turnover of myofibrillar proteins does not decrease with aging. This finding is not consistent with the finding that the fractional rate of synthesis of mixed myofibrillar proteins and of myosin heavy chain is ~30% to 40% slower in the vastus lateralis muscle of older individuals (Welle et al. 1993; Welle et al. 1994b; Welle, Thornton, and Statt 1995Balagopal et al. 1997). Even though there is no reason that synthesis and degradation rates need to match in short term studies, over long periods of time the myofibrillar mass is fairly stable even in older individuals, so that synthesis and degradation must be approximately equal. It is possible that the effect of aging on myofibrillar turnover in vastus lateralis muscle is not representative of the effect on all muscles. Another possible explanation for the apparent discrepancy between age effects on synthesis and degradation of myofibrillar proteins is that nonmuscle sources of 3-methylhistidine might make a larger contribution to 3-methylhistidine excretion in the older subjects.

It is interesting to compare the efflux of 3-methylhistidine from the limbs with the normal values for 24 h 3-methylhistidine excretion. A daily rate of 3-

methylhistidine excretion of 160 μmol/g creatinine corresponds to a rate of about 8 μmol/kg muscle, which corresponds to a rate of about 0.5 μmol/100 g leg or arm tissue per day, or 0.35 $nmol \times min^{-1} \times 100$ g $limb^{-1}$. Mean efflux of 3-methylhistidine from arms or legs of postabsorptive normal subjects generally is at least twice this value (Arfvidsson et al. 1991; Moller-Loswick et al. 1994; Svanberg et al. 1996), although the standard errors of the mean are usually quite large. Thus, it seems that 24 h 3-methylhistidine excretion is significantly less than what would be expected from limb efflux. There are at least two possible explanations for the discrepancy other than measurement errors. One is that limb myofibrillar turnover is greater than that of other muscles. This explanation is unsatisfactory, because the limbs account for about 75% of muscle mass. Even if myofibrillar turnover in the other muscles were zero, the limb efflux in postabsorptive subjects, extrapolated to 24 h, is more than the urinary excretion. The more likely explanation is that feeding significantly reduces the rate of myofibrillar proteolysis. However, neither insulin nor elevated amino acid levels, which are thought to mediate postprandial inhibition of whole body proteolysis (Chapters 5 and 6), have been found to suppress efflux of 3-methylhistidine from the limbs of normal subjects (Svanberg et al. 1996; Moller-Loswick et al. 1994; Arfvidsson et al. 1991). Perhaps the limb balance method is not precise enough to accurately determine the rate of 3-methylhistdine appearance from myofibrillar proteolysis. A possible problem with the limb efflux studies is that 3-methylhistidine concentrations were measured in plasma, whereas whole blood flow was used to calculate efflux. It seems unlikely that 3-methylhistidine concentrations would equilibrate between red blood cells and plasma in the few seconds that it takes for the cells to travel from arterial vessels, through the capillary, and into the venous sampling site. Thus it might be more appropriate to multiply the difference in plasma concentrations of 3-methylhistidine by the plasma flow rather than the blood flow. This calculation would reduce the value obtained for net efflux by ~40%.

The fractional myofibrillar turnover rate based on daily 3-methylhistidine excretion (~1%/d, as discussed above) is slightly less than the estimates of myofibrillar synthesis that we have obtained in healthy young adults (1.3%/d postabsorptive, 2.1%/d fed). Over the long run, these values for degradation and synthesis must match. Several factors may account for the mismatch. One factor is that synthesis was measured in a single muscle, the vastus lateralis. The daily excretion of 3-methylhistidine reflects all muscles. However, all muscles that have been studied in man seem to have fractional synthesis rates similar to vastus lateralis. Another factor is that our crude myofibrillar fraction contained proteins other than actin and myosin, the ones that contain the 3-methylhistidine. Myosin heavy chain synthesis was reported to be only 0.72%/d in postabsorptive subjects (using plasma KIC to estimate leucyl-tRNA enrichment in muscle; 1.06%/d using tissue free leucine as the estimate of precursor enrichment) (Balagopal, Ljungqvist, and Nair 1997). In animals, actin synthesis is slower than myosin heavy chain synthesis, whereas some other myofibrillar proteins have faster turnover rates (Waterlow, Garlick, and Millward 1978). However, there is evidence that turnover of actin in human muscle is more rapid than that of myosin heavy chain (Yarasheski et al. 1998).

Tissue Protein Turnover in Other Mammals

Table 4.3 shows data for in fractional rates of tissue protein synthesis for some other mammals. The pig appears to be quite similar to humans in terms of fractional synthesis rates in muscle and liver. Smaller animals have higher synthesis rates than humans, but relative synthesis rates of the various tissues are generally similar across species.

Turnover of Specific Proteins

Most of the data on turnover of specific protein comes from studies done in young or middle-aged adults. Unless noted otherwise, assume that the values given in this section refer to healthy adults.

Collagen

In spite of the fact that collagen is the most abundant protein, there is little information about its turnover rate in vivo in humans. Perhaps this lack of research reflects the perception that collagen is very inert and is not a very interesting protein from a metabolic standpoint. There is quite a bit of interest in examining collagen degradation products as markers of bone turnover, but these markers do not provide quantitative data on whole-body collagen turnover because of uncertainty about how much collagen must be degraded to produce a certain amount of the degradation products. Another problem with studying collagen turnover is that different types of collagen and different tissues are likely to have different turnover rates. Newly synthesized collagen is more susceptible to degradation than mature collagen that has been fully crosslinked. Thus, there are many problems in studying collagen turnover.

Waterlow, Garlick, and Millward (1978) estimated the turnover rate of collagen in humans by assuming that 10% of the hydroxyproline from collagen degradation is excreted in the urine and that each gram of hydroxyproline represents 7 g collagen. In adults, collagen synthesis should be approximately equal to collagen degradation, but in children collagen synthesis should exceed collagen degradation to allow for growth. Collagen breakdown was estimated to be 1.4 to 3.85 g/d in young adults, which is equivalent to an average fractional turnover rate of only about 0.1%/d. This turnover rate may reflect turnover of procollagen or immature collagen that has not been fully crosslinked, which apparently is more susceptible to degradation, so that turnover of mature collagen fibers may be even slower than 0.1%/d. Older adults have a reduced rate of hydroxyproline excretion. Children under 5 years old excrete as much hydroxyproline as adults, in spite of their much smaller body size, indicating a faster fractional turnover. Children 5 to 15 years old have the highest rates of hydroxyproline excretion on an absolute basis. If these estimates are accurate, collagen turnover accounts for only about 1% of whole-body protein turnover even though it accounts for 20% to 25% of the protein mass of the body.

The fractional synthesis rate of the soluble collagen in human skin has been determined (El-Harake et al. 1998). Collagen that was soluble in 1% Na dodecyl-sulfate had had a fractional synthetic rate of 1.8%/d according to measurements of [^{13}C]hydroxyproline enrichment during [^{13}C]proline infusion. This soluble collagen has not been crosslinked extensively with mature collagen, and probably represents a small fraction of the total skin collagen. No data were presented on the fractional rate of synthesis of the total skin collagen.

Myosin Heavy Chain

Skeletal muscle accounts for about half of body protein mass, myofibrils account for about two thirds of muscle protein mass, and myosin heavy chain accounts for about half of the myofibrillar mass. Thus, myosin heavy chain accounts for about one sixth of the protein mass in the human body. It is the second most abundant protein. Myosin heavy chain synthesis was reported to be 0.72%/d in the vastus laterlais muscle of postabsorptive subjects, when plasma KIC enrichment was used to estimate leucyl-tRNA enrichment in muscle during infusion of [^{13}C]leucine. It was 1.06%/d when tissue free leucine was used as the estimate of precursor enrichment (Balagopal, Ljungqvist, and Nair 1997), which is likely to be a more accurate estimate. Myosin heavy chain synthesis is slower in middle-aged and older adults than in young adults (Balagopal et al. 1997).

Other Myofibrillar Proteins

A preliminary report suggests that the fractional rate of actin synthesis is about twice as rapid as that of myosin heavy chain (Yarasheski et al. 1998). In rabbit and rat skeletal muscle, the specific activity of tracer amino acids in actin is about 30% to 65% of the specific activity of myosin heavy chain in the first few minutes or hours after administration of the tracers, although there was a report that in rat diaphragm incubated in vitro the actin synthesis is more rapid than myosin heavy chain synthesis (Waterlow, Garlick, and Millward 1978; Martin 1981; Bates et al. 1983). In rat skeletal muscle, the ratio of actin to myosin heavy chain specific activities was similar in fed animals, whereas actin specific activity was only 56% of myosin heavy chain specific in fasting animals (Bates et al. 1983).

Because of limitations on the amount of muscle that can be obtained from human subjects, the synthesis rates of muscle proteins that are less abundant (on a mass basis) than myosin heavy chain and actin have not been measured. In animal muscles, tropomyosin synthesis is similar to that of myosin heavy chain, whereas troponin and a-actinin synthesis are more rapid than that of myosin heavy chain (Waterlow, Garlick, and Millward 1978; Martin 1981). Myosin light chain synthesis has been reported to be slower than myosin heavy chain synthesis (Waterlow, Garlick, and Millward 1978; Martin 1981), but some isoforms may have a faster synthesis rate (Morkin et al. 1973). Different troponin isoforms have different synthesis rates in the rat heart (Martin 1981).

Plasma Proteins

Most of the research on the turnover of specific proteins in humans has dealt with plasma proteins. Much of this information is based on decay curves of radioiodinated proteins. Although there is the possibility that iodination of a protein alters its metabolism, generally there is good agreement between this method and results obtained by measuring the incorporation of tracers into proteins. The latter method is hampered by uncertainty about the enrichment of the aminoacyl-tRNA at the site of synthesis of the secreted protein. The liver synthesizes many of the plasma proteins, and ways to estimate aminoacyl-tRNA enrichment in the liver were discussed in Chapter 3. The approximate turnover rates of many plasma proteins are summarized in Table 4.4. The early studies that relied mainly on decay of radioiodinated proteins were reviewed by Waterlow, Garlick, and Millward (1978), and much of the data in Table 4.4 comes from that review and from the data compiled by Pussell et al. (1985) . The rest of this discussion will consider data not included in those summaries.

Albumin

Albumin is the most abundant plasma protein. There are about 4 g albumin per 100 ml plasma in a normal person, accounting for more than half of the total plasma protein mass. There also is a significant extravascular pool of albumin, and plasma albumin accounts for only about 40% of the total body albumin mass (Gersovitz et al. 1980). In brief studies (a few hours), the transfer of labeled albumin to the extravascular compartment is not significant. However, during prolonged tracer administration there is substantial equilibration between plasma and extravascular albumin pools. In this case, the fractional rate of synthesis appears to be much slower than it is in short studies, but the absolute synthesis rate is similar when the extravascular pool is taken into account.

Albumin is synthesized in the liver, so the main problem in determining its synthesis rate is estimating the isotopic enrichment of the aminoacyl -tRNA

TABLE 4.4. Average plasma protein turnover in adult humans.

Protein	Fractional turnover, % of intravenous pool per day	Absolute turnover ($mg \times kg^{-1} \times d^{-1}$)
Albumin	10	160
Transferrin	20	20
Fibrinogen	25	35
Haptoglobin	20	20
Fibronectin	35	6
VLDL apolipoprotein B100	500	8
HDL apolipoprotein A-I	20	11
LDL apolipoprotein B100	40	11
Thyroxine-binding prealbumin	72	10
Thyroxine-binding globulin	20	0.25
Ceruloplasmin	13	0.2
α_1 Antitrypsin (type MM)	38	17
Immunoglobulins: IgG	6	30
IgM	12	5
IgA	20	25
IgD	37	0.4
IgE	89	0.02
Complements: C3	40	20
C1q	67	4
C4	50	11
C5	40	2
B	47	4
H	32	9

in the hepatocytes. One index of hepatic aminoacyl-tRNA enrichment is the plateau enrichment of VLDL-apolipoprotein B100 (see Chapter 3). When normal subjects were fed protein-containing meals, their albumin *FRS* was 13.1%/d when VLDL-apolipoprotein B100 enrichment was used to calculate aminoacyl-tRNA enrichment (Cayol et al. 1997). Under postabsorptive conditions during [^{13}C]leucine infusion, plasma KIC enrichment may be an adequate index of intrahepatic leucyl t-RNA enrichment. Studies using this method have yielded values for postabsorptive albumin *FRS* of 6-11%/d (Ballmer et al. 1990; Olufemi et al. 1990; Lecavalier, De Feo, and Haymond 1991; Cayol et al. 1995; De Feo et al. 1995). The flooding dose method has produced values for mean postabsorptive albumin *FRS* of 5.8-9.9%/d (Ballmer et al. 1990; Ballmer et al. 1995; Hunter et al. 1995). It has been argued that the flooding dose method increases the *FRS* (Smith et al. 1994). When postabsorptive subjects were studied with [^{13}C]valine as the tracer and plasma KIV enrichment as the index of valyl-tRNA enrichment in the liver (analogous to plasma KIC enrichment during [^{13}C]leucine infusion), the albumin *FRS* was 6%/d. When a flooding dose of leucine was given, the value increased to 10.4%/d. However, it is possible that the ratio of plasma KIV enrichment to intrahepatic valyl-tRNA enrichment was altered by the leucine infusion, and that the actual albumin *FRS* did not change in response to the flooding dose of leucine. Albumin synthesis also has been estimated by using the [^{15}N]urea enrichment during [^{15}N]glycine administration as an index of the hepatic arginyl-tRNA enrichment. In young adults with a normal protein intake during a 60 h study, mean daily albumin synthesis was 10% of the intravascular pool (4% of the whole-body albumin pool) according to this method (Gersovitz et al. 1980). These data agree with the radioiodinated albumin decay method. In another study done in postabsorptive women over a 12 h period, the same *FRS* (6%/d) was obtained using the [^{15}N]urea enrichment method and using plasma KIC enrichment to reflect hepatic leucyl-tRNA enrichment (Olufemi et al. 1990).

In summary, although postabsorptive albumin *FRS* generally is less than 10%/d, meals stimulate albumin synthesis (Chapter 5) and there is no reason to dispute the plasma turnover rate of about 10%/d that is based on radioiodinated albumin decay (Table 4.4). Albumin turnover may be slightly slower in the elderly (Gersovitz et al. 1980), although this is not always reported (Fu and Nair 1998). Premature infants have an albumin turnover rate about three times that of young adults, which is consistent with the increased whole-body protein turnover of infants (Yudkoff et al. 1987).

Fibrinogen

Fibrinogen is an abundant plasma protein (2–4 mg/ml) synthesized by the liver. It is converted to fibrin by thrombin activity and is essential for blood clotting. Studies using isotope incorporation methods have yielded values for fibrinogen synthesis of about 20% to 30%/d (Stein et al. 1978; De Feo, Gaisano, and Haymond 1991; De Feo, Horber and Haymond 1992; Cayol et al. 1996), in agreement with earlier estimates based on radioiodinated fibrinogen decay (Table 4.4). In contrast to albumin, there is no difference in fibrinogen synthesis between postabsorptive and fed subjects (De Feo, Horber, and Haymond 1992). Although fibrinogen concentrations tend to increase with age, the fractional synthesis rate of fibrinogen declines by about one third from the age of 25 yr to the age of 70 yr (Fu and Nair 1998).

Fibronectin

Fibronectin is a large glycoprotein that is present on cell surfaces and in plasma. It is an opsonic protein that adheres to fibrin, collagen, actin, and other molecules, assisting in the removal of cellular debris, foreign particulates, and immune complexes by the reticuloendothelial system. Its plasma concentration is about 0.4 mg/ml. A study with radioiodinated fibronectin in normal adults indicated a turnover rate of about 115%/d for plasma fibronectin (Pussell et al. 1985). However, subsequent studies using [^{15}N]glycine infusion, and [^{15}N]hippurate enrichment as the index of hepatic [^{15}N]glycine enrichment, indicated that the *FRS* of fibronectin is only about 35%/d (Thompson et al. 1989a; Carraro, Rosenblatt, and Wolfe 1991). Thus, radioiodinated fibronectin may have a shorter plasma half-life than native fibronectin. Infants have lower plasma fibronectin concentrations than adults, which appears to be the result of reduced fibronectin synthesis (Polin et al. 1989).

Apolipoproteins

Apolipoproteins are essential for the transport and clearance of lipids in the circulation. Most are produced in the liver. Values given in Table 4.4 are the average of the postabsorptive and postprandial data of Cohn et al. (1990). These investigators infused tracer until the VLDL apolipoprotein B100 enrichment reached a steady state, thus ensuring a proper estimate of the aminoacyl-tRNA enrichment in the cells synthesizing the protein. They found that fractional apolipoprotein synthesis was slightly slower in the postabsorptive state than in the fed state. Absolute synthesis of VLDL apolipoprotein B100 was increased during feeding, whereas absolute synthesis of LDL apolipoprotein B100 decreased during feeding. A similar rate of VLDL apolipoprotein B turnover was found in another study using the decay of radioiodinated apolipoprotein (Lewis et al. 1993). *FRS* values for VLDL apolipoprotein B100 of 800 %/d or more have been reported in other studies (Cryer et al. 1986; Parhofer et al. 1991; Arends et al. 1995).

Immunoglobulins

Most of the data on immunoglobulin turnover (Table 4.4) comes from studies of radioiodinated immunoglobulin disappearance over many days. [^{13}C]Leucine incorporation over a few hours in healthy, postabsorptive subjects indicates that IgG synthesis is 2.6%/d, or 12 $mg \times kg^{-1} \times d^{-1}$ (Thornton et al. 1996), which is significantly slower than the value given in Table 4.4. There is some uncertainty about whether or not plasma leucine enrichment reflects the enrichment of leucyl-tRNA in the cells that secrete IgG, but the very rapid rate of synthesis of IgG by these cells suggests that most of the amino acid used for protein synthesis must be of extracellular origin (Thornton et al. 1996). In a similar study with [^{14}C]leucine as the tracer, the *FRS* of IgG was 3%/d when plasma leucine specific activity was used to calculate FRS in postabsorptive subjects (4.9%/d when plasma KIC specific activity was used) (De Feo et al. 1995). IgG synthesis increased by one third after subjects were fed (De Feo et al. 1995).

Hemoglobin

The hemoglobin in an erythrocyte is stable until the erythrocyte is degraded. The mean life span of a human erythrocyte is about 120 days. Thus, fractional turnover is only about 0.8%/d. The hemoglobin mass of an average adult male is about 900 g, so absolute synthesis of hemoglobin must be about 7.5 g/d, or about 100 $mg \times kg^{-1} \times d^{-1}$. Thus hemoglobin synthesis contributes more to whole-body protein synthesis than all other circulating proteins except albumin.

5
Nutritional Influences

The basic aspects of protein synthesis and degradation are considered a subspecialty area of biochemistry and molecular biology. However, more global aspects of bulk protein metabolism in animals, and especially in humans, are more often considered a subspecialty of nutritional sciences. There has been a great deal of research on the nutritional influences on human protein metabolism since the stable isotopic methods became available. Even before then, there was much research on nutritional influences on N balance, especially on the minimum requirements for protein and individual essential amino acids. There is a great deal of overlap between the topic of nutritional effects and that of hormonal and metabolic regulation (Chapters 6 and 7). This chapter describes how what we eat affects protein metabolism, with only brief mention of how hormonal and metabolic signals might be responsible for such effects. The specific effects of these hormonal and metabolic mediators are then described in the following chapters. Nutritional effects will be divided into those associated with each meal, and those associated with changing the diet for periods of a few days or longer. This distinction does not necessarily mean that there are not common mechanisms that mediate short-term and long-term nutritional influences on protein metabolism.

Acute Responses to Meals

In the context of this discussion, the term postabsorptive will refer to situations in which the subjects have not eaten for a few hours, so that circulating amino acids, glucose, and fat are derived from endogenous stores rather than from food being absorbed from the gut. In almost all cases, postabsorptive studies are done after subjects have fasted overnight for 10 to 12 h or more. Studies involving withholding food from the subjects for more than a day will be discussed in the context of responses to fasting. Hence, acute responses to meals refer to responses elicited within the first few hours after food is ingested after a fast of less than a day.

Effect of Meals on Whole-Body Protein Metabolism

A postabsorptive person is in negative protein balance. Even though most of the amino acids derived from protein breakdown are reincorporated into proteins, some are oxidized or removed through other pathways, and therefore protein balance is negative because no amino acids are available from food. After a meal is consumed, protein balance becomes positive, at least if the protein and energy content are adequate. Over a typical day, the negative protein balance of the postabsorptive state is counterbalanced by the positive protein balance of the fed state, if a weight-maintaining diet is consumed. The magnitude of the oscillations between postabsorptive and postprandial protein balance increases as daily protein intake increases (Figure 5.1).

After a protein-containing meal is consumed, protein oxidation rate increases, but the amount of protein oxidized over the first few hours after the meal is less than the amount consumed. The magnitude of the protein accretion after a meal depends on the caloric content and protein content of the meal. The more calories and the more protein, the greater the protein gain. When meals ranging from 800 to 3200 kcal, with 15% of energy as protein, were fed to lean or obese men, the amount of N retained over the subsequent 8 h increased as meal size increased (Owen et al. 1992). The amount of N retained was ~ 40% of the amount of N ingested when 800 kcal meals were consumed and increased progressively to ~ 80% of the ingested amount when

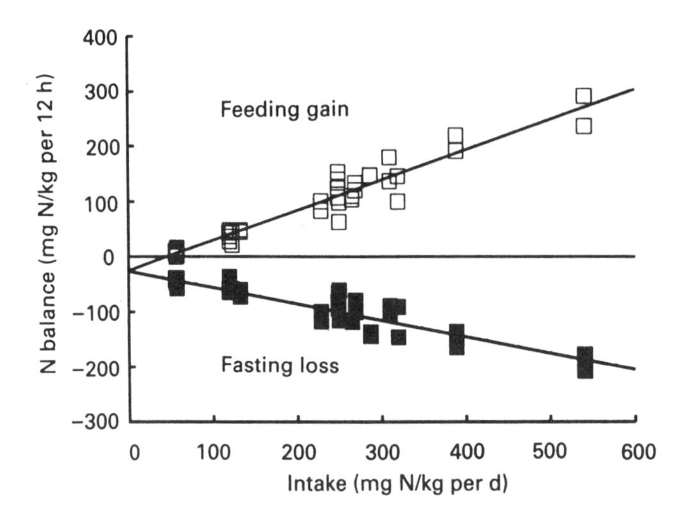

FIGURE 5.1. N balance over 12 hr periods of fasting (closed symbols) and feeding (open symbols) as a function of the daily protein (N) intake. From Price et al. "Nitrogen homeostasis in man influence of protein intake on the amplitude of diurnal cycling of body nitrogen," Clinical Science, Vol 86, pp. 91–102, 1994, permission granted by The Biochemical Society and Portland Press.

3200 kcal meals were consumed. When 1655 kcal were given over 9 h as intermittent small meals, the positive N balance was equivalent to 35% of the ingested N if the meals were 15% protein and was equivalent to 60% of the ingested N if the meals were 70% protein (Robinson et al. 1990). However, with the very high protein intake, the positive N balance probably reflects not only positive protein balance, but also increases in the pools of free amino acids and urea.

The transition from negative protein balance in the postabsorptive state to positive protein balance after meal ingestion could result from suppression of proteolysis with an unchanged rate of protein synthesis, stimulation of protein synthesis with an unchanged rate of proteolysis, or a combination of reduced proteolysis and increased synthesis. There is general agreement that meals inhibit whole-body proteolysis. There is no consensus about whether or not meals stimulate protein synthesis. My interpretation of the literature is that whole-body protein synthesis is increased by meals when both energy and protein intake are adequate.

Proteolysis

The R_a of an essential amino acid, an indicator of whole-body proteolysis in postabsorptive subjects, generally increases when a protein-containing meal is consumed. However, the postprandial R_a reflects not only proteolysis, but also the entry of amino acids from the food being digested in the gut. Proteolysis is determined by subtracting dietary R_a from the total R_a (Chapter 3). Numerous studies have used this method to demonstrate that whole-body proteolysis is slower in fed than in postabsorptive subjects, by up to 65% (Motil et al. 1981b; Hoffer et al. 1985; Melville et al. 1989; Bruce et al. 1990; Biolo et al. 1992a; De Feo, Horber and Haymond 1992; De Feo et al. 1995; Gibson et al. 1996; Boirie, Gachon, and Beaufrere 1997; Forslund et al. 1998). The suppression of proteolysis associated with feeding often is overestimated, because it is assumed that all of the essential amino acids contained in the dietary protein are absorbed as free amino acid. To the extent that the dietary contribution to R_a is overestimated, the suppression of proteolysis by feeding also is overestimated (Figure 5.2). Moreover, there is first-pass splanchnic extraction of the amino acids entering the body from the digestive tract, which causes an underestimation of whole-body R_a after meals (Figure 5.2). This problem also causes an overestimation of the suppression of proteolysis associated with feeding. In studies in which a basal postabsorptive period of several hours is followed by several hours of feeding, the tracer recycling later in the study could lead to some underestimation of $R_{a,}$ and therefore an overestimate of the suppression of proteolysis. These problems cannot completely account for the reduced endogenous R_a of essential amino acids after meals, so there must be some reduction in proteolysis.

Increased insulin and amino acid concentrations account for most, if not all, of the suppression of proteolysis after meals. These factors will be dis-

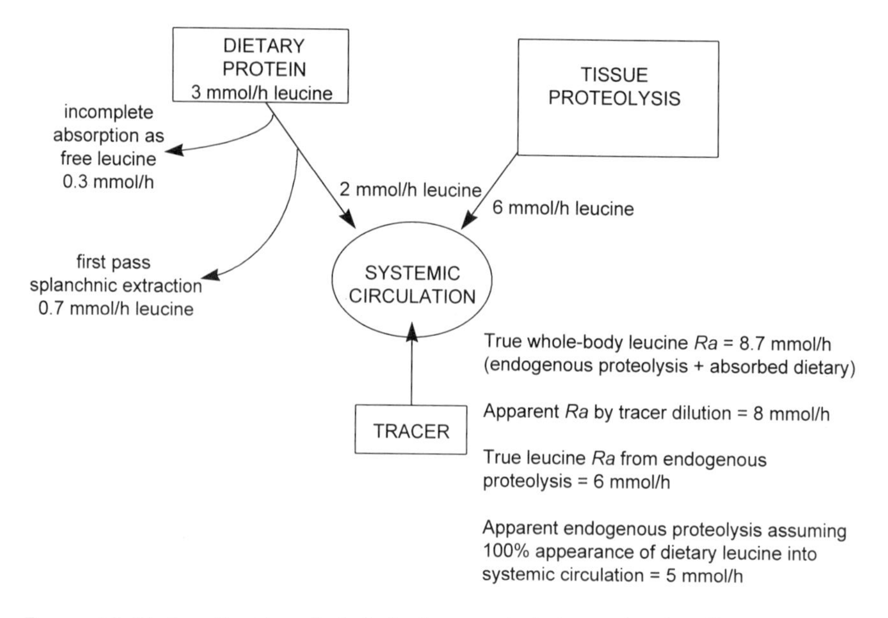

FIGURE 5.2. Underestimation of whole-body proteolysis (tracer leucine dilution method) in a fed subject because of the assumption that all of the leucine in the diet enters the systemic circulation before being oxidized or incorporated into tissue proteins. Magnitude of error depends on the portion of ingested leucine not absorbed and the portion retained in splanchnic tissues.

cussed further in Chapters 6 and 7. Protein-free or low-protein meals, which increase insulin levels but reduce amino acid levels, appear to suppress leucine R_a less than normal or high-protein meals, which increase both insulin and amino acid levels (Gibson et al. 1996; Cayol et al. 1997). However, when a two-pool model was used to estimate splanchnic and extrasplanchnic turnover separately, it appeared that addition of protein to the meal did not suppress proteolysis any more than the protein-free meal (Cayol et al. 1997). A pure protein meal, which raised amino acid levels but caused only a modest increase in insulin levels, produced only a slight, transient suppression of proteolysis (Boirie et al. 1996). The importance of insulin secretion in the suppression of proteolysis was disputed by a study showing that somatostatin did not affect the postprandial suppression of leucine R_a (McHardy et al. 1991). Somatostatin markedly inhibited postprandial insulin secretion, but did not prevent the decrease in whole-body proteolysis even in subjects in whom insulin secretion was almost completely inhibited. Moreover, in insulin-deprived diabetic patients who secreted no insulin after a meal, the postprandial suppression of leucine and phenylalanine R_a was similar to that of nondiabetic control subjects (Biolo et al. 1995a). However, the absence of insulin in these studies was associated with a marked rise in branched-chain amino acid levels, which may account for the suppression of proteolysis in spite of low insulin levels (see Chapter 7).

Protein Synthesis

Some studies have indicated that feeding stimulates whole-body protein synthesis, whereas others have not. Melville et al. (1989) concluded that, based on their own study and research done before 1989, there was no solid evidence that meals increase whole-body protein synthesis in humans, although they cited a few studies that did conclude that meals increase whole-body protein synthesis. Since then, there have been some reports that whole-body protein synthesis is increased by feeding (Tessari et al. 1996e; De Feo, Horber, and Haymond 1992; De Feo et al. 1995) and some that it is not (Bruce et al. 1990; McHardy et al. 1991; Boirie et al. 1996). In my own research, whole-body protein synthesis is consistently greater (~ 25% increase in nonoxidized leucine disposal) in fed subjects than in overnight-fasted subjects (Welle et al. 1994b; Welle and Thornton 1998). Unfortunately, in reviewing the literature I cannot decipher any particular reason (isotope method used, meal size, meal composition) why some studies show a stimulation of protein synthesis and others do not.

Most studies of whole-body protein synthesis have used the nonoxidized portion of leucine R_d as the index of protein synthesis. The first-pass splanchnic extraction of leucine entering the body from food causes an underestimation of whole-body leucine R_a. Because protein synthesis is calculated as the difference between leucine R_a and leucine oxidation, this problem causes an underestimation of protein synthesis during meal absorption. This problem does not explain the discrepancy between studies that do and studies that do not indicate increased protein synthesis during meal absorption, because the same method was used. Nevertheless, some of the studies that showed no effect of meals on protein synthesis may have found a slight increase if they had accounted for leucine that was used for protein synthesis in the first pass through the liver.

There is a difference in tracer dilution between postabsorptive and fed subjects that can cause a slight overestimation of the effect of feeding on protein synthesis. As pointed out in Chapter 3, both plasma leucine enrichment and plasma KIC enrichment are higher than the intracellular leucine enrichment when labeled leucine is infused. This problem causes some underestimation of the true protein synthesis rate. With feeding, the high flux of amino acids from the diet reduces the discrepancy between plasma and intracellular tracer enrichments (Table 3.2), so that protein synthesis is underestimated less during absorption of protein-containing meals. Thus, there is some overestimation of the feeding effect, although the magnitude of the error is minor.

The [^{15}N]glycine end product method does not have the same problems as the [^{13}C]leucine method, but it does have a number of flaws that were discussed in Chapter 3. The [^{15}N]glycine method indicated a 50% increase in the rate of whole-body protein synthesis when subjects were fed, relative to their postabsorptive rate (McNurlan et al. 1987).

An increase in the amino acid concentrations is likely to contribute to any stimulatory effect of meals on protein synthesis. As discussed further in Chapter 7, increasing the amino acid concentrations appears to stimulate protein synthesis even when no other energy source is provided. Whole-body protein synthesis is significantly faster during absorption of protein-containing meals than during absorption of protein-free meals, according to a two-pool (splanchnic and nonsplanchnic) model of leucine kinetics (Cayol et al. 1997). Single-pool models of leucine kinetics also suggest that whole-body protein synthesis is stimulated more as the protein content of the meals increases (Forslund et al. 1998; Gibson et al. 1996).

Effect of Meals on Muscle Protein Metabolism

Protein Balance

There is a negative balance of phenylalanine and many other amino acids in the forearm or leg after an overnight fast. Because phenylalanine cannot be synthesized in muscle, or used for any purpose other than protein synthesis, a negative phenylalanine balance reflects a net loss of protein. Consumption of meals causes a transition to a positive phenylalanine (and therefore protein) balance (Tessari et al. 1996e). This anabolic effect of meals in muscle seems to be mediated by both a suppression of proteolysis and a stimulation of protein synthesis. There is some debate about which of these processes is most important in mediating the protein accretion during meal absorption.

Proteolysis

The method typically used to evaluate proteolysis in muscle involves determination of the R_a of an essential amino acid in the whole forearm or leg. Thus, many studies of "muscle" proteolysis include some contribution of skin, bone, and fat in the limb. However, muscle is likely to account for most of the effects of meals on limb protein metabolism, and for convenience I will equate leg or forearm protein metabolism with muscle protein metabolism.

Most of the research in this field has examined the effects of insulin or of amino acids on muscle proteolysis, rather than the effects of ingesting meals. Although insulin and amino acids may be the primary determinants of the changes in muscle protein metabolism associated with meal ingestion, isolated hyperinsulinemia and isolated hyperaminoacidemia do not necessarily reflect the ordinary postprandial metabolic milieu. The following discussion refers only to investigations using ordinary meals, or infusions of insulin+glucose+amino acids to mimic typical postprandial metabolic changes. There is general agreement that the R_a of phenylalanine, leucine, and tyrosine is suppressed by meals and by insulin+glucose+amino acid infusions. The magnitude of the suppression, relative to postabsorptive values, is ~10-50% (Cheng et al. 1987; Moller-Loswick et al. 1994; Newman et al. 1994; Tessari et al. 1996e). 3-Methylhistidine efflux does not appear to be affected by insulin+glucose+amino acid infusion, suggesting that the effect on limb proteolysis is explained exclusively by nonmyofibrillar proteins (Moller-Loswick et al. 1994). Because of the technical difficulty in obtaining precise arterial-venous differences in 3-methylhistidine concentrations, this conclusion needs to be verified.

After a meal there is increased influx of amino acids into muscle fibers, and therefore the ratio of plasma tracer enrichment to intracellular tracer enrichment is closer to unity. Because the intracellular enrichment reflects tracer dilution at the site of proteolysis, it seems likely that muscle proteolysis is underestimated more in postabsorptive subjects than in fed subjects. Thus, the suppression of muscle proteolysis may be underestimated by tracer dilution methods.

Protein Synthesis

Two methods have been used to study the effects of meals on muscle protein synthesis—the direct isotope incorporation method and the limb tracer balance method. As discussed in Chapter 3, both methods suffer from uncertainty about the tracer enrichment of the aminoacyl-tRNA, the immediate precursor for protein synthesis. Moreover, the limb tracer balance method has the problem of having some contribution of tissues other than muscle. In spite of these problems, there is much evidence that meals increase muscle protein synthesis.

Direct isotope incorporation methods suggest that meals can increase muscle protein synthesis by 10% to 100%. The first study of this kind in humans indicated that muscle protein synthesis increased by 100% when subjects were fed 3.3 kcal$\times$kg$^{-1}\times$h^{-1} (Rennie et al. 1982). A reassessment of this work with a larger number of subjects indicated an increase of about 60% (Halliday et al. 1988). My research indicated that myofibrillar synthesis increased about 50% when subjects were fed frequent small meals (Welle et al. 1994b). These studies used plasma KIC enrichment as the index of intramuscular leucyl-tRNA enrichment during [^{13}C]leucine infusion. The problem with this method is that the meal alters the ratio of intramuscular leucyl-tRNA enrichment to plasma KIC enrichment (Ljungqvist et al. 1997). Thus, the stimulatory effect of feeding is overestimated. An estimate of the degree of overestimation can be made using the tissue free leucine enrichment, which is very close to the leucyl-tRNA enrichment (Ljungqvist et al. 1997). Retrospectively applying the data of Ljungqvist et al. (Ljungqvist et al. 1997) to estimate the tissue free leucine enrichment from plasma KIC in postabsorptive subjects, and our own tissue free leucine enrichment data during feeding, it appears that meals containing ~ 2.6 kcal$\times$kg$^{-1}\times$h^{-1} increase myofibrillar protein synthesis by ~ 25% (Figure 5.3). The protein concentration of the diet did not affect the rate of protein synthesis, over the range of 7% of energy (half the concentration in the typical U.S. diet) to 28% (twice the usual concentration) (Figure 5.3). When meals containing only half as many calories

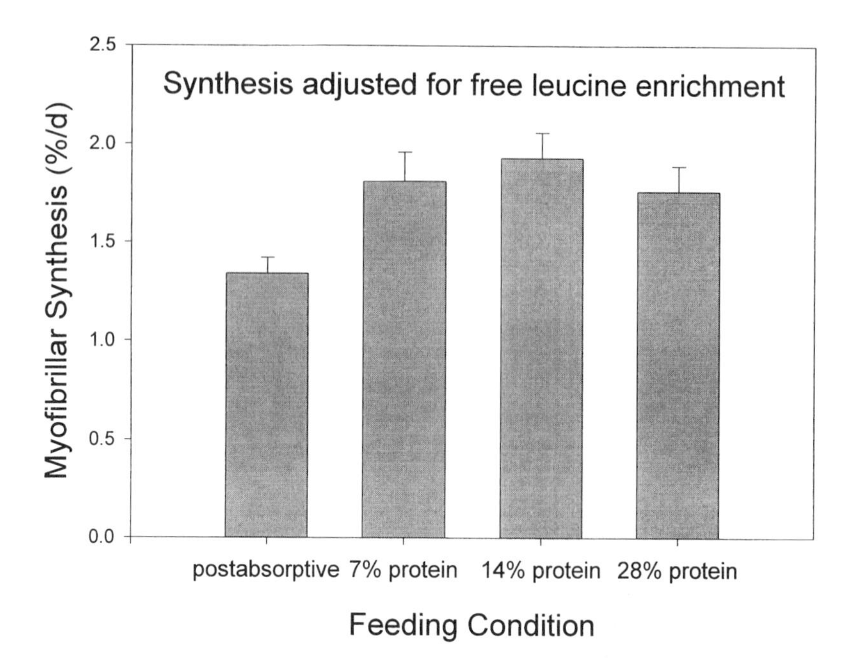

FIGURE 5.3. Stimulation of myofibrillar protein synthesis by meals. Fed subjects received small meals every 30 minutes throughout the protein synthesis determination. Assumed that postabsorptive enrichment of free leucine in muscle is 68% of plasma KIC enrichment (Table 3.2). Postprandial enrichments of free leucine in muscle were directly measured (Welle and Thornton 1998). The stimulation by meals was not affected by the protein concentration of the meals, over the range of 7% to 28% of energy.

were consumed, the increase in muscle protein synthesis, calculated by direct measurement of muscle leucyl-tRNA enrichment, was only 9% and not statistically significant (Ljungqvist et al. 1996).

Another study attempted to circumvent the problem of uncertain aminoacyl-tRNA enrichment by using the phenylalanine flooding dose method (McNurlan et al. 1993). Subjects were studied after an overnight fast, then after consuming hourly meals (~ 3.3 kcal×kg^{-1}×h^{-1}). After the first hour of feeding there was no change in muscle protein synthesis. After 10 h of feeding, the muscle protein synthesis was increased an average of 27%, but the response was quite variable (-27% to + 108%). The values for muscle protein synthesis were quite variable at all time points, suggesting that the method used to measure protein synthesis was not precise enough to give a clear answer to the question of whether or not food intake alters muscle protein synthesis.

Tracer balance studies of the forearm and leg also have examined the effects of meals on muscle protein synthesis. Subjects fed ~ 3.75 kcal×kg^{-1}×h^{-1} had a forearm phenylalanine R_d that was three times greater than that of postabsorptive subjects (Tessari et al. 1996e). However, the variability of the postprandial R_d was very high, and measurements based on plasma enrichments rather than whole-blood enrichments indicated a stimulation by meals of only 31%. A method that accounts for all the routes of leucine disposal in the forearm indicated that incorporation into proteins increased an average of 81% when subjects were fed (Cheng et al. 1987). Intravenous infusion of insulin+glucose+amino acids, which mimics many of the metabolic effects of meal ingestion, increased forearm protein synthesis by 68% in one study (Newman et al. 1994), but reduced forearm and leg protein synthesis by ~ 35% in another study (Moller-Loswick et al. 1994). However, the latter study had somewhat higher insulin levels than those usually observed after ordinary meals, and amino acid administration was not sufficient to increase the amino acid concentrations relative to postabsorptive levels. Hence, the limb tracer balance method also supports the conclusion that meals increase muscle protein synthesis. This method suffers from the same uncertainty about aminoacyl-tRNA enrichment as the direct incorporation method, so that the magnitude of the stimulatory effect probably is overestimated.

Effect of Meals on Protein Metabolism in Other Tissues

Relatively little is known about the effects of meals on protein metabolism in other tissues in humans. Because the splanchnic tissues account for a large proportion of whole-body protein metabolism and because they are exposed to the highest concentrations of nutrients during meal absorption, you might expect that changes in splanchnic protein metabolism account for much of the whole-body response to meals. The best way to study overall splanchnic protein metabolism is to measure tracer and tracee balance across the splanchnic bed, which requires catheterization of the hepatic vein. A further difficulty of this method is that all routes of amino acid disposal and appearance must be accounted for. To my knowledge, there have not yet been any studies examining the effects of ordinary meals on splanchnic protein metabolism using such technology. When amino acids are infused intravenously, splanchnic amino acid balance becomes much more positive and splanchnic proteolysis drops sharply (Gelfand et al. 1986; Gelfand et al. 1988). Net splanchnic uptake accounted for 70% of whole-body amino acid net uptake. If all of the increase in amino acid uptake were used for protein synthesis, then splanchnic protein synthesis could have approximately doubled during amino acid infusion. However, amino acids are used in pathways other than protein synthesis, so it is not possible to calculate the increase in protein synthesis from these data. Moreover, infusion of amino acids intravenously does not necessarily reflect the effects of enteral feeding. During meal absorption, first-pass splanchnic extraction of the essential amino acid leucine is ~ 25%, and that of phenylalanine is ~ 50% (Hoerr et al. 1993; Biolo et al. 1992a). These data cannot be used to quantify protein synthesis rates because not all of the leucine and phenylalanine taken up by the splanchnic tissues is used for protein synthesis, and because first-pass extraction does not account for all of the splanchnic amino acid uptake. Cayol et al. (Cayol et al. 1997) used a two-pool model of splanchnic and nonsplanchnic leucine metabolism to compare protein-free with protein-containing meals. They did not place a catheter in the hepatic vein, but used enrichment of VLDL apolipoprotein B100 to estimate tracer dilution in the splanchnic tissues (which is problematic because this protein is formed in the liver but not the gut). They found that splanchnic leucine uptake during feeding of the protein-containing meals was twice that observed during feeding of protein-free meals. Again, the relation between leucine retention and protein synthesis is uncertain because of leucine deamination and oxidation. Splanchnic proteolysis did not appear to be influenced by the protein in the meals, but no comparison was made with the postabsorptive state.

A number of studies have examined the effect of feeding on the fractional rate of synthesis of albumin, which accounts for most of the protein secreted by the liver. Oral feeding, or combined infusion of insulin+glucose+amino acids increases albumin synthesis (Lecavalier, De Feo, and Haymond 1991; De Feo, Horber, and Haymond 1992; De Feo et al. 1995; Hunter et al. 1995). Most of these studies used the plasma KIC enrichment to estimate hepatic leucyl-tRNA enrichment during [^{13}C]leucine infusion, so there is some uncertainty about the magnitude of the feeding effect because feeding could have changed the ratio of plasma KIC enrichment to hepatic leucyl-tRNA enrichment. The aminoacyl-tRNA enrichment is less likely to be affected by feeding with the phenylalanine flooding dose method, which indicated a 25% increase in albumin synthesis after feeding (Hunter et al. 1995). Albumin synthesis is significantly slower during feeding of protein-free meals than during feeding of protein-containing meals (Cayol et al. 1996; Cayol et al. 1997).

It has been suggested that the increase in albumin synthesis accounts for ~ 20% of the increase in whole-body protein synthesis associated with meals (De Feo, Horber, and Haymond 1992), although this estimate may be inaccurate because the index of hepatic leucyl-tRNA (plasma KIC) is probably not precise. The increase in albumin synthesis determined with the flooding dose method (Hunter et al. 1995) suggests that the stimulation of albumin synthesis accounts for less than 10% of the increase in whole-body protein synthesis.

Meals have no acute effect on fibrinogen synthesis, VLDL apolipoprotein B100 synthesis, and gut mucosal protein synthesis (Cohn et al. 1990; De Feo, Horber, and Haymond 1992; De Feo et al. 1995; Bouteloup-Demange et al. 1998).

Acute Effect of Ethanol on Protein Metabolism

In some individuals, ethanol comprises a significant proportion of total energy intake, often more than the energy provided by protein. In alcoholism, ethanol may provide more energy than all other nutrients. When given alone, ethanol slightly increases whole-body protein breakdown (leucine R_a) and synthesis (nonoxidized leucine R_d), and slightly reduces leucine oxidation (Berneis, Ninnis, and Keller 1997). Although leucine oxidation does not necessarily reflect overall protein oxidation, leucine oxidation does reflect the difference between protein synthesis and breakdown when there is no dietary leucine available. Thus, ethanol has a modest protein-sparing effect acutely, at the whole-body level. When the same amount of ethanol was given during insulin+glucose+amino acids, which mimics the postprandial state, no detectable effect of ethanol on protein metabolism was observed (Berneis, Ninnis, and Keller 1997). Adding ethanol to a mixed meal does not appear to affect whole-body protein breakdown or synthesis, but it reduces the synthesis of both albumin and fibrinogen in a dose-dependent fashion (Volpi et al. 1997; De Feo et al. 1995). The inhibition of hepatic protein synthesis may be related to a more reduced intracellular redox status (Volpi et al. 1998). In rats, large doses of ethanol suppress protein synthesis in muscle, intestine, skin, bone, and heart (Preedy et al. 1994; Preedy et al. 1996; Reilly et al. 1997). Addition of ethanol to a meal reduces leucine oxidation (De Feo et al. 1995), but it is unclear whether or not ethanol is more or less effective than other macronutrients in this regard. With chronic alcohol abuse, there is a reduction in muscle protein synthesis (Chapter 9), but this effect may be related to chronic malnutrition or toxic effects of ethanol, and may not reflect the acute response to modest amounts of ethanol.

Starvation

Cessation of food intake for more than a day has profound effects on protein, carbohydrate, and fat metabolism, all of which are interrelated. The most urgent problem during starvation is maintaining an adequate supply of glucose or ketone bodies for the brain, which does not use other nutrients for energy. Early in starvation most of the glucose comes from glycogen stored in the liver. Glycogen stored in muscle cannot be released as glucose into the circulation (because of the lack of glucose-6-phosphatase), but lactate formed from glycolysis of muscle glycogen can be converted to circulating glucose by the liver. Gluconeogenesis from amino acids derived from proteolysis is a relatively minor source of glucose early in starvation. As starvation continues, glycogen stores are depleted rapidly and the brain relies on glucose derived from the carbon skeletons of amino acids that are derived from protein breakdown, as well as from the glycerol released by lipolysis. Life ends when the protein stores of the body decrease by ~ 50%.

Protein balance always is negative during starvation, because there is no protein intake and there is always some obligatory protein oxidation. Daily protein oxidation, as reflected by total N excretion, decreases sharply immediately after starvation begins, mainly because no dietary protein is being oxidized (Owen et al. 1969). Compared with the early postabsorptive rate of protein oxidation, there is not much change in whole-body protein oxidation for the first few days after starvation begins (Gelfand and Sherwin 1986; Nair et al. 1987c; Hoffer and Forse 1990; Owen et al. 1998). Leucine oxidation increases after 2 to 4 days of fasting (Nair et al. 1987c; Frexes-Steed et al.

1990; Fryburg et al. 1990; Carlson, Snead, and Campbell 1994), compared with the early postabsorptive leucine oxidation. Thus, branched chain amino acid oxidation may be selectively increased early in starvation. After about a week of starvation, N excretion progressively diminishes until stabilizing at its minimal value of ~ 5 g/d in a typical obese adult (Owen et al. 1969; Hoffer and Forse 1990; Owen et al. 1998). Part of this minimal rate of protein oxidation appears to be related to the requirement for renal NH_3^+ excretion to maintain pH while ketoacid production is elevated (Hannaford et al. 1982).

These changes in protein oxidation seem to be mediated, at least in part, by parallel changes in whole-body protein breakdown. After 2 to 4 days of starvation, the leucine R_a is increased by 20% to 50% relative to the early postabsorptive rate (Nair et al. 1987c; Jensen et al. 1988; Frexes-Steed et al. 1990; Fryburg et al. 1990; Carlson, Snead, and Campbell 1994). After 1 to 3 weeks of starvation in obese patients, leucine R_a is reduced ~ 10-30% (Henson and Heber 1983; Nair, Halliday, and Griggs 1988; Hoffer and Forse 1990; Vazquez, Morse, and Adibi 1985). It is unclear whether or not lean subjects would have such a decrease in proteolysis after more prolonged fasting, because they have less body fat to supply energy.

Forearm R_a of leucine and phenylalanine increases significantly after 60 h of fasting compared to values after overnight fasting (Fryburg et al. 1990). If muscle accounts for all of the increase in forearm proteolysis associated with starvation, and forearm muscle is representative of all muscle, then muscle could account for ~ 90% of the whole-body increase in leucine R_a. In rats, the increase in muscle proteolysis after a day of starvation is much greater in myofibrillar than in nonmyofibrillar proteins, which is the result of increased ubiquitination and proteasomal degradation of myofibrils (Wing, and Goldberg 1995; Wing, Haas, and Goldberg 1995; Medina, Lowell, Ruderman, and Goodman 1986). The decrease in whole-body proteolysis after the first week of starvation is associated with a decrease in muscle proteolysis. Whereas there is an increase in muscle efflux of amino acids after 2 days of starvation (Fryburg et al. 1990), there is a reduction in forearm and leg amino acid efflux after 10 days of starvation (Albert et al. 1986). There is an initial increase in 3-methylhistidine excretion (an index of myofibrillar proteolysis) during the first 2 to 3 days of fasting (Giesecke et al. 1989), but after 3 weeks of starvation, urinary excretion of 3-methylhistidine is reduced to the same extent as whole-body proteolysis (Hoffer and Forse 1990).

During starvation, a greater fraction of the leucine coming from proteolysis is oxidized rather than reincorporated into proteins. Thus, the whole-body protein synthesis, as reflected by the nonoxidized portion of leucine R_d, increases substantially less than the increase in R_a after 2 to 4 days of starvation (Nair et al. 1987c; Jensen et al. 1988; Frexes-Steed et al. 1990; Fryburg et al. 1990; Carlson, Snead, and Campbell 1994). After 3 weeks of starvation, the whole-body protein synthesis rate is reduced (Hoffer and Forse 1990). Muscle protein synthesis declines even in the early stages of starvation (2 to 4 d), as reflected by stable isotope incorporation studies and polyribosome analyses

(Wernerman, von der Decken, and Vinnars 1985; Essen et al. 1992a), although forearm phenylalanine R_d was unchanged after only 60 h of fasting (Fryburg et al. 1990). Muscle protein synthesis after more prolonged starvation has not been determined, but is likely to decrease further.

The effect of starvation on protein metabolism in tissues other than muscle has not been examined in humans, but has been studied in animals. Most of the research has been done in rats, and it is important to consider that a day of starvation in a rat is much more of a metabolic challenge than a day of starvation in a human, because of the much faster metabolic rate and protein turnover in the rat. The animal studies generally indicate that starvation reduces the fractional rate of synthesis at least modestly in all of the tissues that have been studied (brain, heart, gut, liver, kidney, skeletal muscle, skin), but that skeletal muscle protein synthesis tends to decline more than that of other tissues (Millward 1970; Garlick et al. 1975; McNurlan, Fern, and Garlick 1982; Cherel et al. 1991). The response varies somewhat from one muscle to another (Baillie and Garlick 1991a), and starvation inhibits synthesis of myofibrillar proteins more than synthesis of sarcoplasmic proteins (Millward 1970; Bates et al. 1983). Older rats decrease their muscle protein synthesis in response to starvation much less than young rats (Baillie and Garlick 1991b).

The increase in proteolysis and reduction in protein synthesis early in starvation may be mediated to a large extent by the decline in insulin levels and by acidosis. The decrease in proteolysis later in starvation reflects the diminished need for amino acids for energy (because of an overall decrease in metabolic rate and increase in fat oxidation) without an increased demand for amino acids for gluconeogenesis (because of increased use of glycerol for gluconeogenesis and reduced oxidation of glucose by the body). Increased free fatty acid levels and reduced triiodothyronine (T_3) production may be involved in the reduction in proteolysis from early starvation to long-term starvation. The further reduction in protein synthesis could be mediated by reduced T_3 and reduced amino acid levels, and possibly by an increased sensitivity to the protein-sparing actions of insulin. However, Nair et al. (1989) were unable to reverse the decrease in whole-body protein turnover with short-term T_3 replacement after 2 to 3 weeks of fasting in obese patients. This finding does not rule out the possibility that starvation reduces sensitivity to the effect of T_3 on protein metabolism.

Very Low Energy Diets

Starvation leads to rapid loss of body fat, which is beneficial to obese patients. However, it also leads to a significant loss of body proteins, which is a health risk and which slows the metabolic rate because of the decrease in active cell mass. Therefore, obese patients are often placed on very low energy diets (less than 30% of the energy requirement for weight maintenance) that have a normal supply of protein but very little carbohydrate and fat.

Obese patients can survive on such diets for many weeks if they contain adequate high-quality proteins and micronutrients.

The loss of body proteins is much less on very low energy diets than with total starvation. Hence, these diets are sometimes referred to as "protein sparing" diets. Generally, negative N balances of ~ 1-2 g/d (6 to 12 g protein/d) can be eventually be achieved with long term administration of such diets (DeHaven et al. 1980; Fisler et al. 1982; Pasquali et al. 1992; Davies et al. 1989), which is much less than the minimum N loss of ~ 5 g/d (30 g protein/d) during total starvation. There have even been some claims of positive N balance on such diets (Apfelbaum 1976; Amatruda et al. 1983; Hoffer et al. 1984a; Vazquez, Morse, and Adibi 1985), but these have not been verified with independent body composition analyses. Positive N balance is very unlikely as long as subjects are losing weight, but the loss of some protein during weight loss in obese subjects is not necessarily harmful because lean body mass is elevated in obesity (Forbes 1987a). When protein is provided by the diet, less endogenous protein breakdown is needed for energy and gluconeogenesis than is needed during starvation. Thus, a larger fraction of the amino acids derived from proteolysis can be retained via protein synthesis. Substitution of some carbohydrate for protein in isocaloric very low energy diets generally results in somewhat less sparing of body proteins (DeHaven et al. 1980; Hoffer et al. 1984a; Vazquez, Morse, and Adibi 1985; Pasquali, Casimirri, and Melchionda 1987; Piatti et al. 1994). Even though protein balance is less negative during consumption of very low energy diets than it is during total starvation, the daily protein oxidation is greater because much of the dietary protein is oxidized.

Adaptation to very low energy diets includes a reduction in the rate of whole-body proteolysis and protein synthesis. According to measurements made over 60 h with the [^{15}N]glycine end product method, both whole-body protein degradation and synthesis were reduced ~ 30% after 3 weeks of a very low energy diet (Bistrian, Sherman, and Young 1981). Another study verified this finding with a shorter (14 hr) [^{15}N]glycine protocol (NH_3 end product method) when the low energy diet was protein-free, but not when it contained adequate protein (Garlick, Clugston and Waterlow 1980). The leucine R_a and nonoxidized R_d decline 10% to 30% after 1 to 3 weeks on various very low energy diets of 350 to 800 kcal/d, indicating that both whole-body protein breakdown and whole-body protein synthesis are slower (Garlick, Clugston, and Waterlow 1980; Hoffer et al. 1984a; Vazquez, Morse, and Adibi 1985; Hendler and Bonde 1990; Vazquez and Adibi 1992). The combination of energy and protein deficiency appears to inhibit protein turnover more than energy deficiency alone (Garlick, Clugston, and Waterlow 1980). Extending the very low energy diet from 3 to 8 weeks does not diminish whole-body protein turnover any further in obese patients (Hoffer et al. 1984a).

The excretion of 3-methylhistidine has been used to determine the effect of very low energy diets on myofibrillar protein breakdown. Reductions in 3-methylhistidine excretion of ~ 25% to 50% occur after consumption of very low energy diets for several weeks (Marliss, Murray, and Nakhooda 1978; Hoffer et al. 1984b; Pasquali, Casimirri, and Melchionda 1987; Piatti et al. 1994), although not all investigators have reported this (Garlick, Clugston, and Waterlow 1980). 3-Methylhistidine excretion does not decline as much if protein intake is inadequate (Pasquali, Casimirri and Melchionda 1987; Piatti et al. 1994). It is important to remember that some of the 3-methylhistidine comes from turnover of actin in tissues other than skeletal muscle, and that 3-methylhistidine is present in only two muscle proteins, so that changes in 3-methylhistidine excretion do not necessarily reflect changes in overall muscle protein degradation.

In summary, the metabolic adaptations to very low energy diets are similar to the adaptations to long-term total starvation. The reduction in protein degradation serves to spare body proteins under both circumstances, but the provision of dietary protein with the very low energy diets results in much less protein wasting. The slowing of protein synthesis may be secondary to the reduction in proteolysis. Slower protein synthesis contributes to the decline in energy expenditure, and allows more of the amino acids coming from the diet and from protein breakdown to be used for energy and glucose production.

Malnutrition

There are many forms of malnutrition. The term refers to any diet that is incompatible with optimal health. Even overeating, which leads to obesity and its attendant problems, can be considered malnutrition. However, the current discussion will be limited to situations in which energy intake, protein intake, or both energy and protein intake are below the minimum required for good health. One problem in assessing the effects of malnutrition on protein metabolism is that malnutrition often is the result or cause of other diseases that may have independent effects on protein metabolism. This section is limited, to the extent possible, to the effects of the malnutrition per se. The effects of diseases will be covered in Chapter 9. Another problem is in defining a "suboptimal" intake. There is abundant evidence that restricting food intake over most of the life span of rodents and many other animal species can extend longevity and reduce morbidity (Weindruch and Sohal 1997). Thus, a diet that optimizes body mass, strength, and physical performance may not be the optimal diet for longevity. It is beyond the scope of this book to expound on this issue. I will define an inadequate diet as one that does not have enough protein and energy to sustain what is considered an "ideal" body weight in affluent societies.

Reduction in Both Energy and Protein Intake

The effects of severe restriction of protein and energy (total starvation), and of severe energy deficiency with adequate protein intake have already been discussed. Such diets cannot sustain life for more than a few weeks in lean individuals. This section describes the effects of deficient diets that have enough protein and energy to sustain life for much longer periods of time, but not enough to sustain the ideal body weight.

The classic experiment of Keys et al. (1950) described the effect of 24 weeks of food deprivation on N balance in healthy, nonobese men. The average energy intake of these subjects before semistarvation was ~ 50 kcal×kg^{-1}×d^{-1}, and protein intake was ~ 1.6 g×kg^{-1}×d^{-1}. During semistarvation they initially consumed an average of ~ 23 kcal×kg^{-1}×d^{-1}, with a protein intake of ~ 0.7 g×kg^{-1}×d^{-1}. Although the absolute amount of energy and protein consumption was fairly constant during semistarvation, by the end of the study the average intake per kg had increased to ~ 30 kcal×kg^{-1}×d^{-1} and ~ 0.9 g protein×kg^{-1}×d^{-1}. While the energy intake clearly was deficient, the protein intake should have been enough to maintain a normal lean body mass if energy intake had been adequate. In spite of the adequacy of the protein intake, N (protein) balance was negative throughout the period of food deprivation. After the first 12 weeks they still were losing an average of 2 g N (12 g protein) per day. After 23 weeks of semistarvation the rate of protein loss had diminished to 1 g N (6 g protein) per day. The cumulative loss of protein during the first 12 weeks of food deprivation averaged 2 kg. During the second 12 weeks, it was only 0.7 kg. About 20% to 25% of the initial total body protein was lost during the 24 weeks of semistarvation. Muscle wasting was apparent from a 38% reduction in urinary creatinine excretion and reductions in thigh (19%), upper arm (24%), and calf (12%) circumferences. Concentrations of plasma proteins fell 10%, although the total pool of albumin barely changed because of an increase in plasma volume. The total amount of circulating hemoglobin fell an average of 29%. This study clearly showed that protein intakes that are adequate when weight-maintenance energy is provided are not sufficient to maintain protein balance when energy intake is deficient. Although there is evidence of adaptation of protein metabolism to food deprivation, even after 24 weeks the adaptation could not restore protein balance. Had the experiment continued, it is likely that net protein losses would eventually have ended, but only after further emaciation so that the energy and protein intake per kg would be increased.

Muscle wasting does not account for all of the whole-body protein loss during severe food deprivation. Autopsy data reviewed by Keys et al. (1950) indicate that the liver, heart, kidneys, spleen, and endocrine glands all shrink during severe malnutrition. The brain and lungs are relatively spared. Reductions in organ weights may underestimate protein losses, because the protein concentration of the tissues may decline. Adipose tissue is depleted more than any other tissue during energy deprivation. However, the protein content of adipose tissue (~ 3% of its mass) is so low that loss of adipose tissue explains only a trivial fraction of the total body protein loss. With less severe food deprivation, there is evidence that most of the protein loss comes from muscle rather than internal organs.

The effect of total starvation and very severe energy deprivation on human protein metabolism was discussed earlier. I am aware of only one controlled study of the effect of several weeks of more moderate food deprivation on human protein turnover. Stein et al. (1991) examined a 50% reduction in food intake over a 4 week period in moderately overweight men (~ 30% of weight as fat). The mean energy intakes before and during the food deprivation were similar to those of the men in the Keys et al. (1950) experiment, although somewhat less per kg. Protein intake during food deprivation should have been adequate to maintain protein mass if weight-maintenance energy requirements had been met. The [^{15}N]glycine end product method indicated that both whole-body protein breakdown and whole-body protein synthesis were reduced ~ 20%.

Other studies of food deprivation on protein turnover were not controlled experiments, but comparisons of normal subjects to patients with malnutrition associated either with poverty or diseases. Probably the most uncomplicated disease, from a metabolic standpoint, that causes malnutrition is anorexia nervosa. In severe cases there is marked emaciation, but there is no evidence of a primary systemic metabolic or hormonal imbalance that leads to this disease. A group of young women with anorexia nervosa, weighing an average of 78% of the ideal for height, was compared with a group of normal young women by the [^{15}N]glycine end product method (Vaisman et al. 1992). The authors of this study concluded that whole-body protein synthesis and breakdown, per kg body weight, were the same in the patients and controls. However, it is important to note that the total rate of protein turnover per subject was ~ 30% less in the anorexia nervosa patients, because they weighed much less than the normal subjects. Per kg fat free mass, the anorexia nervosa patients had a 24% reduction in the rate of protein synthesis and a 15% reduction in the rate of protein breakdown. This reduction in protein turnover per kg fat free mass probably underestimates the true effect of malnutrition on protein turnover, because skeletal muscle, which has a relatively slow rate of protein turnover (Chapter 4), is generally depleted more than visceral organs, which have a high rate of protein turnover. Thus, if the rate of protein turnover per g of tissue had remained the same in each organ, there should have been an increase in the whole-body protein turnover per kg fat free mass. Another factor to consider in interpreting these data is that the anorexia nervosa patients were eating the same number of calories and protein, per kg body weight, as the normal subjects. If the study had been done early in the course of the disease, when energy intake per kg is reduced, the decline in protein turnover could have been greater.

Studies of undernourished Indian men did not appear to support the idea that long-term malnutrition is associated with slower protein turnover (Soares et al. 1991; Soares et al. 1994). Affluent men were compared with poor manual laborers with a low body mass index (body mass index is weight/height^2 and is a surrogate measure of fatness). In one study (Soares et al. 1991), the affluent group was subdivided into those with a low body mass index and those with a normal body mass index. Protein turnover was evaluated with [^{15}N]glycine as the tracer, with the urea and NH_3 end product averaging method. The total rate of whole-body protein synthesis was 24% less in the undernourished poor subjects than in the normal-weight affluent men and 18% less than in the underweight affluent men. However, per kg body weight, the underweight groups had a higher rate of protein synthesis. This finding is not surprising, because the normal-weight men had much more body fat, which does not participate in protein turnover (although the stromal component of adipose tissue does contribute). More important was the finding that the protein synthesis per kg of fat free mass was not significantly different among the groups, although it tended to be higher in underweight men. The problem of adjusting the whole-body values for fat free mass was addressed above in the context of anorexia nervosa. Muscle mass was reduced much more than visceral mass, which should lead to a higher rate of protein turnover per kg of fat free mass, because of the higher rate of protein turnover in visceral organs than in muscle. Another problem in interpreting this study is that the underweight men consumed more energy, per kg of total or fat free mass, than the normal-weight men consumed during the study period. A follow-up study was done to address this last problem (Soares et al. 1994). A group of normal weight subjects was fed the same amount of energy and protein (per kg total or fat free weight) as an undernourished group. Whole-body protein turnover, per subject, was slower in the malnourished men, but was similar per kg fat free mass. Once again, there was the problem that the distribution of the fat free mass between muscle and other tissues was not the same in normal-weight and malnourished subjects. Table 5.1 uses the data from this study to emphasize that there may have been a reduction in the fractional rate of protein turnover in the tissues of the malnourished men, which would have been masked by expressing the data per kg fat free mass. If the fractional rate of visceral protein turnover were the same, the fractional rate of muscle protein turnover could have been reduced by as much as 40% in the malnourished subjects. If the fractional rate of muscle protein synthesis had remained the same, nonmuscle protein turnover may have been ~ 10% slower in malnourished subjects.

It is unclear whether the underweight men in these studies had been consuming any less energy or protein, per kg fat free mass, before participating in the experiments. Once body weight has stabilized, each kg of active tissue may be receiving as much food energy in the malnourished person as in the well-nourished person. Thus, the effect of chronic malnutrition may appear to be different than the effect of short-term malnutrition, even though the chronically malnourished person would have had an initial reduction in protein turnover if he had previously been on an adequate diet.

Malnutrition is most devastating to infants and young children. They require much more energy and protein (per kg) than adults both for maintenance and growth. The malnourished child might not even have a loss of body protein, but the prevention or reduction in protein accretion by malnutrition

TABLE 5.1. Comparison of protein synthesis and body composition in normal-weight and undernourished Indian men.

	Normal average	Undernourished average
Weight (kg)	60.2	43.2
Fat free mass (kg)	49.4	38.6
Muscle mass (kg)	26.0	16.2
Nonmuscle fat free mass (kg)	23.4	22.4
Whole-body protein synthesis (g/12 h)	151.3	118.1
Protein synthesis per kg fat free mass ($g \times 12\ h^{-1} \times kg^{-1}$)	3.1	3.1
Muscle protein synthesis (g/12 h)	45.4[a]	
Nonmuscle protein synthesis (g/12 h)	105.9[a]	
Expected muscle protein synthesis if no effect of malnutrition (g/12 h)		28.4[b]
Expected nonmuscle protein synthesis if no effect of malnutrition (g/12 h)		101.5[b]
Expected whole-body protein synthesis if no effect of malnutrition (g/12 h)		129.9[b]
Difference between observed and expected whole-body protein synthesis (g/12 h)		−11.3 (−8.7%)
% reduction in fractional rate of muscle protein synthesis required to explain discrepancy		40%[c]
% reduction in fractional rate of nonmuscle protein synthesis required to explain discrepancy		11%[d]

Source: Data of Soares et al. (1994).

[a]Based on assumption that muscle accounts for 30% of whole-body value in normal subjects.

[b]Derived by multiplying normal value, per kg, by measured mass of muscle or nonmuscle tissues. Whole-body expected value obtained by adding expected values for muscle and nonmuscle components.

[c]100×(11.3/28.4)

[d]100×(11.3/101.5)

prevents normal development and can have lifelong consequences, even after nutritional rehabilitation. Because the resynthesis of proteins from amino acids after proteolysis is never 100% efficient, with some of the essential amino acids being oxidized rather than reincorporated into proteins, it would seem to be beneficial for malnourished infants to reduce their rate of protein turnover. Moreover, protein turnover requires much energy, and reducing protein turnover would benefit the malnourished infant by conserving energy. Thus, it was surprising that the earliest study to examine protein turnover in malnourished infants indicated that their rates of whole-body protein breakdown and synthesis were increased, relative to infants who had recovered from malnutrition (Picou and Taylor-Roberts 1969). However, it now is apparent that this unexpected result is explained by the fact that the children were studied while in the recovery phase, which is associated with increased protein turnover (see later). Later studies have demonstrated, with both the [^{15}N]glycine end product method and the [^{13}C]leucine dilution method, that malnourished infants have slower rates of whole-body protein synthesis and breakdown (per kg body weight) than infants who have recovered from malnutrition (Golden, Waterlow, and Picou 1977b; Manary et al. 1997). Some of the initial increase in the ratio of protein turnover to body weight in children recovering from malnutrition could be attributed to the initial reduction in body weight as edema subsides, but this factor cannot explain the effect entirely. The fact that myofibrillar proteins participate in this response is suggested by a marked reduction in 3-methylhistidine excretion in malnourished infants compared to its excretion after recovery, even though the rehabilitation diet did not contain sources of 3-methylhistidine (Rao and Nagabhushan 1973). Albumin turnover also is reduced in malnourished children (Waterlow, Garlick, and Millward 1978).

In contrast to the general finding of reduced protein turnover during malnutrition in humans, there is evidence that long-term restriction of food intake increases the fractional rate of protein turnover in various tissues of rodents, including liver, kidney, heart, gut, and muscle (Ricketts et al. 1985; Ward 1988; Merry, Lewis and Goldspink 1992; Sonntag, Lenham, and Ingram 1992). Restriction of food intake to 40% to 75% of the ad libitum level (with adequate micronutrient intake) after weaning leads to marked increases in longevity and reductions in the incidence and severity of various pathologies in old animals. Hence, this regimen cannot be considered to be malnutrition. In fact, this type of food restriction significantly slows most of the physiological changes associated with aging. Because young animals have much higher rates of tissue protein turnover than adult animals, these results may reflect the slowing of the aging process rather than a stimulation of protein turnover by food restriction. As is the case with chronic malnutrition in humans, the deprived animals are smaller, so that food intake per g of body weight is only slightly less than that of animals fed ad libitum.

Protein or Essential Amino Acid Deprivation

Protein oxidation never is zero, even with protein-conserving physiological adaptations. Therefore, it is necessary to consume some protein to maintain the protein stores of the body, no matter how generous energy intake might be. Some of the amino acids can be synthesized by humans de novo, but the essential amino acids cannot be synthesized and must be provided by the diet. Methionine, threonine, tryptophan, isoleucine, leucine, lysine, valine, and phenylalanine are essential for adults. Histidine also is an essential amino acid for infants, and probably for adults as well (Laidlaw and Kopple 1987). Tyrosine is synthesized from the essential amino acid phenylalanine, so if phenylalanine intake is marginal, tyrosine can be an essential amino acid. Other amino acids may become essential under various conditions (Laidlaw and Kopple 1987). Synthesis of the nonessential amino acids requires N. Therefore, the requirement for essential amino acids is greater when total protein intake is low than the requirement when total protein intake is high.

Thus, both protein quality and total N intake are important determinants of the requirement for dietary protein. Energy intake also is an important determinant of the protein requirement. Reduced energy intake raises the level of protein needed to maintain the protein stores of the body, whereas a surfeit of energy lowers the protein requirement. This section discusses the effect of diets that are deficient in protein quantity or quality, but which contain enough energy that they could maintain protein and energy stores if protein intake were adequate.

In affluent countries, it is rare for an individual to have an adequate energy intake without also having an adequate protein intake. Usually, the amino acid composition of the protein sources is such that enough of the essential amino acids are provided. However, in some regions of the world there may not be enough high-quality protein in the diet to supply enough essential amino acids for tissue maintenance or growth. Because the protein requirement for growth is greater than the requirement for maintenance, protein deficiency is tolerated better by adults than by infants and children. Kwashiorkor refers to a syndrome that occurs in impoverished areas where the protein intake of children is inadequate. Breast milk supplies enough protein for normal development during the period of breast feeding, but after weaning the diet may not contain enough high-quality protein. Kwashiorkor is marked by stunted growth, poor immunity against infections, poor wound healing, edema, and other problems. The edema may prevent weight loss even if protein stores are depleted.

The World Health Organization (WHO) and the National (USA) Research Council (NRC) have attempted to define safe levels of protein intake to avoid protein deficiency (Table 5.2). It is difficult to arrive at a single value, because protein requirements depend on protein quality, energy intake, antecedent protein intake, and individual variations in efficiency of protein utilization. The levels established by these organizations allow for individual variations, so they are higher than the average protein requirement. The WHO values are closer to the average. The NRC has adopted higher levels of recommended protein intake, which may reflect the fact that it is easier to achieve these levels of protein intake in the United States. The WHO values specifically refer to ideal proteins (egg or milk protein, or proteins with equivalent biological value), and higher levels are recommended when protein quality is lower.[1] Most of the research on protein requirements is based on the amount of protein required to maintain N balance over several days in healthy subjects. The problems with N balance studies were mentioned in Chapter 3. Another problem with this approach is the assumption that the protein stores of the subjects in such studies, often college students, are at optimal levels. It is unclear whether some protein loss in these subjects, followed by reestablishment of protein balance on a lower protein intake at a smaller lean body

---

[1] Protein quality refers to the essential amino acid content of the protein. If a protein lacks one or more of the essential amino acids, it cannot sustain protein synthesis.

TABLE 5.2. Safe levels of protein intake ($g \times kg$ ideal body weight $^{-1} \times d^{-1}$).

	NRC	WHO
Infants		
to 6 months	2.2	
6–12 months	2.0	1.5
Children		
1–3 yr	1.8	1.2
4–6 yr	1.5	1.0
7–9 yr	1.2	0.9
Males		
10–12 yr	1.0	0.8
13–15 yr	0.9	0.7
16 yr-adult	0.8	0.6
Females		
10–12 yr	1.0	0.8
13–15 yr	0.9	0.6
16–19 yr	0.8[a]	0.6[b]
Adult	0.8[a]	0.5[b]

Source: Adapted from Whitney and Cataldo (1983). NRC, National Research Council (USA). WHO, World Health Organization. Values rounded to nearest 0.1 $g \times kg$ ideal body weight $^{-1} \times d^{-1}$.

[a]Add 30 g protein per day for pregnancy and 20 g protein per day for lactation.

[b]Add 9 g protein per day for pregnancy and 17 g protein per day for lactation.

mass, would make them any less healthy. Nevertheless, the WHO and NRC recommendations are useful approximations of the minimum protein intake compatible with maintenance of protein stores in adults, and compatible with normal growth in children.

The adult requirements for essential amino acids that have been estimated by the WHO and the NRC are considerably lower than the amounts that would be consumed by someone eating recommended amounts of high quality proteins (Young, Bier, and Pellett 1989). Young and his colleagues have argued that the current recommendations for intake of essential amino acids are too low (Young, Bier, and Pellett 1989). They abandoned the traditional N balance approach and used stable isotopic tracers of individual amino acids to determine the minimum amount of essential amino acids needed to ensure that irreversible metabolism of the essential amino acids does not exceed the amount consumed. It is not entirely clear why the N balance approach indicates lower values for essential amino acid requirements. Both approaches are imperfect, and the issue can be settled definitively only by showing a difference in the health or performance between groups maintained for a long time on either the lower or higher levels of intake of essential amino acids. Someone eating the amount of protein recommended by the WHO (Table 5.2) would be consuming the minimum amounts of essential amino acids recommended by Young, but only if almost all of the protein were high quality (milk, eggs, meat).

Protein oxidation declines as protein intake is reduced, but the daily protein oxidation exceeds the protein intake when intake falls below the requirement. In healthy adults consuming weight-maintenance amounts of energy with no protein intake, the average daily obligatory N loss is ~ 50 mg/kg after a week of adaptation (Scrimshaw et al. 1972; Young and Marchini 1990). A 70 kg man would lose about 1.3% of his protein every week on such a regimen. If protein oxidation were to proceed at the rate associated with a typical daily protein intake (1.2 g/kg), he would lose about 5% of his protein stores each week on a protein-free diet. The major reason for a reduction in protein oxidation during protein deficiency is simply the lack of substrate provided by the diet, leading to a reduction in the levels of amino acids available for oxidation. A modest reduction in the rate of whole-body protein breakdown further reduces the availability of amino acids for oxidation. A reduction in whole-body protein breakdown was observed with the $[^{15}N]$alanine end product method as soon as the first day after switching subjects from a conventional diet to a low protein diet (Taveroff, Lapin, and Hoffer 1994). After a week on a low protein diet (0-0.3 $g \times kg^{-1} \times d^{-1}$), young adults have a lower postabsorptive rate of whole-body protein breakdown than they do after consuming generous amounts of protein (1.5 $g \times kg^{-1} \times d^{-1}$) (Motil et al. 1981b; Yang et al. 1986; Garlick, McNurlan, and Ballmer 1991). However, the suppression of proteolysis by meals is diminished in subjects adapted to a low protein diet, so that absolute postprandial rates of whole-body proteolysis are not diminished by a low protein diet (Motil et al. 1981b; Garlick, McNurlan, and Ballmer 1991). Postprandial protein synthesis is significantly slower in subjects on a protein-deficient diet than in those consuming adequate protein (Motil et al. 1981b; Garlick, McNurlan, and Ballmer 1991; Hoerr et al. 1993; Taveroff, Lapin, and Hoffer 1994). In spite of more efficient reutilization of proteolysis-derived amino acids for protein synthesis than for oxidation during periods of protein deficiency, the protein synthesis rate remains below the protein breakdown rate, and protein loss is inevitable. When the level of protein intake is near the average maintenance requirement of adults (0.4 $g \times kg^{-1} \times d^{-1}$), rather than being obviously deficient, significant changes in whole-body protein turnover may not be observed (Gersovitz et al. 1980; Motil et al. 1994a; Castaneda et al. 1995). In very young children consuming adequate energy, N retention and protein synthesis are much less when protein intake is 0.7 $g \times kg^{-1} \times d^{-1}$ than when it is 1.7 $g \times kg^{-1} \times d^{-1}$ (Jackson et al. 1983).

Protein malnutrition leads to low serum albumin concentrations. Both breakdown and synthesis of albumin are reduced by protein deprivation, but the reduction in synthesis is greater so that albumin levels decline (Gersovitz et al. 1980). Even a marginal protein deficiency (0.4 $g \times kg^{-1} \times d^{-1}$ for 2 weeks) can reduce albumin production in healthy subjects, although elderly subjects may be less sensitive in this regard (Gersovitz et al. 1980).

Some of the urea produced by the liver is salvaged by the microflora of the lower gut, and the proportion increases as protein intake decreases (Jackson 1993). Some of the N salvaged by the microflora is used for synthesis of essential amino acids, which can be absorbed and used by the host. However, the capacity to meet the protein requirement by the urea salvage pathway is limited, and cannot sustain protein balance when protein intake is below the minimum requirement.

Animal research has provided some insights into the mechanisms whereby protein deficiency alters protein metabolism. In rats, protein deficiency reduces both lysosomal and ATP-dependent proteolysis in muscle, but increases the capacity for Ca^{2+}-dependent proteolysis (Tawa, Kettelhut, and Goldberg 1992). This response may be mediated in part by reduced thyroid hormone levels, which also could participate in the reduction in protein synthesis. The efficiency of protein synthesis is enhanced by protein deficiency, in that a larger proportion of the essential amino acids derived from protein breakdown or the diet are used for protein synthesis rather than oxidation. However, because the flux of amino acids is severely limited by protein deficiency, the rate of protein synthesis is reduced. Because the K_m values for the aminoacyl-tRNA synthetases are well below the amino acid levels within amino acid-deprived cells, aminoacyl-tRNA availability does not appear to be rate-limiting for protein synthesis. Rat livers perfused with an amino acid-deficient medium have a reduced initiation of translation and reduced formation of the 40S initiation complex (Flaim et al. 1982a). Extracts from amino acid-deprived livers can inhibit initiation of translation in a reticulocyte lysate system, by increasing phosphorylation of eIF-2α (Everson et al. 1989) and reducing the activity of eIF-2B (formerly called GEF) (Kimball et al. 1989). Amino acid availability also may regulate levels of mRNAs for some ribosomal proteins (Kilberg, Hutson, and Laine 1994). It has been suggested that the essential amino acid tryptophan may have hormone-like actions on the activity of ribosomes, which may be involved in the reduced protein synthesis associated with tryptophan (or overall protein) deficiency (Lin, Smith, and Bayley 1988). Protein deficiency also reduces the abundance of specific mRNAs in rat liver, including those encoding albumin, transferrin, transthyretin, fibrinogen (β-chain), and apolipoprotein E (Young and Marchini 1990). The reduced concentration of muscle and liver ribosomal RNA associated with protein deficiency is probably mediated at the level of transcription (Young and Marchini 1990). The precise mechanisms whereby multiple aspects of protein synthesis are regulated by concentrations of one or more amino acids are unclear.

Recovery from Malnutrition

Repletion of body proteins occurs rapidly when adequate amounts of protein and energy are provided to the malnourished person. In the Minnesota semistarvation experiment (Keys et al. 1950), positive N balance was obtained during recovery on an intake of as little as 3,000 kcal/d in men who had been malnourished for 24 weeks (about 1600 kcal/d). When energy intake was increased to ~ 4,000 kcal/d, N accretion was as high as 6.3 g/d (~ 40 g protein/d). However, the degree of positive N balance in this study is an overestimate because fecal losses were not recorded during recovery. With protein intakes of 60 g/d or more, additional protein supplementation did not improve N balance further, as long as energy intake was constant. Recovery of protein mass appears to occur more rapidly than recovery of fat mass. When a group of anorexia nervosa patients had been rehabilitated to ~ 90% of the ideal body weight, the total body N was restored to ~ 90% of the normal level and body fat was restored to only ~ 80% of the normal level (Russell et al. 1994). The proportion of weight gain accounted for by fat free mass (and therefore protein) is much higher during recovery from malnutrition than it is during weight gain in subjects who are already well nourished (Forbes 1987a).

In infants and children recovering from malnutrition, the rates of whole-body protein breakdown and synthesis, per kg body weight, are much higher than the rates observed after recovery (Picou and Taylor-Roberts 1969; Golden, Waterlow, and Picou 1977b). As long as protein intake meets the minimum requirement, this effect appears to be mediated primarily by the increase in energy intake and is not enhanced by additional protein (Golden, Waterlow, and Picou 1977b). The rate of protein synthesis correlates with the energy intake in children recovering from malnutrition (Golden, Waterlow, and Picou 1977a). The rate of protein breakdown also is positively correlated with energy intake, but the strength of this association is much weaker than the association between energy intake and protein synthesis. The rate of weight gain and N accretion can be predicted from the rate of protein synthesis, but is not significantly related to the rate of whole-body protein breakdown.

It seems to be energetically wasteful to increase protein breakdown at a time when tissues are trying to increase their protein content. The same thing happens during normal growth or the protein accretion associated with excessive energy intake. There could be some advantage to increasing protein degradation during nutritional rehabilitation. It seems likely that not only the total amount of protein, but also the pattern of relative concentrations of different proteins, would be affected by malnutrition. Increasing both protein degradation and protein synthesis would allow a more rapid restoration of the normal pattern than an increase in protein synthesis only.

High Energy Intake

The result of prolonged periods of high energy intake is obesity, which can influence protein metabolism itself. This section will deal with the effect of the elevated energy intake per se. The effect of obesity on protein metabolism is discussed in Chapter 9.

Whenever energy intake is enough to cause weight gain, protein balance is positive (Figure 5.4). In overfeeding studies of normal-weight volunteers, lasting up to 100 days, ~ 30% to 40% of the weight gained is fat free tissue, so

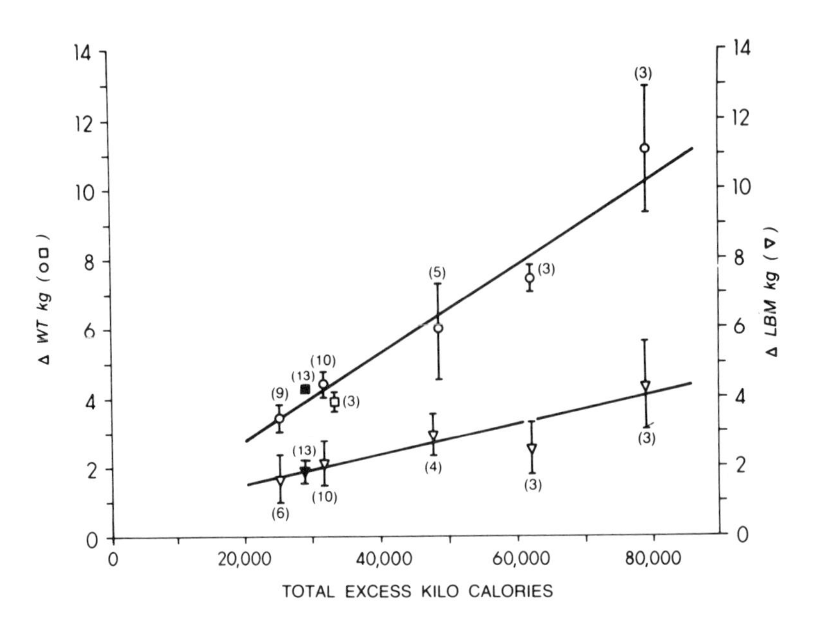

FIGURE 5.4. Accretion of lean body mass (LBM)and total body weight associated with surfeit energy intake. Data taken from a number of studies. From Forbes (1987a), with permission, Springer-Verlag.

as much as 8% of the weight gained is protein if 20% of fat free tissue is protein (Forbes 1987a). Thin individuals gain more protein per kg of weight gain than those who are already overweight (Forbes 1987b). Computed tomography suggested that increased muscle volume accounted for all of the increase in the total volume of fat free tissue observed after 100 days of overfeeding in young men (Deriaz et al. 1992). However, volumes may not precisely reflect protein content, and some accretion of protein in viscera should not be discounted. About 25% of the excess weight of obese persons is fat free tissue, suggesting that as weight gain proceeds, less protein is added per unit of weight gain. This finding might be explained by the fact that the energy intake per kg of fat free mass declines as weight is gained, and is not different in lean and obese persons. Obese persons have an increased visceral mass (Naeye and Roode 1970), indicating that not all of the protein accretion during weight gain occurs in muscle. The accumulation of protein during overfeeding does not require an increase in protein intake, either in absolute amounts or as a proportion of energy, as long as protein intake meets the minimum requirement. For example, we provided an extra 400 to 500 g of carbohydrate each day to healthy young men without changing their protein or fat intake. There was an immediate reduction in N excretion, which remained about 4 g/d (25 g protein) less throughout the 20 days of carbohydrate supplementation than it was before supplementation (Figure 5.5).

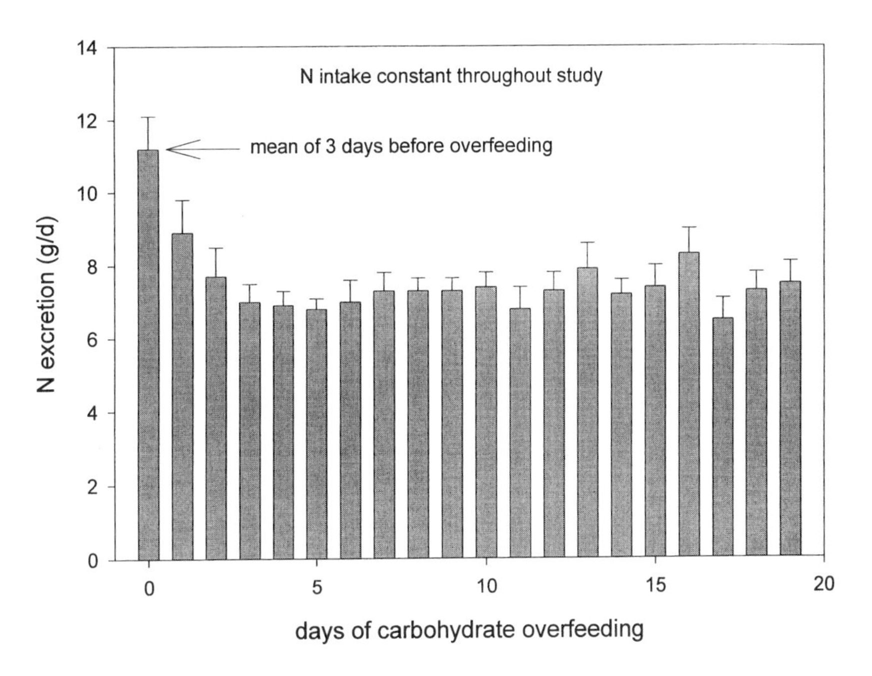

FIGURE 5.5. N accretion associated with excessive energy intake (as carbohydrate), in the absence of any change in protein intake, in normal men. Mean N intake from protein was 13 g/d throughout study. Baseline energy intake was 2200 kcal/d. Mean energy intake during overfeeding was 4000 kcal/d. Author's unpublished data.

The efficiency of protein retention is enhanced by excess energy intake. In other words, a greater proportion of the available amino acids (from protein intake and breakdown of body proteins) are used for protein synthesis rather than oxidation. Whether there is an increase the absolute level of amino acid oxidation during excessive energy intake depends on whether or not protein intake increases. If protein intake is constant or reduced, amino acid oxidation declines. If protein intake increases just enough to meet the amino acid requirement for protein accretion, then total amino acid oxidation does not change. If the increase in protein intake is greater than the increase in protein retention, then amino acid oxidation increases, but is less than it would if energy intake were not excessive.

There is very little information about the effect of excessive energy intake on human protein turnover, and no information about effects on individual tissues. As discussed above, the [^{15}N]glycine end product method revealed an increase in the amount of endogenous protein degraded as energy intake increased in previously-malnourished children, and an even greater increase in protein synthesis (Golden, Waterlow, and Picou 1977a). However, modest increases (25% to 30%) in energy intake in well-nourished adults do not appear to increase whole-body protein turnover, as reflected by leucine or lysine R_a, and the nonoxidized portion of leucine R_d (Motil et al. 1981a; Yang et al. 1986). A greater energy excess can increase postabsorptive protein turnover. Young men overfed by 1600 kcal/d with carbohydrates for 10 days (~ 60% above maintenance), with no change in protein intake, had a 13% increase in proteolysis and a 12% increase in protein synthesis after overnight fasting, according to leucine kinetics (Welle et al. 1989). This response may be mediated by hormonal changes associated with overfeeding, which include elevated plasma concentrations of T_3, insulin, glucagon, insulin-like growth factor-I, and possibly other hormones (Forbes et al. 1989; Welle et al. 1989).

High Protein Intake

It is difficult to define what level of protein intake is "high." Some early humans may have lived on diets containing five times the proportion of energy from protein than that of the typical U.S. diet (Eaton and Konner 1985). Except for certain disease states (e.g., nephropathy, phenylketonuria), a high protein diet is not necessarily incompatible with good health, although a large intake of meat may increase the risk of atherosclerosis and cancer. Thus, a "high" protein intake can only be defined in relative terms. This section will review studies in which two or more levels of protein intake above the minimum requirement (~ 0.5 $g \times kg^{-1} \times d^{-1}$) were examined.

N balance becomes more positive as protein intake increases (Oddoye and Margen 1979; Cheng et al. 1978; Motil et al. 1981b; Yang et al. 1986; Campbell et al. 1994a; Price et al. 1994; Campbell et al. 1995; Pannemans et al. 1995; Pannemans, Halliday, and Westerterp 1995). For the first few days after a change in the level of protein intake, a significant fraction of the change in N balance may be the result of a change in the mass of urea, free amino acids, or other nitrogenous compounds. After a new steady state level of these compounds is achieved, almost all of the difference in N balance should be related to protein gains or losses. There may be a pool of "labile" proteins whose concentrations are very sensitive to changes in protein intake, and an increase in this pool may account for some of the N retention associated with a high protein intake (Oddoye and Margen 1979). However, the N balance method generally tends to overestimate protein gains and may overestimate the gains the most during periods of high protein intake. A study of N balance and body composition during 12 weeks of resistance training at two levels of protein intake illustrates this problem (Campbell et al. 1995). The subjects with a high protein intake (1.6 $g \times kg^{-1} \times d^{-1}$) had an apparent positive N balance of ~ 25 $mg \times kg^{-1} \times d^{-1}$ throughout the experimental period. Over the 12 week study, this degree of N retention should have increased total body protein mass by ~ 1 kg, which would have increased total fat free mass by ~ 5 kg. The measured increase in fat free mass was only 1.8 kg by underwater weighing. Although the body composition assays also are subject to some error, these data suggest that the N balance method overestimated protein accretion by ~2 to 3-fold. Subjects on the lower protein diet appeared to be retaining only one fourth the amount retained by the high protein group. Their increase in fat free mass was 1 kg, which is close to the value predicted by N balance.

Even though the apparent N balance is more positive with high levels of dietary protein, most of the excess protein is oxidized with a corresponding increase in urea production and N excretion. Increased oxidation of leucine also is evident when protein intake is increased, both in absolute terms and as a fraction of the total leucine R_a (Motil et al. 1981b; Tarnopolsky et al. 1992; Campbell et al. 1995).

For protein accretion to occur, protein synthesis must be more rapid than protein breakdown. This could occur at a higher or lower turnover rate. Some evidence suggests that protein accretion is superimposed on a higher whole-body protein turnover rate when protein intake is increased. After a period of adapting to a high protein intake, the postabsorptive protein turnover is elevated. When protein intake was increased from 0.6 to 1.5 $g \times kg^{-1} \times d^{-1}$ in normal subjects, the postabsorptive R_a of leucine and lysine increased ~ 20% (Motil et al. 1981b; Yu et al. 1985; Yang et al. 1986). An increase in protein intake from 0.77 to 1.59 $g \times kg^{-1} \times d^{-1}$ increased leucine and phenylalanine R_a ~10-15% (Pacy et al. 1994). Postabsorptive N flux increased by about one third when protein intake was increased from 12% of the energy intake to 21% (Pannemans et al. 1995; Pannemans, Halliday, and Westerterp 1995). Postabsorptive protein synthesis, estimated from the nonoxidized leucine R_d, increased 27% when protein intake was changed from 0.6 to 1.5 $g \times kg^{-1} \times d^{-1}$(Motil et al. 1981b), although it did not change in another study when protein intake was increased from 0.77 to 1.59-2.07 $g \times kg^{-1} \times d^{-1}$ (Pacy et al. 1994). Postabsorptive protein synthesis increased 12% to 42%, according to the [^{15}N]glycine end product method, when protein intake was increased from 12% of energy to 21% of energy (Pannemans et al. 1995; Pannemans, Halliday, and Westerterp 1995). In elderly women, increasing the protein intake from 10% to 20% of energy intake did not influence postabsorptive protein metabolism according to [^{13}C]leucine kinetics, but increased postabsorptive protein synthesis 45% and protein breakdown 39% according to the [^{15}N]glycine end product method (Pannemans et al. 1997). This discrepancy could reflect the limitations of one or both of the techniques, or it could be related to the fact that the [^{15}N]glycine study commenced much sooner after the latest meal than the [^{13}C]leucine study.

Postprandially, high protein diets can inhibit endogenous proteolysis more than lower protein diets, but this effect is not always found (Motil et al. 1981b; Tarnopolsky et al. 1992; Pacy et al. 1994; Quevedo et al. 1994; Campbell et al. 1995). Protein synthesis increases ~10% to 30% when protein intake increases from levels near the maintenance requirement (0.6-0.8 $g \times kg^{-1} \times d^{-1}$) to generous levels (1.4 $g \times kg^{-1} \times d^{-1}$) (Motil et al. 1981b; Tarnopolsky et al. 1992; Pacy et al. 1994; Campbell et al. 1995). According to the [^{15}N]glycine method, daily protein breakdown is unaffected by increasing protein intake over the range of 0.6 to 1.2 $g \times kg^{-1} \times d^{-1}$, whereas daily protein synthesis either does not change

(Hou et al. 1986) or increases (Meredith et al. 1989). Not all of the effects discussed above were reported as statistically significant, because of small numbers of subjects, but it is a fairly consistent finding that higher protein intakes increase the average postabsorptive protein turnover and postprandial protein synthesis in healthy adults.

Once a person has adapted to a high protein diet, he temporarily may be unable to maintain his protein mass upon changing to a lower protein diet, even if the lower level of protein usually is adequate (Oddoye and Margen 1979). For example, healthy men who were consuming 1.82 $g \times kg^{-1} \times d^{-1}$ protein had positive N and leucine balances (Quevedo et al. 1994). When their diet was switched to one containing 0.77 $g \times kg^{-1} \times d^{-1}$, their N and leucine balances were transiently negative, but within 7 days they became neutral (not significantly different from zero). It is interesting that the major mechanism of the greater protein retention on the high protein diet in this study, in contrast to those cited above, was a greater postprandial suppression of proteolysis. Postprandial protein synthesis was increased modestly.

In children who had recovered from malnutrition, increasing the protein intake above an adequate level did not appear to affect the rate of protein synthesis (Golden, Waterlow, and Picou 1977a,b). This research is a bit difficult to interpret, because the different levels of protein intake were not always given with exactly the same energy intakes. However, two groups were studied at the same energy intake with very different protein intakes. At 1.2 $g \times kg^{-1} \times d^{-1}$ protein intake, protein balance was +0.7 $g \times kg^{-1} \times d^{-1}$, and at 5.2 $g \times kg^{-1} \times d^{-1}$ protein intake, protein balance was +2.2 $g \times kg^{-1} \times d^{-1}$. The major difference in protein metabolism between groups was a 20% lower rate of protein breakdown in the high protein group, although this difference was not statistically significant with the small number of children studied. The rate of protein synthesis in these children was much faster than that of adults and might have been difficult to increase further. The very high rate of protein synthesis might be related to the very high energy intake (120 $kcal \times kg^{-1} \times d^{-1}$) relative to the usual adult energy intake (40 $kcal \times kg^{-1} \times d^{-1}$).

Because many athletes believe that a high protein intake increases muscle mass, there has been interest in how a high protein intake affects muscle protein mass and protein metabolism (see Chapter 8). Increased protein intake could increase muscle protein mass either by reducing the rate of proteolysis or by stimulating protein synthesis. In short-term studies, high amino acid levels (such as those associated with a high-protein diet) can inhibit muscle protein degradation (Louard, Barrett, and Gelfand 1990; Nair, Schwartz, and Welle 1992; Louard, Barrett, and Gelfand 1990) and stimulate protein synthesis (Bennet et al. 1989; Svanberg et al. 1996). However, these studies examined the effects of isolated increases in amino acid levels for a few hours and might not reflect the effect of elevated amino levels in the normal postprandial metabolic milieu. There is no evidence that a high protein diet for several weeks inhibits the daily myofibrillar protein degradation, as reflected by 3-methylhistidine excretion (Hickson and Hinkelmann 1985; Campbell et al. 1995). In fact, there was a gradual trend toward increasing 3-methylhistidine excretion when healthy young men consumed a high protein diet (2.4 $g \times kg^{-1} \times d^{-1}$) for a month (Hickson and Hinkelmann 1985). We did not find any difference in the postprandial myofibrillar synthesis rate over a dietary protein concentration from 7% to 28% of energy (Figure 5.2), but the effect of long-term feeding of a high protein diet on muscle protein synthesis has not been studied. In 56 to 80-year-old men and women who were performing resistance exercises, increasing the protein intake from 0.8 to 1.6 $g \times kg^{-1} \times d^{-1}$ resulted in positive N balance, but without any evidence of muscle hypertrophy (Campbell et al. 1995). Evidence for a beneficial effect of high protein intake on muscle mass (Lemon 1996) relies mainly on N balance data, which may exaggerate the effect of a high protein diet on whole-body protein retention and provides no specific information about muscle protein mass.

In malnourished children and children recovered from malnutrition, a high protein intake (2–4.8 $g \times kg^{-1} \times d^{-1}$) promoted albumin synthesis relative to rates observed when the children consumed marginal amounts of protein (0.7-1 $g \times kg^{-1} \times d^{-1}$) (Waterlow, Garlick, and Millward 1978). Albumin half-life was not significantly influenced by the dietary protein concentration.

Summary of Effects of Energy and Protein on Whole-Body Protein Metabolism

Most of the research on dietary effects on protein metabolism in humans has been done at the whole-body level, using N balance or stable isotopic methods to evaluate whole-body protein breakdown and synthesis. It is important to remember that such effects may not be uniform in all tissues. Nevertheless, such information has been useful in evaluating the nutritional requirements of humans.

When body weight is being lost because of an energy deficit, part of the loss is protein. Protein mass is defended to some extent by a reduction in the rate of protein breakdown, but protein synthesis declines even more and some protein loss is probably unavoidable during periods of weight loss. Muscle appears to account for most of the whole-body protein losses. During periods of weight gain, the protein mass of the body increases, as long as protein intake meets the minimum requirement for tissue growth. Even in adults, protein stores increase during weight gain, although not as much as they do in growing children. The increase in protein stores during periods of excess energy intake is mediated by increased protein synthesis, and protein breakdown can even increase somewhat if energy intake is excessive enough. When dietary protein intake is constant, more dietary protein is used for protein synthesis and less is used for oxidation when energy intake is high; less protein is used for synthesis and more is used for oxidation when energy intake is low. Figure 5.6 schematically illustrates the relative whole-body protein breakdown, synthesis, and balance as a function of energy intake.

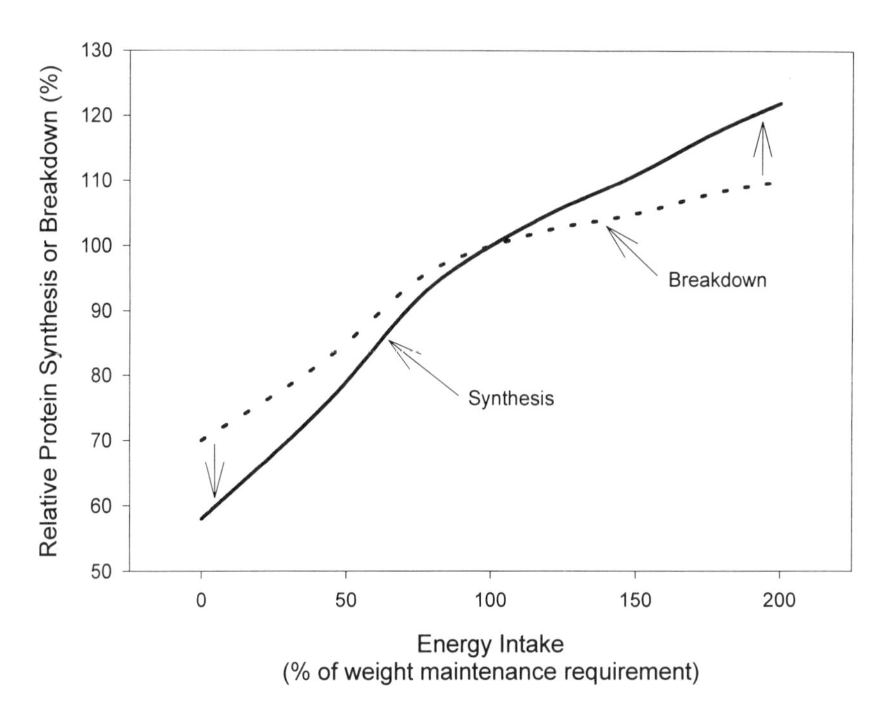

FIGURE 5.6. Hypothetical relation between energy intake and protein metabolism. Overall turnover is proportional to level of energy intake, with total daily proteolysis and protein synthesis being equal when energy expenditure equals energy intake. Synthesis (*solid line*) is less than breakdown (*dotted line*) when energy intake is insufficient to maintain energy balance, leading to protein depletion (*arrow pointing down*). Synthesis exceeds breakdown when energy intake exceeds expenditure, leading to protein accretion (*arrow pointing up*).

Protein intake often goes hand in hand with energy intake, and the effects of energy intake on protein metabolism are exacerbated by the simultaneous effects of protein intake. Even at a constant energy intake, the level of protein intake influences protein metabolism. Dietary protein deficiency is associated with reduced protein degradation, but protein synthesis declines even more and protein mass is lost. Increasing protein intake above the minimum requirement usually seems to induce protein accretion (positive N balance), but there generally is a lack of verification of this finding with independent body composition methods. I am skeptical that you could increase your protein mass very much simply by increasing the dietary protein concentration without increasing energy intake. At protein intakes above the minimum requirement, protein turnover is faster than turnover at marginal or deficient levels of protein intake. However, at very high levels of protein intake, the rate of synthesis plateaus and any further gain in protein mass appears to be

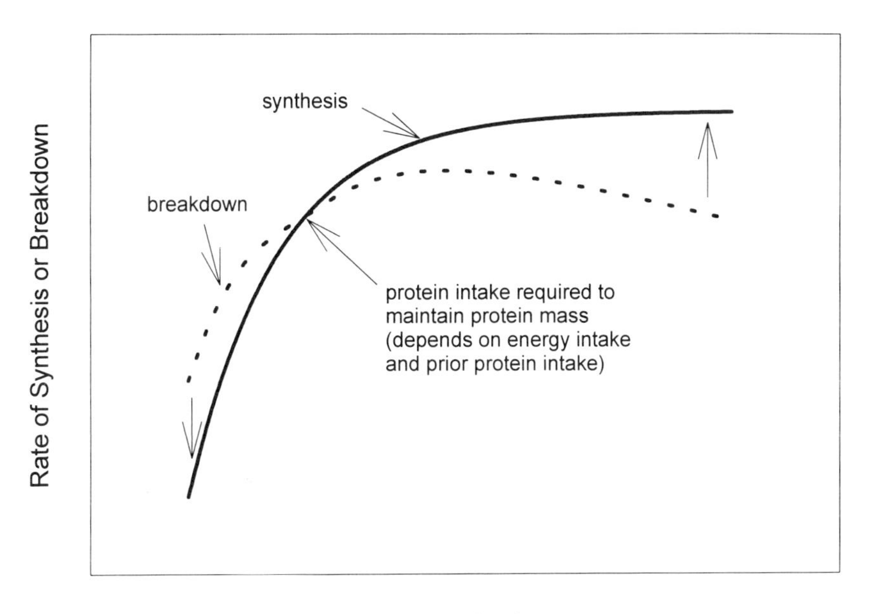

FIGURE 5.7. Hypothetical relation between protein intake and protein metabolism. The minimum requirement to maintain protein stores is the point at which the synthesis (*solid line*) and breakdown (*dotted line*) curves intersect. This value varies according to protein quality, energy intake, and recent protein intake. Although reduced proteolysis compensates to some extent for low protein intake, below a certain level of protein intake the breakdown must exceed synthesis and protein depletion occurs (*arrow pointed down*). The level of protein intake at which this occurs declines when recent protein intake is low, but breakdown always exceeds synthesis when protein intake is less than 0.4-0.5 $g \times kg^{-1} \times d^{-1}$ in an adult. At chronically high levels of protein intake, the point of intersection between synthesis and breakdown moves to the right (higher protein intake needed to maintain N balance). The gap between synthesis and breakdown, reflecting protein accretion (*arrow pointing up*), probably narrows with chronic intake of large amounts of protein, although N balance studies do not always show such an effect.

mediated by reduced proteolysis. Figure 5.7 schematically illustrates the relative whole-body protein breakdown, synthesis, and balance as a function of protein intake at a constant energy intake. The positive N balances associated with elevated protein consumption are likely to be transient. The amount of protein needed to maintain N balance depends on recent protein intake. When recent protein intake is higher, the protein requirement for N balance is higher. When recent protein intake is lower, the protein requirement for N balance is lower. Below a certain level of protein intake, adaptation is inadequate and the body loses protein.

6 Endocrine, Paracrine, and Autocrine Regulation

As with most human metabolic processes, the regulation of protein metabolism involves redundant control mechanisms to maintain homeostasis. Many hormones can influence protein metabolism (Table 6.1). Some hormones, such as thyroid hormones, do not fluctuate much from day to day or from hour to hour in healthy, well-nourished individuals, but their presence may be essential for maintaining normal rates of protein turnover. Other hormones, such as insulin, have concentrations that vary over a wide range from hour to hour and are involved in acute regulation of metabolism. Hormones may have direct effects on protein metabolism through specific signal transduction systems, or indirect effects related to effects on levels of metabolic substrates (Chapter 7). Not only the concentrations of the hormones, but also the sensitivity to specific actions of hormones can vary to maintain metabolic homeostasis. In addition to hormones that are produced in endocrine glands and act at tissues remote from the glands, there are many peptides released by nonendocrine cells that act on nearby cells (paracrine effects) or on the cells producing them (autocrine effects). These effects are hormone-like in that they require interaction of the peptide with specific receptors on the cell surface. Although this chapter will focus on the effects of each hormone separately, many situations are associated with changes in multiple catabolic and anabolic hormones and ultimate effect of all of these influences may be somewhat upredictable.

Table 6.1 categorizes hormones as generally anabolic or generally catabolic. Every hormone may increase the levels of certain proteins and decrease the levels of others. The term anabolic in this context means that the hormone tends to increase the total protein mass, whereas the term catabolic means that the hormone tends to reduce the total protein mass. The categorization of hormones as anabolic and catabolic should not be taken to imply that more of the anabolic hormones or less of the catabolic hormones is good for a person. Even though the term catabolic usually has a negative connotation, a protein catabolic effect may have an overall beneficial effect under some conditions. Normal levels of catabolic hormones (e.g., thyroid hormones, cortisol) may be required to maintain a normal rate of protein turnover, and reduced levels

TABLE 6.1. Categorization of hormones and as anabolic or catabolic.

Anabolic	Catabolic
Insulin	Cortisol
Growth hormone	Thyroid hormones
Insulin-like growth factor-I	Glucagon
Testosterone	Cytokines
Estradiol?	Progesterone?
Epinephrine	

of the catabolic hormones may have many negative physiological effects. Supraphysiological concentrations of anabolic hormones can have many adverse consequences.

Insulin

Insulin probably has been studied more than any other hormone with respect to the regulation of protein metabolism. The long-standing interest in insulin probably derives from the fact that one of the hallmarks of insulin deficiency (insulin-dependent diabetes, before insulin therapy was available) is emaciation, including both the fat and the protein stores of the body. Thus, insulin is essential for maintaining a normal protein mass. Insulin concentrations vary over a ~ 10-fold range in a normal person, from ~ 50 pmol/L after an overnight fast to ~ 500 pmol/L shortly after a meal. Although postabsorptive insulin levels are much lower than postprandial levels, even the low postabsorptive insulin concentration can have major metabolic effects. Insulin probably is one of the key factors in determining the acute response of protein metabolism to meals, and a decrease in insulin levels may be one of the major determinants of the initial increase in proteolysis during starvation (Chapter 5).

Effect of Insulin on Whole-Body Protein Metabolism

Insulin Deprivation of Diabetic Patients

Diabetes mellitus is divided into two categories—insulin-dependent (IDDM or type 1) and non-insulin-dependent (NIDDM or type 2). The primary problem with patients in the latter category is resistance to insulin. They secrete enough insulin to maintain glucose homeostasis in a nondiabetic person, but their response to insulin is inadequate. Protein metabolism in these patients is discussed in Chapter 9 and will not be considered here. Patients with IDDM do not secrete enough insulin because of destruction of the pancreatic beta cells that produce insulin. They afford the opportunity to examine the importance of maintaining normal levels of insulin. When insulin is withdrawn from patients with IDDM, their protein balance becomes markedly negative. Restoration of insulin therapy reverses this effect. A study published shortly after insulin became available to treat IDDM illustrates this point very well (Atchley et al. 1933). Figure 6.1 shows data on N balance obtained in that study, in which two patients with IDDM were deprived of their insulin for several days under very controlled conditions. N balance was maintained near equilibrium when they were treated with insulin, became negative very

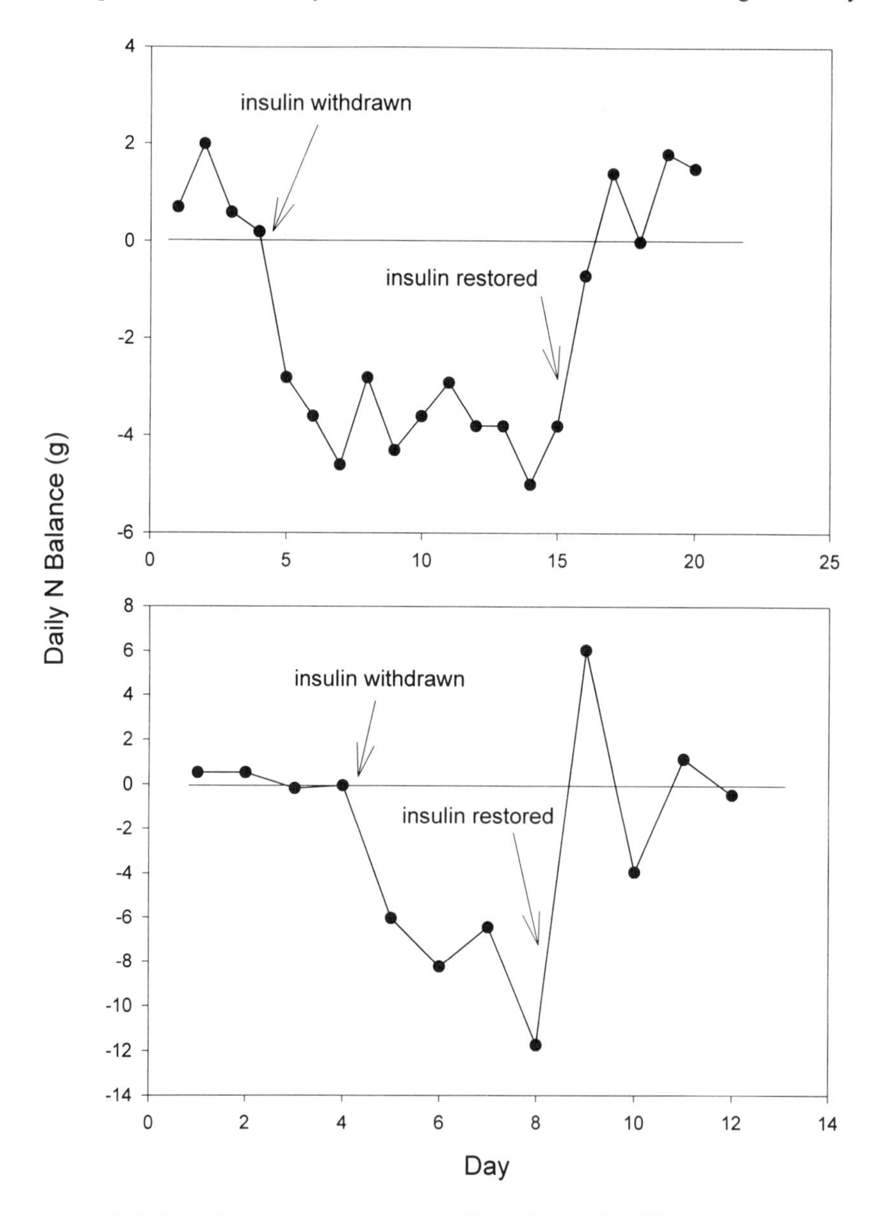

FIGURE 6.1. Effect of insulin treatment and insulin withdrawal on N balance in 2 patients with insulin-dependent diabetes mellitus. Data of Atchley et al. (1933).

rapidly after treatment was withdrawn and became neutral again very soon after insulin treatment was restored. The magnitude of the N losses during insulin deficiency was similar to what is observed during total starvation.

The protein catabolic effect of insulin deprivation is mediated primarily by an increase in whole-body proteolysis. Compared with nondiabetic subjects, patients with IDDM have postabsorptive rates of leucine R_a that are ~ 20% to 50 % faster when insulin is withdrawn or is inadequate to control glucose levels well (Nair et al. 1983; Umpleby et al. 1986; Luzi et al. 1990; Carlson and Campbell 1993; Biolo et al. 1995a). IDDM patients who are given enough insulin to maintain normal glucose concentrations have significantly slower rates of leucine R_a than those given no insulin, or not enough insulin to prevent hyperglycemia (Robert et al. 1985; Umpleby et al. 1986; Nair, Ford, and Halliday 1987; Luzi et al. 1990; Pacy, Thompson, and Halliday 1991; Carlson and Campbell 1993; Nair et al. 1995; Vogiatzi et al. 1997). Phenylalanine R_a also is increased in insulin-deprived IDDM patients (Pacy, Thompson, and Halliday 1991; Biolo et al. 1995a; Nair et al. 1995). The fact that adequate insulin can normalize whole-body protein breakdown indicates that the effect of IDDM is mainly the result of insulin deficiency rather than some other secondary effect of chronic IDDM. A somewhat surprising finding was that whole-body protein synthesis is faster in postabsorptive, insulin-deprived IDDM patients than in normal individuals or IDDM patients given adequate insulin. The nonoxidized portion of leucine R_d, an index of whole-body protein synthesis, was 5% to 50% more rapid in the IDDM patients when they were not given insulin, or when they were not given enough insulin to prevent marked hyperglycemia (Nair et al. 1983; Robert et al. 1985; Umpleby et al. 1986; Nair, Ford, and Halliday 1987; Luzi et al. 1990; Pacy, Thompson, and Halliday 1991; Nair et al. 1995; Vogiatzi et al. 1997). The portion of phenylalanine R_d not converted to tyrosine, another index of protein synthesis, was 15% to 24% more rapid when insulin was withdrawn than when it was given in a dose that normalized glucose levels (Pacy, Thompson, and Halliday 1991; Nair et al. 1995). The increase in whole-body protein synthesis during insulin deprivation of IDDM patients appears to be accounted for by increased splanchnic protein synthesis, with no increase in muscle protein synthesis (Nair et al. 1995). A finding of more rapid protein synthesis was unexpected in light of previous cell culture and in vivo animal studies indicating that insulin deprivation reduces protein synthesis. However, the insulin deprivation causes many other metabolic effects, and in vivo these effects could offset the tendency of insulin deficiency to slow protein synthesis. The increase in amino acid concentrations because of increased proteolysis, for example, could help to maintain elevated rates of protein synthesis in spite of insulin deficiency (see section on amino acid effects in Chapter 7). If leucine incorporation into proteins is expressed as a fraction of the leucine R_a, then insulin deficiency clearly inhibits the use of leucine for protein synthesis. Leucine oxidation is increased relative to leucine R_a during insulin deficiency in IDDM patients (Nair et al. 1983; Robert et al. 1985; Umpleby et al. 1986; Nair et al. 1995). The conversion of leucine to its ketoacid, KIC, is accelerated even more than leucine oxidation (Nair et al. 1995).

In spite of the fact that insulin deficiency can increase whole-body protein synthesis, insulin deficiency is catabolic because proteolysis increases more than protein synthesis.

Insulin Effects on Whole-Body Protein Metabolism in Nondiabetic Subjects

Insulin concentrations can be reduced from the normal postabsorptive level by somatostatin, a hormone that inhibits secretion of insulin and several other hormones. After infusion of somatostatin into postabsorptive subjects, the average leucine R_a increases slightly (Nair et al. 1987b; McHardy et al. 1991). The increases in proteolysis during somatostatin infusion in normal subjects is much less than the increases in protein breakdown in insulin-deprived IDDM patients, but the degree of insulin deficiency is less, and even a minimal amount of insulin may be sufficient to restrain proteolysis. The somatostatin studies are somewhat difficult to interpret, because secretion of glucagon, growth hormone, and other hormones also is inhibited by somatostatin. Nevertheless, the trend toward elevated proteolysis during somatostatin-induced insulin deficiency in normal subjects is consistent with the effect of insulin deficiency in IDDM.

Many studies have demonstrated that insulin infusion inhibits whole-body proteolysis, as reflected by the R_a of leucine or other essential amino acids. In all of these studies, hypoglycemia was prevented by simultaneous infusion of glucose. Because insulin also reduces plasma concentrations of many amino acids, some investigators have infused amino acids during insulin administration to prevent hypoaminoacidemia. However, none of these studies resulted in ideal amino acid replacement. Some amino acids were more concentrated during insulin+glucose+ amino acid administration than they were under basal conditions, whereas some were less concentrated. Generally, total amino acid levels have been elevated above the basal level in such studies. Whether or not amino acid levels decrease or increase in studies of hyperinsulinemia, the whole-body R_a of leucine and other essential amino acids declines in response to insulin infusion. The magnitude of the effect is greater when insulin infusion is combined with amino acid replacement, as expected from the independent effect of amino acids (discussed in Chapter 7).

When insulin is given to normal adult volunteers without amino acid replacement, the leucine R_a declines rapidly, falling by as much as ~ 40% from postabsorptive values within 2 to 3 h (Castellino et al. 1987; Shangraw et al. 1988; Flakoll et al. 1989). Increases in insulin concentrations of as little as 30 to 40 pmol/L (cf. postabsorptive levels of ~ 50 pmol/L) can induce detectable decreases in whole-body proteolysis (Jensen and Haymond 1991). At insulin concentrations similar to those observed after ordinary meals (~ 200–600 pmol/L), the suppression of leucine R_a is ~ 15% to 30% (Fukagawa et al. 1985; Tessari et al. 1986b, Tessari et al. 1987; Caballero and Wurtman 1991;

Tessari et al. 1991; Welle et al. 1994b; Luzi, Castellino, and DeFronzo 1996). Several dose-response studies have been done (Figure 6.2). Three of these suggest that half-maximal suppression of leucine R_a occurs at insulin levels of ~ 250 pmol/L or even less (Fukagawa et al. 1985; Tessari et al. 1986b; Shangraw et al. 1988), whereas another suggests that insulin levels as high as 1000 pmol/L might be required for half-maximal suppression (Flakoll et al. 1989). Even when insulin levels exceed 4,000 pmol/L, suppression does not exceed ~ 40% (Fukagawa et al. 1985; Tessari et al. 1986b; Shangraw et al. 1988; Flakoll et al. 1989). The suppression of proteolysis persists for at least 12 h during continuous insulin administration, but is diminished after 24 h of insulin infusion (Petrides, Luzi, and DeFronzo 1994).

Raising insulin levels from postabsorptive to typical postprandial levels, without infusing amino acids, reduces total amino acid oxidation, as reflected by urea excretion and the change in total body urea, by > 50% (Tappy, Owen, and Boden 1988; Ang, Halliday, and Powell-Tuck 1995). However, NH_3 excretion is increased during insulin administration (Ang, Halliday, and Powell-Tuck 1995). Leucine oxidation typically decreases when insulin is infused without amino acid replacement (Tessari et al. 1986b; Castellino et al. 1987; Tessari et al. 1987;

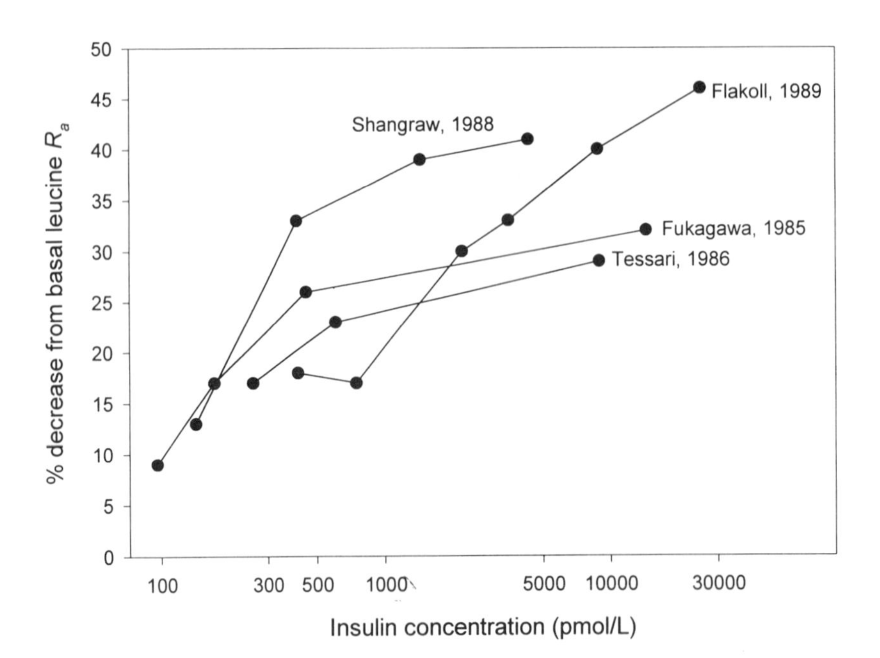

FIGURE 6.2. Suppression of whole-body proteolysis, as reflected by leucine R_a, by insulin. Curves show the mean suppression of proteolysis during insulin infusion (with glucose to maintain euglycemia), compared to rate at the postabsorptive insulin concentration, as a function of the plasma insulin concentrations in 4 studies in which multiple insulin levels were examined. Figure includes only studies in which amino acids were not infused.

Shangraw et al. 1988; Flakoll et al. 1989; Luzi, Castellino, and DeFronzo 1996), but this effect is not always found, particularly at lower doses of insulin (Fukagawa et al. 1985; Caballero and Wurtman 1991; Petrides, Luzi, and DeFronzo 1994). The drop in proteolysis undoubtedly is a major factor in the suppression of amino acid oxidation that accompanies insulin administration.

The rate of whole-body protein synthesis, as reflected by the nonoxidized portion of leucine R_d, declines during insulin infusion when no amino acids are infused (Tessari et al. 1986b; Castellino et al. 1987; Tessari et al. 1987; Flakoll et al. 1989; Petrides, Luzi, and DeFronzo 1994; Luzi, Castellino, and DeFronzo 1996). Nevertheless, protein synthesis is suppressed less than protein breakdown, so that the net effect of insulin is anabolic even when there is no amino acid replacement. Ordinarily, insulin is secreted after meals that usually contain enough protein to maintain or raise the plasma amino acid concentrations. Thus, the reduction in protein synthesis during insulin infusion in these studies must be considered in the context of the reduced amino acid levels and does not disprove a role for insulin in postprandial stimulation of protein synthesis.

Most of the studies of insulin on protein kinetics have involved tracing the carbon of leucine or other essential amino acids. A study using the [^{15}N]glycine-urea end product method also indicated that insulin suppresses whole-body protein breakdown and synthesis (Ang, Halliday, and Powell-Tuck 1995). However, when NH_3 rather than urea was used as the traced end product, insulin appeared to increase protein breakdown and synthesis. It was suggested that the NH_3 end product method was invalid during insulin infusion because insulin stimulates renal NH_3 production.

When amino acids are infused along with insulin and glucose, the suppression of whole-body proteolysis is even greater than it is during infusion of insulin and glucose only (Castellino et al. 1987; Tessari et al. 1987; Flakoll et al. 1989; Bennet et al. 1990a) (Figure 6.3). One study even suggested that complete suppression of whole-body proteolysis could be achieved with high insulin and amino acid levels (Flakoll et al. 1989), although others have not reported such a dramatic effect. The rate of oxidation of particular amino acids may be increased, unchanged, or reduced during insulin and amino acid administration, depending on the degree to which concentrations of the amino acids are maintained. When both insulin and amino acids are infused and the amino acid infusion raises the amino acid concentrations above the basal levels, the rate of whole-body protein synthesis is increased above the basal level (Castellino et al. 1987; Tessari et al. 1987; Fukagawa et al. 1989; Bennet et al. 1990a). However, amino acids can stimulate protein synthesis even when no insulin is infused (Chapter 7), and it is not entirely clear whether the increase in insulin or the increase in amino acid levels is responsible for the stimulation of whole-body protein synthesis in these studies. Whole-body protein synthesis does not appear to be stimulated by insulin when the amino acid replacement regimen leads to only small changes in the plasma concentrations of most amino acids (Castellino et al. 1987; Flakoll et al. 1989;

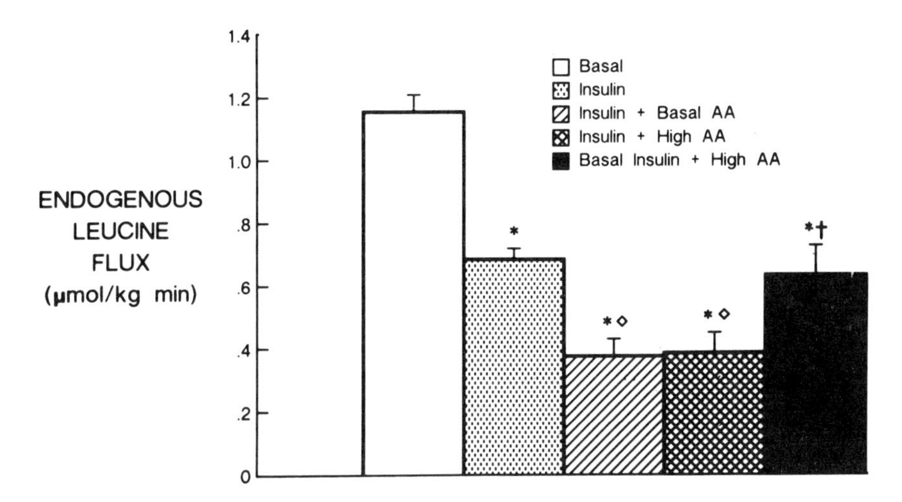

FIGURE 6.3. Enhancement of the insulin-induced suppression of whole-body proteolysis (endogenous leucine flux) by maintenance of plasma amino acid concentrations. Insulin alone significantly reduces proteolysis and plasma amino acid concentrations. Preventing the decline in amino acid concentrations is associated with more suppression of proteolysis at the same insulin concentration. High amino acid levels also suppress proteolysis even if insulin levels do not increase. Reproduced from The Journal of Clinical Investigation, 1987, Vol 80, pp. 1784–93 by copyright permission of The American Society for Clinical Investigation.

Heslin et al. 1992a; Flakoll, Hill, and Abumrad 1993; Tauveron et al. 1995). The problem in interpreting this research is that no studies have accomplished perfect amino acid replacement during insulin administration. In other words, there have been no studies in which insulin has been given systemically without some change in the concentrations of at least some amino acids. Even if perfect replacement of plasma amino acids could be achieved, the intracellular concentrations would not be the same because insulin inhibits proteolysis. Thus, disentangling insulin effects from amino acid effects is very difficult. At present, there is no convincing evidence that insulin stimulates whole-body protein synthesis when amino acid levels are not elevated simultaneously. However, it is quite possible that insulin has a permissive role in allowing amino acids or other stimuli to enhance protein synthesis.

Effect of Insulin on Muscle Protein Metabolism

Three approaches have been used to examine the effect of insulin on muscle protein metabolism in humans—direct measurement of isotope incorporation into muscle proteins, limb amino acid balance studies (with or without tracer), and 3-methylhistidine excretion. The limitations of each of these methods were discussed in Chapter 3. The general consensus from these approaches is that insulin promotes protein anabolism in human muscle by inhibiting proteolysis rather than by stimulating protein synthesis. However, there is not unanimous agreement about this conclusion.

Direct Measurement of Isotope Incorporation

In patients with IDDM, insulin deficiency is not associated with any change in the fractional rate of muscle protein synthesis, as determined by [^{13}C]leucine incorporation with plasma KIC enrichment as the index of intramuscular leucyl-tRNA enrichment (Pacy et al. 1989; Bennet et al. 1990c; Charlton, Balagopal, and Nair 1997). Myosin heavy chain synthesis in skeletal muscle also is unaffected by insulin deficiency (Charlton, Balagopal and Nair 1997). Any inhibitory effect that insulin deficiency might have on muscle protein synthesis might have been offset by a stimulatory effect of high intracellular amino acid concentrations during insulin deficiency. In nondiabetic subjects, raising insulin from postabsorptive levels (~ 50 pmol/L) to typical postprandial levels (~ 200–250 pmol/L) by systemic insulin administration (with glucose to maintain euglycemia), did not affect the incorporation of [$^{2}H_{5}$]phenylalanine into muscle proteins according to the flooding dose method (McNurlan et al. 1994a). Insulin administration reduced the levels of many amino acids, which could have offset any stimulatory effect of insulin. Biolo, Fleming, and Wolfe (1995) reported that intraarterial infusion of insulin, which raised local venous insulin concentrations from 60 to 460 pmol/L, increased phenylalanine incorporation into muscle proteins by 69%. Although use of an intraarterial infusion to raise local insulin levels minimized the increase in systemic insulin, and therefore minimized the decrease in arterial amino acid concentrations, it did not prevent a decrease in the concentrations of free amino acids in muscle. It seems unlikely that the rather small difference in insulin concentrations between these latter two studies completely accounts for the dramatic difference in results. The greater arterial delivery of amino acids to muscle, because of local rather than systemic insulin infusion, also could contribute to the discrepancy. Both studies relied on small numbers of subjects (6 in each), and the 95% confidence intervals for the effect of insulin in these two studies overlap. The weight of evidence from limb tracer balance studies is that physiological increases in insulin levels do not stimulate muscle protein synthesis in adult humans, as described below. Thus, the Biolo, Fleming, and Wolfe (1995) study seems to be the only evidence that insulin can stimulate muscle protein synthesis in humans, when there is no simultaneous infusion of amino acids.

Limb Tracer Balance Studies

Muscle is likely to be the major determinant of amino acid balance, uptake, and release from the limbs, but it is important to remember that there is some protein metabolism in the skin, bone, and adipose tissue within the limbs. Almost all studies agree that insulin promotes a more positive (or less negative) balance of essential amino acids in the limbs, reflecting a protein anabolic effect of insulin. Most of the research indicates that this effect is mediated by reduced proteolysis, but Biolo, Fleming, and Wolfe (1995) suggested that increased protein synthesis is the primary effect of insulin. This issue is examined more closely below.

In IDDM patients, raising insulin to normal postprandial values (250–600 pmol/L) reduces forearm and leg proteolysis, as reflected by leucine and phenylalanine R_a (Tessari et al. 1990; Bennet et al. 1991). Even a small increment in insulin, from 20 to 70 pmol/L, reduces leucine and phenylalanine R_a in the leg by ~ 40% (Nair et al. 1995). There was no evidence for stimulation of protein synthesis by insulin in these studies, as reflected by unchanged or reduced limb phenylalanine and leucine R_d during insulin infusion. Similarly, in nondiabetic subjects insulin administration generally has inhibited proteolysis in the forearm or leg (Table 6.2). Even studies that reported no effect of insulin usually have had a statistically nonsignificant reduction in proteolysis. When there is no amino acid infusion during insulin administration, the effect on limb protein synthesis has been quite variable, ranging from small decreases to large increases (Table 6.2). Except for one study, none of the changes in limb protein synthesis[1] during insulin administration was statistically significant, and the overall average effect of insulin on limb protein synthesis is close to zero.

The study of Biolo, Fleming, and Wolfe (1995) stands out as giving different results from the others and is worthy of further discussion. They argued that the suppression of limb proteolysis is overestimated in the other studies, because the other studies did not adequately model the amino acid kinetics. Their model required an estimate of the free amino acid tracer enrichment in the muscle tissue, which they obtained by biopsy. Had they used the usual tracer balance calculations (described in Chapter 3), they would have concluded that insulin did suppress proteolysis (phenylalanine and leucine R_a, although lysine R_a still would not be suppressed). Their model attempts to account for the unlabeled amino acid that dilutes the intracellular free amino acid tracer pool, but is reincorporated into protein without diluting the tracer in the venous efflux from the limb. Because they found that protein synthesis was markedly increased by insulin, more of this unlabeled amino acid would be used for protein synthesis than for efflux from the tissue. Their model, as well as direct incorporation of tracer into muscle proteins, indicated that insulin increased protein synthesis by 36% to 69%, depending on the method used. When they used the calculations employed by other investigators, they also found increased protein synthesis during insulin administration—59% with phenylalanine as the tracer and 77% with lysine as the tracer. These increases in protein synthesis are much greater than those found by any other investigators using similar methods and calculations.

[1] Leucine R_d in the limb is not a valid measure of protein synthesis, because leucine can be oxidized or deaminated by muscle. Phenylalanine, tyrosine, and lysine R_d are valid indices of protein synthesis in the limb.

TABLE 6.2. Effect of insulin on limb amino acid R_a and R_d in non-diabetic subjects.

Study	Limb	~Insulin concentration (pmol/L)	Amino acid	Δ% Ra	Δ% Rd
Insulin infusion without amino acid replacement					
Gelfand and Barrett 1987	Forearm	700	phe	-25	-2
			leu	-50	-2
Fryburg et al. 1990	Forearm	200	phe	-19	+19
			leu	-37	+9
Arfvidsson et al. 1991	Leg	600	tyr	-13	-15
Denne et al. 1991	Leg	16,000	phe	-41	-21
Tessari et al. 1991	Forearm	450	phe	-23	-10
			leu	-26	-7
Heslin et al. 1992a	Forearm	400	leu	-37	+18
Louard et al. 1992	Forearm	200	phe	-33	0
			leu	-32	-3
		250	phe	-26	+10
			leu	-41	+9
		400	phe	-44	0
			leu	-28	+34
		750	phe	-42	-2
			leu	-50	+10
Biolo, Fleming and Wolfe 1995	Leg	450	phe	-7	+48
			leu	+6	+39
			lys	+10	+36
			phe[a]	-28	+59
			leu[a]	-17	
			lys[a]	+4	+77
Insulin + amino acid infusion					
Bennet et al. 1990a	Leg	600	phe	-11	+55
			leu	-30	+48
Newman et al. 1994	Forearm	450	phe	-17	+68
			leu	-42	+5
Moller-Loswick et al. 1994[b]	Leg	650	phe	-48	-32
			tyr	-39	-25
			phe	-65	-52
			tyr	-48	-39
Hillier et al. 1998	Forearm	50,000	phe	-17	+100-150[d]

Values in last 2 columns are mean values; many of these changes were not statistically significant at $P < 0.05$.

[a]Calculated using model employed by other investigators rather than the more complex model used by these authors.

[b]There was a slight reduction in amino acid levels during insulin + amino acid infusion in this study, because amino acids also were infused in baseline period, and amino acid infusion rate was not increased during insulin infusion.

[d]Degree of stimulation of protein synthesis dependent on rate of amino acid infusion.

When amino acids are infused during insulin administration, limb proteolysis is inhibited, although the magnitude of the effect does not appear to be any greater than the suppression of proteolysis induced by insulin alone (Table 6.2). The main additional effect of combining amino acid with insulin administration is a stimulation of limb protein synthesis, as reflected by phenylalanine R_d. However, this effect is observed only when plasma amino acids levels increase during the combined infusion, or when insulin levels are extremely high. When amino acids have already been infused in the basal period before insulin administration, and amino acid levels decline during the combined insulin+amino acid infusion, then limb protein synthesis can be inhibited rather than stimulated (Moller-Loswick et al. 1994). In studies in which both insulin and amino acids levels increase (Bennet et al. 1990a; Newman et al. 1994), there is some question about whether the insulin or the amino acids are the primary factor in mediating the increase in protein synthesis. In patients with IDDM, leg protein synthesis was about twice as fast with simultaneous administration of insulin and amino acid than with insulin only (Bennet et al. 1991). However, amino acids alone were almost as effective as amino acids+insulin in stimulating leg protein synthesis in these patients.

Efflux of 3-methylhistidine from a limb is an index of the degradation of myofibrillar proteins (see Chapter 3). In two studies, insulin administration failed to significantly inhibit 3-methylhistidine efflux from the leg or forearm, even though tyrosine and phenylalanine R_a were inhibited (Arfvidsson et al. 1991; Moller-Loswick et al. 1994). If these results are accurate, insulin selectively inhibits degradation of proteins other than actin and myosin. However, small changes in myofibrillar degradation could easily be missed by this method, because of the very small arterial-venous differences in 3-methylhistidine concentrations. In one of the studies, 3-methylhistidine efflux from the leg was undetectable during insulin administration, but the suppression of efflux from baseline still did not achieve statistical significance (Arfvidsson et al. 1991).

Cardiac muscle appears to be similar to skeletal muscle in the response of protein metabolism to insulin. Phenylalanine R_a in the heart was suppressed 80% by insulin administration (insulin levels of ~ 1,600 pmol/L) in men with coronary artery disease, even though the subjects were insulin resistant (McNulty et al. 1995). Insulin did not stimulate phenylalanine R_d.

Daily 3-Methylhistidine Excretion

Patients with IDDM, when they do not receive enough insulin to control their hyperglycemia very well, excrete more 3-methylhistidine than nondiabetic subjects. One group of IDDM patients excreted 42% more 3-methylhistidine per kg body weight than a group of nondiabetic subjects (Huszar et al. 1982). Although they received insulin during the study, their average glucose concentration was over 11 mmol/L, indicating that the dose was inadequate.

When they received 15% or 25% less insulin than usual, 3-methylhistidine did not change significantly. The ratio of 3-methylhistidine to creatinine excretion was not affected by diabetes in that study because of the unexpected finding that the IDDM patients excreted significantly more creatinine than the controls. It is very unlikely that the diabetics had 40% more muscle than weight-matched controls. Thus, either the diabetic patients were more careful than the controls about collecting all urine (the creatinine excretion values in the controls seem too low), or diabetes somehow alters the relation between muscle mass and creatinine excretion. Increased creatinine production per kg muscle mass in diabetic patients is supported by another study showing that diabetic subjects (both insulin-dependent and noninsulin-dependent) with severe fasting hyperglycemia (15.6 mmol/L) excreted 20% more creatinine than they did when their glycemia was fairly well controlled (6.3 mmol/L) (Marchesini et al. 1982). Their 3-methylhistidine excretion was increased by 35% when they were more hyperglycemic. The IDDM patients excreted 67% more 3-methylhistidine when they were more hyperglycemic, whereas the NIDDM patients excreted only 14% more. There was no difference in 3-methylhistidine excretion between diabetic and nondiabetic subjects when the diabetic patients were less hyperglycemic.

Effect of Insulin on Other Tissues and Individual Proteins

Splanchnic proteolysis was ~ 30% to 50% more rapid in insulin-deprived IDDM patients (free insulin level of 18 pmol/L) than it was when they received enough insulin to normalize their glucose concentrations (insulin level of 72 pmol/L) (Nair et al. 1995). The increase in splanchnic protein synthesis was even greater than the increase in proteolysis during insulin deprivation. Thus, insulin deprivation actually resulted in a more positive splanchnic protein balance, which is quite different from the effect on muscle. The elevated splanchnic protein synthesis during insulin deprivation could account for all of the increase in whole-body protein synthesis. There have not been any studies of insulin infusion on splanchnic protein metabolism in nondiabetic subjects. Thus, it is unclear whether or not raising the insulin concentrations above the usual postabsorptive levels would have any further effect on splanchnic protein metabolism.

Albumin synthesis in IDDM patients was reported to decrease 29% when their insulin was withdrawn for 12 h, whereas fibrinogen synthesis increased 50% (De Feo, Gaisano, and Haymond 1991). However, in another study there was no effect of insulin withdrawal on albumin synthesis of IDDM patients (Pacy, Read, and Halliday 1990). Both of these studies relied on plasma KIC enrichment as an index of hepatic leucyl-tRNA enrichment during labeled leucine infusion. Because there is no guarantee that insulin does not alter the relation between plasma KIC and hepatic leucyl tRNA enrichment, the quantitative aspects of these studies are suspect. However, this problem does not invalidate the dissociation between insulin's effect on albumin synthesis and its effect on fibrinogen synthesis. Insulin administration in normal subjects, resulting in insulin levels of ~ 400 pmol/L, reduced apolipoprotein B production by 53%, according to measurements of apolipoprotein concentrations and clearance of radiolabeled apolipoprotein (Lewis et al. 1993).

Raising insulin levels from 40 to 210 pmol/L reduced proteolysis (phenylalanine R_a) in subcutaneous abdominal adipose tissue by 55% (Coppack, Persson, and Miles 1996). The hyperinsulinemia was induced by hyperglycemia, but it is very likely that the high insulin rather than the high glucose levels inhibited adipose tissue proteolysis. Protein synthesis (phenylalanine R_d) also tended to decline by a similar percentage, but the effect was not statistically significant. This level of hyperinsulinemia usually causes reduced forearm proteolysis (Table 6.2), but this effect was not observed in this study.

Possible Mechanisms of Insulin Effects on Protein Metabolism

In cell culture and animal studies, the reduction in proteolysis seems to be related to reductions in both lysosomal and proteasomal protein degradation. Insulin reduces the number of autophagic vacuoles in hepatocytes and cardiac myocytes, and inhibitors of lysosomal proteolysis attenuate the insulin-induced suppression of proteolysis (Lee and Marzella 1994; Long et al. 1984; Seglen and Bohley 1992). In skeletal muscle of rats made insulin-deficient with streptozotocin, the elevated proteolysis was not blocked by inhibition of lysosomal or calcium-dependent proteases, but was blocked by inhibiting ATP synthesis and proteasome activity (Price et al. 1996). Moreover, insulin deficiency increased the levels of mRNAs encoding ubiquitin and proteasome subunits (Price et al. 1996). Insulin suppressed ubiquitin mRNA levels in goat muscle, but not in other tissues (Larbaud et al. 1996). There is little evidence for a stimulatory effect of insulin on protein synthesis in adult humans, but there is some debate about this conclusion as discussed above. In various cell culture and animal experiments, there has been evidence that insulin promotes protein synthesis by increasing RNA production (ribosomal and total mRNA), inhibiting ribosomal degradation, and promoting translation initiation (Flaim, Copenhaver, and Jefferson 1980; Hsu et al. 1992; Antonetti et al. 1993). Stimulation of protein synthesis in muscle may depend on increased prostaglandin $F_{2\alpha}$ production (Reeds and Palmer 1983; Reeds et al. 1985). The effect of insulin on translation initiation may be mediated by the activity of eIF2-B (Kimball and Jefferson 1988; Karinch et al. 1993) and eIF-4E (O'Brien 1994). The effect on eIF-4E appears to be caused by phosphorylation of the protein PHAS-I, which releases it from eIF-4E so that the latter protein is free for association with eIF-4G and subsequent cap binding (O'Brien 1994; Kimball et al. 1997). Whether or not changes in amino acid transport induced by insulin can influence either proteolysis or protein synthesis is uncertain. Insulin was reported to cause a modest increase in system A amino acid transport into human muscle (Bonadonna et al. 1993), but another study suggested that the effect of insulin on amino acid transport was not an important factor in regulating muscle protein metabolism (Biolo, Fleming, and Wolfe 1995). More information is needed about the regulatory role of intracellular concentrations or transmembrane uptake of specific amino acids before any conclusions can be made about the importance of amino acid transport to the regulation of protein metabolism.

Growth Hormone

Growth hormone is the primary regulator of the production of insulin-like growth factor-I (IGF-I), which may mediate most of the effects of growth hormone on protein metabolism. There is no way to raise or lower growth hormone levels *in vivo* without affecting IGF-I production, so this issue is difficult to address. In this section, the effects of growth hormone deficiency and growth hormone administration are discussed without any attempt to distinguish between direct effects and those mediated by IGF-I.

Effect of Growth Hormone on Body Composition and N Retention

Growth hormone deficiency causes a reduction in the fat free mass of the body and an increase in the fat mass, effects that are reversed by growth hormone replacement (De Boer, Blok, and Van der Veen 1995). Excessive secretion of growth hormone from the pituitary gland, or exogenous administration of growth hormone to normal subjects, has the opposite effect (Rudman et al. 1990; Leger et al. 1994; O'Sullivan et al. 1994; Vaisman et al. 1994; Thompson et al. 1995;). The protein concentration of the fat free mass could be altered by derangements of growth hormone levels, so some caution is required in equating changes in fat free mass with changes in protein mass. Nevertheless, measurement of total body N, probably the best index of total body protein mass, indicates a role for growth hormone in regulating the protein mass of the body. After 2 yr of growth hormone replacement therapy, adults who had been growth hormone deficient had an increase in total body N of ~ 10% to 15% (Bengtsson et al. 1993; Johannsson et al. 1997). Studies with imaging techniques and urinary creatinine excretion suggest that much of the increase in protein mass associated with growth hormone administration occurs in skeletal muscle (Cuneo et al. 1991; Bengtsson et al. 1993; Leger et al. 1994; Jorgensen et al. 1996; Welle et al. 1996a). Children with growth hormone deficiency or excess have much larger changes in fat free mass than adults, because growth hormone deficiency markedly inhibits linear growth and growth hormone excess markedly accelerates it.

N balance studies also support an anabolic role for growth hormone under various conditions. Growth hormone administration for several days promotes N retention in healthy adults (Horber and Haymond 1990), in growth hormone-deficient adults (Valk et al. 1994), in obese patients during energy deprivation (Clemmons et al. 1987), in HIV-positive men (Mulligan et al. 1993), in subjects on short-term or chronic glucocorticoid treatment (Horber and Haymond 1990; Giustina et al. 1995), in patients recovering from surgery (Hammarqvist et al. 1992) or multiple trauma (Petersen, Holaday, and Jeevanandam 1994), in patients on a hypocaloric parenteral nutrition regimen (Manson, Smith, and Wilmore 1988), in normal men on a very-low-protein diet (Lundeberg et al. 1991), in malnourished elderly men (Kaiser, Silver, and Morley 1991), and in healthy elderly men and women (Marcus et al. 1990; Butterfield et al. 1997).

Effect of Growth Hormone on Whole-Body Protein Turnover and Amino Acid Oxidation

In contrast to the anabolic effect of insulin, the anabolic effect of growth hormone does not include an inhibition of whole-body proteolysis. Several studies have demonstrated that the R_a of leucine does not change during administration of growth hormone (Horber and Haymond 1990; Russell-Jones et al. 1993; Copeland and Nair 1994; Mauras 1995b; Welle et al. 1996a; Bowes et al. 1997). The [^{15}N]glycine end product method suggested that whole-body protein breakdown increased an average of 8% (per kg body weight) in elderly women treated with growth hormone for a month, but it is unclear how much of this effect can be attributed to the change in body composition (Butterfield et al. 1997). This method did not indicate any significant effect of growth hormone replacement on the whole-body protein breakdown of growth hormone-deficient patients (Binnerts et al. 1992).

Growth hormone diverts more of the amino acids coming from breakdown of endogenous and dietary proteins into the protein synthetic pathway, with a smaller proportion being oxidized. Total N excretion, an index of whole-body protein oxidation, is reduced by growth hormone under a number of conditions as noted above. The oxidation of leucine also is reduced by growth hormone (Russell-Jones et al. 1998; Fryburg and Barrett 1993; Russell-Jones et al. 1993; Copeland and Nair 1994; Mauras 1995b; Bowes et al. 1997). The increase in leucine oxidation induced by glucocorticoid administration is abolished by growth hormone treatment (Horber and Haymond 1990; Oehri et al. 1996). The nonoxidized portion of leucine R_d, an index of whole-body protein synthesis, usually increases in response to growth hormone administration. In healthy subjects, this effect often has been small and sometimes does not reach statistical significance (Yarasheski et al. 1992; Fryburg and Barrett 1993; Copeland and Nair 1994; Mauras 1995b; Yarasheski et al. 1995; Welle et al. 1996a). In growth hormone-deficient patients, whole-body protein synthesis increased ~ 20% after two months of hormone replacement (Russell-Jones et al. 1993; Russell-Jones et al. 1998). There is some dissociation between the minimum dose of growth hormone required to normalize IGF-I levels and the minimum dose needed to stimulate protein synthesis in growth hormone-deficient patients (Lucidi et al. 1998). Normal subjects and cancer patients treated with large doses of growth hormone for 3 to 7 days had rates of nonoxidized leucine R_d that were > 20% faster than those of placebo treated subjects (Horber and Haymond 1990; Wolf et al. 1992a; Wolf et al. 1992b). Growth hormone also increased whole body protein synthesis in glucocorticoid-treated subjects, and enhanced the stimulation of protein synthesis induced by insulin + glucose + amino acid infusion in these subjects (Berneis et al. 1997). The [^{15}N]glycine end product method indicated that whole-body protein synthesis increased ~ 15% after one month of hormone replacement in growth-hormone deficient patients (Binnerts et al. 1992). This method revealed a similar increase in protein synthesis in healthy young subjects who received growth hormone for 3 months along with a resistance exercise program, although the exercise program itself had no effect on protein synthesis (Yarasheski et al. 1992). This method also indicated that growth hormone increased whole-body protein synthesis in healthy elderly women (+9%) (Butterfield et al. 1997) and in trauma patients (+28%) (Petersen, Holaday and Jeevanandam 1994). Overall, the evidence is overwhelmingly in favor of the concept that growth hormone promotes protein anabolism by reducing amino acid oxidation and promoting protein synthesis, and not by inhibiting proteolysis.

Effect of Growth Hormone on Muscle Protein Metabolism

An anabolic effect of growth hormone on muscle is indicated by reduced muscle mass in growth hormone-deficient patients (De Boer, Blok, and Van der Veen 1995), and by increased muscle volume (by imaging methods) and creatinine excretion (an index of muscle mass) after growth hormone administration to growth hormone-deficient patients and healthy volunteers (Jorgensen et al. 1989; Cuneo et al. 1991; Bengtsson et al. 1993; Leger et al. 1994; Welle et al. 1996a). There is no evidence that this effect is mediated by reduced protein breakdown in muscle. Excretion of 3-methylhistidine does not decrease with growth hormone administration, suggesting that myofibrillar degradation is unaffected (Yarasheski et al. 1995; Welle et al. 1996a; McNurlan et al. 1997). Moreover, there is no evidence from limb protein turnover studies that proteolysis is inhibited by growth hormone (Table 6.3). Thus, any anabolic effect of growth hormone in muscle must be mediated by increased protein synthesis. Several studies have demonstrated increased muscle protein synthesis after growth hormone administration, but others have not (Table 6.3). Some of the failures to show increased muscle protein synthesis can be explained. For example, Copeland et al. (1994) used a shorter period of growth hormone infusion and a smaller dose of hormone than other investigators. Yarasheski et al. (1992, 1995) combined growth hormone treatment with a resistance exercise program. The exercise program itself caused a substantial increase in muscle protein synthesis, which may have prevented any further effect of growth hormone. Another factor in these studies is that

TABLE 6.3. Effect of growth hormone on muscle protein metabolism.

Study	Subjects	GH concentration (ng/ml)	Dose and duration	Method	%Δ synthesis	%Δ degradation
Fryburg, Gelfand and Barrett 1991	Healthy young adults	35, deep venous	6 h local (forearm) 14 ng/kg/min	Forearm phe Rd. Ra	+71	16
Wolf et al. 1992a	Healthy adults	59	3d, 100–200 μg/kg/d	Forearm phe Rd, Ra	+148	+64
Wolf et al. 1992b	Cancer patients	10	3d, 100–200 μg/kg/d	Forearm phe Rd, Ra	+18	-5
Fryburg and Barrett 1993	Healthy young adults	32	6 h, 60 ng/kg/min	Forearm phe Rd, Ra	+68	11
Copeland and Nair 1994	Healthy young adults	12–20	3.5 h, 33 ng/kg/min	Leg phe Rd, Ra	-43	-35
Garibotto et al. 1997	Malnourished hemodialysis patients	Not reported	6 wk, 5 mg 3 times/wk	Forearm phe Rd, Ra	+25	0
Yarasheski et al. 1992	Healthy young adults, during resistance training	Not reported	12 wk 40 μg/kg/d	[^{13}C]leucine incorporation into vastus lateralis proteins	+6 (relative to effect of exercise alone)	
Yarasheski, Zachwieja and Bier 1993	Experienced weight lifters, young adults	Not reported	2 wk, 40 μg/kg/d	[^{13}C]leucine incorporation into vastus lateralis proteins	0	
Yarasheski et al. 1995	Healthy elderly, during resistance training	Not reported	16 wk, 12–24 μg/kg/d	[^{13}C]leucine incorporation into vastus lateralis proteins	+22 (relative to effect of exercise alone)	
Welle et al. 1996a	Healthy elderly	18, peak	single sc injection, 30 μg/kg	[^{13}C]leucine incorporation into vastus lateralis	+2	
			3 month, 30 μg/kg 3 times/wk	proteins (myofibrillar only)	0 (excludes one outlier in placebo group)	
Butterfield et al. 1997	Healthy elderly women	Not reported	4 wk, 25 μg/kg/d	[^{13}C]leucine incorporation into vastus lateralis proteins	+49	
McNurlan et al. 1997	Healthy adults	Not reported	2 wk, 6 mg/d	[$^{2}H_{5}$]phe incorporation	+25	
	HIV infected			into vastus lateralis	+21	
	AIDS			proteins	-17	
	AIDS with weight loss				-41	

Values in last 2 columns are mean values; many of these changes were not statistically significant at $P < 0.05$.

growth hormone levels may have declined to normal by the time the protein synthesis measurements were made, even though IGF-I levels were still elevated. Studies done after a few months of injections may have been too late to detect an increase in the fractional rate of myofibrillar synthesis, because the anabolic effect of growth hormone can wane with chronic treatment (Snyder, Clemmons, and Underwood 1988; Binnerts et al. 1992).

Effect of Growth Hormone on Protein Metabolisms in Other Tissues

Growth hormone promotes growth of bone and connective tissue. Growth hormone deficient patients have low levels of markers of collagen turnover, and growth hormone administration increases the levels of these markers. Indices of collagen turnover that are affected by growth hormone deficiency and administration include urinary hydroxyproline and pyridinoline excretion, and serum levels of the C-terminal propeptide of type I collagen, the C-terminal pyridinoline cross-linked telopeptide of type I collagen, and the N-terminal propeptide of type III collagen (Marcus et al. 1990; Binnerts et al. 1992; Bengtsson et al. 1993; Holloway et al. 1994; Giustina et al. 1995; Bollerslev et al. 1996; Johansson et al. 1996; Mauras, Doi, and Shapiro 1996; Burman et al. 1997;). These studies, collectively, suggest that both bone and non-bone collagen synthesis and breakdown are stimulated by growth hormone.

Growth hormone does not seem to affect albumin or fibrinogen synthesis in healthy adults when plasma KIC enrichment is used to estimate hepatic leucyl-tRNA enrichment during tracer leucine infusion (De Feo, Horber, and Haymond 1992; Zachwieja, Bier, and Yarasheski 1994). However, the phenylalanine flooding dose method suggested that growth hormone stimulates albumin synthesis by ~25% (McNurlan et al. 1998).

In steers in which growth hormone stimulated muscle protein synthesis, there was no effect on the fractional rate of protein synthesis in the liver and gut (Eisemann, Hammond, and Rumsey 1989). In rats recovering from surgery, growth hormone stimulated the fractional rate of protein synthesis in skeletal and jejunal smooth muscle, but not in liver or jejunal mucosa (Lo and Ney 1996). However, growth hormone does stimulate hepatic protein synthesis in growth hormone-deficient mice (Pell and Bates 1992). Growth hormone promotes both collagen and noncollagen protein deposition in muscle, skin, head, and viscera of growing pigs (Caperna, Gavelek, and Vossoughi 1994).

Mechanisms of Growth Hormone Effects on Protein Metabolism

Growth hormone may promote protein synthesis via its stimulation of IGF-I production, although more direct effects cannot be excluded. In vitro and animal studies indicate that growth hormone increases amino acid uptake, RNA content, RNA polymerase activity, and the initiation of translation (Fryburg and Barrett 1995; Umpleby and Russell-Jones 1996). In human muscle, growth hormone can increase the abundance of myosin heavy chain mRNA, relative to total RNA, within 6 h (Fong et al. 1989). The reduction in amino acid oxidation caused by growth hormone could be mediated in part by increased lipolysis and fat oxidation, which reduces the demand for amino acid oxidation.

Insulin-Like Growth Factor-I (IGF-I)

Many of the effects of growth hormone may be mediated by its stimulation of IGF-I production, so studies discussed in the previous section may be relevant to the discussion of IGF-I effects. Most of the circulating IGF-I probably comes from the liver, although many tissues express IGF-I and could contribute to circulating IGF-I to some extent. An autocrine or paracrine effect could be significant in some tissues. There are several IGF-binding proteins that influence the activity of IGF-I, and production of these proteins also is influenced by growth hormone. Thus, a change in circulating IGF-I concentrations during exogenous IGF-I administration is probably not equivalent to the same change induced by growth hormone administration. At very high IGF-I concentrations, this hormone can bind to insulin receptors enough to exert insulin-like effects that may not be elicited by more physiological levels of IGF-I.

Effect of IGF-I on Whole-Body Protein Metabolism

The importance of IGF-I in maintaining normal growth and protein accretion in children is demonstrated by the fact that long-term IGF-I therapy in children who are insensitive to growth hormone can normalize growth and body composition (Backeljauw, Underwood, and The GHIS Collaborative Group 1996). Treatment with IGF-I for a few days to a few weeks induces positive N balance in healthy young adults (Mauras and Beaufrere 1995), elderly women (Thompson et al. 1995), and patients on weight-reducing diets (Clemmons, Smith-Banks, and Underwood 1992). A transient improvement in N balance was noted in AIDS patients treated with IGF-I, although a higher dose was not as effective as a lower dose in this regard (Lieberman et al. 1994). In elderly women, the N balance data were supported by measurements of fat free mass, extracellular water, and total body water (Thompson et al. 1995). Fat free mass also increases in myotonic dystrophy patients treated with IGF-I for 4 months (Vlachopapadopoulou et al. 1995). Protein oxidation, as reflected by urea or total N excretion, is reduced by IGF-I in healthy young adults (Mauras et al. 1997), growth hormone-deficient adults (Hussain et al. 1993), and patients recovering from surgery (Leinskold et al. 1995).

A number of studies have examined the effect of IGF-I on whole-body protein turnover in humans. Acute studies in normal volunteers, in which IGF-I is infused at high doses for up to 28 h, have generally indicated that IGF-I inhibits whole-body proteolysis (leucine R_a) and protein synthesis (nonoxidized portion of leucine R_d) (Mauras, Horber, and Haymond 1992; Turkalj et al. 1992; Elahi et al. 1993; Laager, Ninnis, and Keller 1993), although the changes were not always statistically significant. In these studies the circulating total IGF-I concentrations were increased by 100% to 400%. These doses of IGF-I significantly reduced amino acid concentrations, which might have inhibited whole-body protein synthesis. When amino acids were given intravenously to prevent the decline in amino acid concentrations during a 3 h infusion of IGF-I, which raised IGF-I levels by ~ 200%, there was no change in proteolysis and a 12% increase in protein synthesis (Russell-Jones et al. 1994). When IGF-I was given for 5 days or more, there was no effect on whole-body proteolysis in normal subjects (Mauras and Beaufrere 1995; Mauras et al. 1997), patients with myotonic dystrophy (Vlachopapadopoulou et al. 1995), or patients with AIDS (Lieberman et al. 1994). There was a 14% increase in the whole-body protein synthesis in normal subjects treated with enough IGF-I to raise their IGF-I concentrations by ~ 250% after a week (Mauras and Beaufrere 1995), although this effect was not evident when the hormone was given continuously with a minipump (Mauras et al. 1997). A more modest increase in IGF-I levels (~80%) after 4 months of IGF-I treatment in myotonic dystrophy patients stimulated whole-body protein synthesis by 7% (Vlachopapadopoulou et al. 1995). In AIDS patients, leucine kinetic studies did not reveal any significant change in postabsorptive whole-body protein synthesis after 10 days of IGF-I treatment, which raised IGF-I levels by ~100% (low dose) or ~250% (high dose) (Lieberman et al. 1994). However, there was a trend toward increased (mean increase of 19%) whole-body protein synthesis over 24 hours with the high dose of IGF-I, according to the [^{15}N]glycine end product method. This method also indicated a stimulation of both whole-body protein synthesis and whole-body proteolysis in elderly women treated with IGF-I for a month, with the stimulation of synthesis exceeding the stimulation of breakdown only slightly (Butterfield et al. 1997). Thus, in long term studies the anabolic effect of IGF-I, like that of growth hormone, appears to be mediated by increased protein synthesis rather than reduced protein breakdown.

Effect of IGF-I on Muscle

When IGF-I was infused directly into a brachial artery, at doses that raised deep venous IGF-I levels by 55% to 315%, the forearm phenylalanine balance shifted from being negative to being positive (Fryburg 1994) This effect was mediated by a 50% to 70% increase in forearm protein synthesis, as reflected by the phenylalanine R_d. At the lowest dose, which raised deep venous IGF-I levels by an average of 55%, the protein synthesis increased 62% and proteolysis was unaffected. This response is similar to that elicited by growth hormone. At higher IGF-I doses the forearm proteolysis declined by ~40%, which is similar to the effect of insulin. It is interesting that the higher IGF-I doses raised the deep venous IGF-I concentrations in the contralateral forearm more than the low dose infusion raised them in the infused forearm, yet there was no effect on protein metabolism in the contralateral forearm. Perhaps the free IGF-I concentration was much higher in the forearm into which the IGF-I was infused, because it could take some time for the newly infused hormone to interact with its binding proteins. Thus, the physiological relevance of the suppression of proteolysis at the higher doses is questionable. The intermediate IGF-I dose was later shown to enhance forearm protein synthesis by 180% when combined with systemic hyperaminoacidemia, much more than the 44% increase induced by hyperaminoacidemia alone (Fryburg et al. 1995b). Inhibiting nitric oxide production can prevent the stimulation of forearm protein synthesis by IGF-I (Fryburg 1996). The inhibition of proteolysis at high doses of IGF-I might be mediated to some extent by increased degradation of an enzyme involved in conjugating ubiquitin to proteins (Wing and Bedard 1996).

In elderly women, IGF-I treatment for a month raised muscle protein synthesis ~60% (Butterfield et al. 1997). This effect was similar to that of growth hormone. However, increases in IGF-I levels (50% to 160%) associated with growth hormone treatment were not associated with increased muscle protein synthesis in healthy elderly men (Welle et al. 1996a) and in young weightlifters (Yarasheski et al. 1993) and did not enhance the exercise induced stimulation of muscle protein synthesis (Yarasheski et al. 1992; Yarasheski et al. 1995). In postoperative patients, myofibrillar degradation as reflected by3-methylhistidine excretion was unaffected by IGF-I administration (Leinskold et al. 1995).

Effect of IGF-I on Collagen Metabolism

IGF-I administration increases serum levels of the carboxyterminal propeptide of type I procollagen, an index of type I collagen production, in growth hormone-deficient adults (Bianda et al. 1998), osteoporotic men (Johansson et al. 1996), postmenopausal women (Ebeling et al. 1993; Ghiron et al. 1995), normal young adults (Mauras, Doi, and Shapiro 1996; Bianda et al. 1997), anorexia nervosa patients (Grinspoon et al. 1996), and fasting young women (Grinspoon et al. 1995). These results generally have been interpreted as reflecting increased bone formation, although type I collagen is not limited to bone. IGF-I stimulates proliferation and collagen production in cultured human osteoblastic cells (Jonsson et al. 1993; Kudo et al. 1996; D'Avis et al. 1997), and also increases collagen accumulation relative to total protein accumulation in cultured human foreskin fibroblasts (Bird and Tyler 1994). Collagen degradation, as reflected by urinary pyridinoline, deoxypyridinoline, hydroxyproline, or N-telopeptide excretion, was also increased in most of the studies showing increased collagen formation (Ebeling et al. 1993; Ghiron et al. 1995; Grinspoon et al. 1996; Johansson et al.

1996; Mauras, Doi and Shapiro 1996; Bianda et al. 1998). IGF-I administration did not increase these markers of collagen degradation in fasting young women, even though it markedly increased the marker of collagen synthesis (Grinspoon et al. 1995). In elderly women, a low dose of IGF-I did not increase collagen degradation, although higher doses did (Ghiron et al. 1995). In general, the consensus is that IGF-I has an anabolic effect on bone, stimulating bone formation more than resorption. Whether it promotes collagen accumulation in other tissues is uncertain.

Effect of IGF-I on Other Tissues

Because IGF-I promotes overall growth in children insensitive to growth hormone, it clearly has an overall anabolic effect on protein metabolism in most tissues of growing children. Whether this effect reflects a direct effect of IGF-I or some secondary effect of accelerated linear growth velocity is uncertain. IGF-I is a mitogenic hormone in many cell culture systems, suggesting that at least part of its effect in growing children is mediated by cellular proliferation. In mice, an acute stimulatory effect of IGF-I on protein synthesis was observed in cardiac and skeletal muscles, but not in liver, kidney, slpeen, small intestine, colon, lung, or brain (Bark et al. 1998).

Other Growth Factors

There are many other growth factors besides IGF-I, including IGF-II, epidermal growth factor, the transforming growth factors, nerve growth factor, fibroblast growth factor, erythropoietin, platelet-derived growth factor, endothelin, and others. These factors generally are autocrine or paracrine hormones that regulate proliferation and gene expression in cells that express the appropriate receptors. Growth factors often are regulated by hormones secreted from endocrine glands (MacGillivray 1995). They are critical for normal growth, developmental changes, and tissue repair and maintenance. Not all growth factors are anabolic. For example, myostatin (GDF-8, a member of the transforming growth factor-β family) inhibits skeletal muscle growth (McPherron and Lee 1997a, 1997b). Thus, growth factors promote or inhibit protein synthesis or degradation, directly or indirectly, as part of the regulation of mitogenesis and growth. Because the protein synthesis associated with these processes comprises a small fraction of the total protein metabolism of most human tissues in vivo, except perhaps during rapid growth or tissue repair, local under- or overexpression of any individual growth factor probably would not have much influence on whole-body protein metabolism. However, we do not know enough about the effects of all of these factors to rule out an important role in overall protein metabolism, similar to that of IGF-I. The extent to which these growth factors influence synthesis and breakdown of specific tissues and proteins in vivo has not been studied in humans.

Androgens

Testosterone

The major circulating androgen in humans is testosterone. Many tissues convert testosterone to dihydrotestosterone (DHT), which mediates many of the effects of testosterone. Men have about 10 times higher plasma testosterone levels than women. The most obvious manifestation of the anabolic effect of testosterone is the marked difference in lean body mass, especially muscle, between men and women. Men who cannot convert testosterone to DHT have the usual masculine pattern of muscle development (Randall 1994), indicating that testosterone itself rather than DHT mediates the anabolic effect of testosterone in muscle.

The anabolic effect of testosterone is evident not only from the greater lean body mass in men than in women, but also from the reduced lean body mass in hypogonadal men. Early studies of testosterone administration indicated that it produced a positive N balance (Kountz 1951). Testosterone replacement in hypogonadal men increases lean body mass (Wang et al. 1996; Bhasin et al. 1997), and high doses of testosterone increase lean body mass in normal men (Welle et al. 1992c; Bhasin et al. 1996). Growth velocity, including growth of the lean body mass, is accelerated by testosterone in prepubertal boys (Arslanian and Suprasongsin 1997). Most of the increase in lean body mass associated with testosterone administration is muscle mass (Griggs et al. 1989; Welle et al. 1992c; Bhasin et al. 1996; Brodsky, Balagopal, and Nair 1996; Bhasin et al. 1997).

Testosterone stimulates protein synthesis in muscle in normal men and myotonic dystrophy patients given supraphysiological doses (Griggs et al. 1986; Griggs et al. 1989). No detectable effect on whole-body protein synthesis or breakdown was noted in these studies, probably because muscle accounts for a relatively small portion of whole-body protein synthesis (Chapter 4). The muscle mass stabilizes after a few months of testosterone treatment (Welle et al. 1992c), even though the fractional rate of protein synthesis in muscle remains elevated (Griggs et al. 1989). Thus, the fractional degradation of muscle protein also may be increased by high testosterone levels. It has been reported that a single injection of testosterone in young men can increase muscle protein synthesis more than 2-fold, while increasing the fractional degradation rate by about one third (Ferrando et al. 1997a). Hypogonadal men have a marked increase (~50%) in muscle protein synthesis after testosterone treatment (Brodsky, Balagopal, and Nair 1996). Myosin heavy chain synthesis increases along with synthesis of other muscle proteins with testosterone replacement, and whole-body protein synthesis increases without a change in protein breakdown (Brodsky, Balagopal, and Nair 1996). Testosterone replacement in elderly men nearly doubled the fractional rate of protein synthesis in muscle (Urban et al. 1995). Together, these studies clearly indicate that stimulation of muscle protein synthesis is the major mechanism of the anabolic effect of testosterone.

In boys with delayed puberty, the whole-body protein synthesis did not change after testosterone treatment, whereas protein breakdown declined slightly (Arslanian and Suprasongsin 1997). In boys with short stature, testosterone increased whole-body protein synthesis by 34%, while it increased protein breakdown by 18% (Mauras et al. 1994). Muscle protein synthesis was not measured in these studies. Based on studies in adults, as discussed above, it seems likely that testosterone increased muscle protein synthesis in these boys, even when there was no detectable effect on whole-body protein synthesis.

Suppression of testosterone production in normal men with a gonadotropin releasing hormone analogue significantly reduced both whole-body protein breakdown and synthesis, suggesting that a normal testosterone level is required for normal protein turnover in men (Mauras et al. 1998a). It is interesting that women, with one tenth the testosterone levels of men, have the same rate of muscle and whole-body protein turnover as men, when data are adjusted for lean body mass. Perhaps women are more sensitive to testosterone than men, or are less dependent than men on androgens in the regulation of their protein metabolism.

In hypogonadal men, testosterone replacement increased serum type I procollagen and reduced excretion of type I collagen crosslinked N-telopeptides, suggesting an anabolic effect of testosterone on bone (Wang et al. 1996). However, an effect on collagen balance in other tissues cannot be excluded.

Testosterone may promote protein anabolism not only by direct effects on androgen receptors, but also by increasing growth hormone and IGF-I levels (De Feo 1996; Rooyackers and Nair 1997) or local production of IGF-I in muscle (Urban et al. 1995; Mauras et al. 1998a).

Dehydroepiandrosterone (DHEA)

The most abundant circulating androgen is DHEA sulfate (DHEAS), which is 200-fold more abundant than testosterone even in men (Miller and Tyrell 1995). DHEAS levels are similar in men and women, but DHEAS is a very weak androgen. Relatively little research has been done on the anabolic effect of DHEA or DHEAS in humans. Because DHEAS levels decline markedly with aging, in parallel with the loss of muscle mass, it is intriguing to speculate that DHEA(S) is an important anabolic hormone. Nestler et al. (1988) reported that high oral doses of DHEA for 4 weeks increased the lean body mass of normal men by an average of 4.5 kg, although the effect was not statistically significant. We were unable to verify this finding (Welle, Jozefowicz, and Statt 1990). We also failed to observe any effect of DHEA on whole-body protein breakdown or synthesis, or on muscle protein synthesis. Moreover, no anabolic effect of DHEA was observed in obese men (Usiskin et al. 1990) or in postmenopausal women (Mortola and Yen 1990). Replacement doses of DHEA, which restored the serum DHEA and DHEAS levels to the youthful range in elderly subjects, did not increase lean body mass (Morales et al. 1994). Thus, DHEA(S) does not appear to be an important anabolic hormone.

Ovarian Steroids

Very little research has been done on the effect of estrogens and progestins on protein metabolism. The major estrogen in humans is estradiol, and the major progestin is progesterone. Although present in men and women, their concentrations are much higher in women at certain stages of the menstrual cycle (Figure 6.4). The much smaller muscle mass in women than in men indicates that these steroids are much less anabolic than testosterone. Progesterone has antiestrogenic properties in some tissues and may antagonize any effect of estradiol on protein metabolism. Progesterone also is a weak androgen. Whether or not antiestrogenic, androgenic, or independent effects of progesterone have any significant role in regulating protein metabolism has not been resolved.

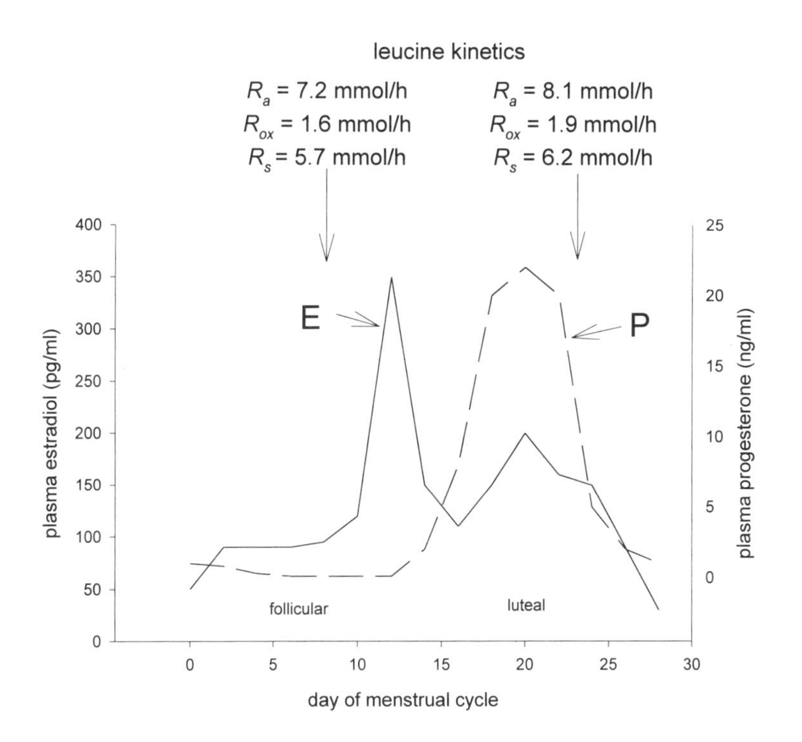

FIGURE 6.4. Evidence for modest increase in protein turnover during luteal phase of menstrual cycle, when estradiol (E) and progesterone (P) levels are high, relative to follicular phase when levels of these hormones are low. Values for leucine kinetics are from Lariviere et al. (1994). Values for hormone levels are from Erickson (1995).

In animal husbandry, large doses of estrogens are considered anabolic (Heitzman 1979), but this may be an indirect or pharmacological effect rather than a physiological one. At puberty, girls have a growth spurt, but the role of ovarian steroids is not clear. Their main effect on linear growth may be mediated by promotion of growth hormone secretion. At menopause, lean body mass declines more than would be expected from aging per se (Poehlman, Toth, and Gardner 1995), suggesting that the monthly surges in estradiol and/or progesterone may have an anabolic effect on protein metabolism. There is some evidence that replacement of ovarian steroids can increase the muscle or total lean body mass of postmenopausal women (Jensen, Christiansen, and Rodbro 1986). However, hormone replacement usually has not been found to increase lean body mass (Hassager and Christiansen 1989; Kohrt, Landt, and Birge, 1996; Gambacciani et al. 1997).

Studies done several decades ago indicated that very high doses of estradiol caused positive N balance, whereas smaller doses had little or no effect (Knowlton et al. 1942; Kountz 1951; Ackermann et al. 1954). Large doses of progesterone also promoted N retention (Ackermann et al. 1954). However, Landau and Poulos (1971) concluded that progesterone was a catabolic hormone with respect to protein metabolism. This conclusion was based mainly on the fact that it increased N excretion in subjects on a constant diet. The effect was maximal when the dose corresponded to the amount of progesterone excreted during the first trimester of pregnancy. The catabolic effect was associated with reduced serum amino acid concentrations. Thus, if the catabolic effect is mediated by increased protein degradation, the amino acids must be metabolized or cleared from the circulation rapidly.

In the late luteal phase of the menstrual cycle, when both estradiol and progesterone levels are high, postabsorptive whole-body proteolysis and protein synthesis are about 10% more rapid than they are during the follicular phase (Figure 6.4) (Lariviere, Moussalli, and Garrel 1994). Leucine oxidation also increases along with the increase in proteolysis, but total N excretion is unaffected. These changes may be related to changes in free T_3 or other hormones rather than to changes in estradiol or progesterone levels (Lariviere, Moussalli, and Garrel 1994; De Feo 1996). In prepubertal girls, estradiol administration for 4 weeks slightly increased the average rate of whole-body protein turnover, but the effect did not achieve statistical significance (Mauras 1995a). The effect of progesterone on protein turnover in humans apparently has not been investigated. Although it is clear that estradiol and progesterone must influence the synthesis or degradation of specific proteins related to development and maintenance of the female reproductive system, more global effects on protein metabolism are probably minor.

Estradiol is important for maintaining normal bone mass in women. Research dealing with bone mass and turnover often include markers of type I collagen turnover, but there is no way to be certain that changes in these markers are restricted to bone collagen. Estrogens increase serum levels of the propeptide of type I collagen in hypogonadal girls, indicating increased type I collagen synthesis (Mauras, Doi, and Shapiro 1996). Excretion of deoxypyridinoline, an index of collagen degradation, also increases in estrogen-treated hypogonadal girls. In contrast, ovarian hormone replacement reduces collagen formation and degradation in postmenopausal women, as reflected by reduced serum type I procollagen carboxy-terminal propeptide and excretion of hydroxyproline, pyridinoline and deoxypyridinoline crosslinks, and cross-linked teloppeptides (Hassager et al 1991; Prestwood et al. 1994).

In summary, except for effects on bone, breast, and reproductive organs, physiological variations in ovarian steroids probably have little importance as regulators of bulk protein metabolism. However, so little research has been done in this area that this conclusion is only tentative until more definitive research is completed.

Epinephrine

Epinephrine usually is considered a catabolic hormone because its levels are highest during periods of stress, when the body is generally in a catabolic state. However, there is evidence that epinephrine is anabolic with respect to protein metabolism while being catabolic with respect to fat and carbohydrate metabolism.

Except during periods of stress, circulating epinephrine levels are quite low, usually less than 0.5 pmol/ml. Certain stimuli, such as fear, hypoglycemia, trauma, or other stresses can elevate epinephrine levels 10 fold or more. Epinephrine has a very short half-life in the circulation, so that its concentrations fall very rapidly when the stressful stimulus is removed. Most stressful situations raise levels of cortisol and other hormones along with those of epinephrine. Consequently, the net effect of the stressful situation on protein metabolism usually is catabolic even though epinephrine may be anabolic.

Epinephrine infusion reduces plasma amino acid levels (Shamoon, Jacob, and Sherwin 1980; Miles et al. 1984; Kraenzlin et al. 1989). Even though epinephrine raises plasma glucose concentrations, its inhibitory effect on insulin secretion prevents or attenuates the increase in insulin levels that normally would accompany the hyperglycemia. Thus the hypoaminoacidemia is not explained by increased insulin levels, occurring even when insulin levels are not increased (Shamoon, Jacob, and Sherwin 1980; Kraenzlin et al. 1989). The decline in amino acid levels is mediated by reduced whole-body and muscle proteolysis and by increased amino acid clearance (Miles et al. 1984; Kraenzlin et al. 1989; Castellino et al. 1990; Fryburg et al. 1995a). Protein synthesis also is reduced by epinephrine when proteolysis is inhibited, but the net effect is anabolic because proteolysis declines more than protein synthesis in muscle (Fryburg et al. 1995a) and the whole body (Kraenzlin et al. 1989). Epinephrine infusions reduce muscle ribosome concentrations, which is consistent with a slower rate of muscle protein synthesis (Wernerman et al. 1989).

One study failed to confirm an antiproteolytic effect of epinephrine after a more prolonged infusion (8.5 h) than those used by other investigators, even though amino acid levels decreased as expected (Matthews, Pesola, and Campbell 1990). One possible explanation for this discrepant finding is that prolonged hypoaminoacidemia offset the anabolic effect of epinephrine. This explanation is supported by a study demonstrating that epinephrine suppressed proteolysis when amino acids were infused to prevent hypoaminoacidemia, but not when amino acid levels declined (Castellino et al. 1990). As pointed out by Matthews et al. (1990), the efficiency of utilizing amino acids for protein synthesis and oxidation must be enhanced by epinephrine, because protein synthesis and amino acid oxidation do not decline along with the fall in amino acid levels. Epinephrine infusion does not appear to enhance the suppression of proteolysis or stimulation of protein synthesis associated with parenteral amino acid administration (Schiefermeier et al. 1997).

Do the low basal epinephrine concentrations have any impact on protein metabolism? This question has been addressed by examining the effect of propranolol on basal protein turnover. Propranolol is a nonselective beta adrenergic antagonist that prevents the effects of infused epinephrine on protein metabolism (Shamoon, Jacob, and Sherwin 1980; Kraenzlin et al. 1989; Del Prato et al. 1990). Lamont et al. (1989) found that propranolol increased leucine concentrations, leucine oxidation, and urea excretion, and increased the average leucine R_a by 25%, although the latter effect was not statistically significant. These data suggest that basal epinephrine secretion (or norepinephrine released from sympathetic neurons) has an inhibitory effect on protein breakdown and oxidation. However, when we placed healthy volunteers on high doses of propranolol for several days, there was no effect on postabsorptive whole-body proteolysis or protein synthesis as determined by leucine kinetics (Welle et al. 1989). These data suggested that basal sympathetic nervous system activity is not important in the regulation of protein metabolism.

Cortisol

Glucocorticoids are steroid hormones that have a wide range of effects in many tissues. Although the name implies that the primary role is related to glucose metabolism, glucocorticoids also have profound effects on protein metabolism and other metabolic processes. The primary glucocorticoid secreted by the human adrenal cortex is cortisol. Its concentrations vary throughout the day in a circadian rhythm and also are elevated by various stressful conditions. Much of the research related to glucocorticoid effects relates to the use of synthetic glucocorticoids rather than to variations in cortisol levels. This section focuses on variations in cortisol levels. Pharmacologic effects of synthetic glucocorticoids are discussed in Chapter 9.

Several studies have examined the effect of infusing cortisol for several hours into healthy subjects. Whole-body proteolysis increases ~10% to 20%

within 12 h of raising plasma cortisol levels by 3 to 7 fold (Simmons et al. 1984; Darmaun, Matthews, and Bier 1988; Brillon et al. 1995; Garrel et al. 1995). When cortisol was infused for 60 h, urea excretion markedly increased from 12 to 36 h of the infusion, but then returned to normal (Darmaun, Matthews, and Bier 1988). However, the increase in proteolysis was maintained for the full 60 h. Insulin levels usually do not increase with very short-term cortisol infusion, but eventually increase when the infusion is prolonged enough. This hyperinsulinemia could help to counteract the catabolic effect of cortisol. However, Brillon et al. (1995) did not observe much of an increase in proteolysis during cortisol infusion when somatostatin was used to reduce insulin levels to normal. The effect of cortisol infusion on the protein synthesis and breakdown in muscle of normal subjects has not been reported. There is much evidence that synthetic glucocorticoids can both stimulate muscle proteolysis and inhibit muscle protein synthesis (Chapter 9), and it seems likely that cortisol would have a similar effect at sufficiently high doses. The finding that a 6 h infusion of cortisol reduced muscle ribosome concentrations by 27% is consistent with an inhibition of muscle protein synthesis (Wernerman et al. 1989).

Although studies of patients with chronic insufficiency or excess of cortisol might be expected to yield some insights into the role of cortisol in protein metabolism, metabolic abnormalities in these patients are difficult to interpret because of the multiple disturbances in many organ systems. Patients with primary adrenocortical deficiency often have loss of lean body mass, muscle wasting, and muscle weakness, but they also have anorexia and multiple medical problems. Therefore, it is difficult to ascribe the protein catabolism to cortisol deficiency. Patients with cortisol excess (Cushing's syndrome) usually are overweight, but they have reduced muscle mass and weakness. When these patients are treated to normalize cortisol secretion, total body potassium (an index of body cell mass) increases by as much as 40% (Lamberts and Birkenhager 1976). Although it is possible that there is some increase in the ratio of potassium to protein, these results suggest that chronic hypercortisolemia is catabolic. With chronic cortisol excess, there is little evidence for increased whole-body proteolysis. Tessari et al. (1989) found that postabsorptive leucine R_a was normal, per kg body weight, in patients with Cushing's syndrome. Moreover, chronic hypercortisolemia did not cause resistance to the antiproteolytic effect of insulin infusion. Whole-body protein synthesis (nonoxidized leucine R_d) tended to be slower per kg body weight, by ~15%, in these patients. Unfortunately, data were expressed per kg total body weight rather than per kg fat free mass, and therefore might have underestimated protein turnover relative to body cell mass in the Cushing's syndrome patients. A later study found that whole-body protein synthesis per kg lean body mass was 45% slower in patients with Cushing's syndrome than in control subjects and that reducing cortisol production resulted in stimulation of whole-body protein synthesis (Bowes et al. 1993). Muscle of patients with Cushing's syndrome does not overexpress the mRNAs encoding various proteolytic proteins, including ubiquitin, proteasome subunits, and a ubiquitin conjugating enzyme (Ralliere et al. 1997). Together with the evidence that synthetic glucocorticoids inhibit muscle protein synthesis (Chapter 9), these data suggest that reduced protein synthesis rather than increased proteolysis mediates the reduced muscle mass associated with chronic hypercortisolemia. It is worth noting that muscle wasting with synthetic glucocorticoids is more severe when there is also muscle disuse and is more prominent in the type 2 muscle fibers, which may be activated less frequently than the type 1 fibers. Thus, the effect of cortisol on protein metabolism in a particular muscle may depend on its pattern of use.

The effect of hypercortisolemia on collagen turnover is unclear. Patients with Cushing's syndrome can have normal (Hermus et al. 1995) or reduced (Osella et al. 1997) formation of type III collagen, without a change in type I collagen production, as reflected by serum levels of N-terminal propeptides. The ratio of hydroxyproline to creatinine in urine is increased in Cushing's syndrome (Hermus et al. 1995), whereas the serum levels of the crosslinked telopeptide of type I collagen are not elevated (Hermus et al. 1995; Osella et al. 1997). Markers of collagen formation (types I and III) and degradation increase markedly after surgical treatment of the hypercortisolemia, then gradually return to baseline levels over many months (Hermus et al. 1995). The endocrine outcomes of the surgical treatment are complex, and it is difficult to interpret the role of the reduced cortisol levels in the postsurgical response.

Thyroid Hormones

Thyroxine (T_4) is the most abundant thyroid hormone in the circulation. T_4 can be converted to triiodothyronine (T_3) in tissues by enzymatic deiodination, and some T_3 is secreted by the thyroid gland. T_3 is the active hormone—the one that binds to nuclear receptors. These hormones are mostly bound to proteins and have a relatively long half-life in the circulation. Consequently, T_4 and T_3 levels are fairly constant throughout the day. Thyroid hormone levels are affected by nutrition, illness, and specific disorders that affect the pituitary or thyroid gland itself.

Thyroid hormones are required for maintaining normal rates of most metabolic cycles, and protein turnover is no exception. Children with thyroid hormone deficiency do not grow normally and can exhibit permanent developmental disabilities. Although adults with thyroid hormone deficiency do not necessarily have an altered lean body mass (Forbes 1987a), they clearly have a subnormal rate of protein turnover (Crispell et al. 1956; Morrison et al. 1988). When hypothyroid patients are treated with thyroid hormones, their protein turnover returns to normal (Crispell et al. 1956; Morrison et al. 1988). The magnitude of the reduction in whole-body protein turnover in untreated hypothyroidism is quite substantial, averaging approximately 25% to 50%.

Net efflux of tyrosine, phenylalanine, and other essential amino acids from the leg is reduced in hypothyroid patients, suggesting reduced muscle proteolysis (Morrison et al. 1988). However, net efflux of 3-methylhistidine is not reduced, suggesting that myofibrillar degradation is not reduced. Both albumin and LDL turnover are reduced in hypothyroid patients (Walton et al. 1965; Lewallen, Rall, and Berman 1959). Thyroidectomized rats have a 50% reduction in hepatic production of albumin and total secreted proteins and a 20% reduction in the synthesis of hepatic intracellular proteins (Peavy, Taylor, and Jefferson 1981). The decline in hepatic protein synthesis is associated with a similar reduction in the tissue RNA concentration. Hypothyroid rats also have slower protein turnover in muscle, kidney, heart, and intestine (Umpleby and Russell-Jones 1996).

Several studies have examined the effect of experimental hyperthyroidism in normal subjects. Although an early study of a single subject showed an 81% decline in whole-body protein synthesis after 10 days of T_3 treatment (100 μg/d) (Crispell et al. 1956), most of the recent studies have demonstrated a significant increase in whole-body protein turnover, as reflected by leucine kinetics (Table 6.4). The T_3 doses in Table 6.4 should be compared to the daily T_3 production rate of about 30 μg/d in a 70 kg man, keeping in mind that the exogenous hormone suppresses endogenous T_4 and T_3 production. The net protein balance generally is negative in experimental hyperthyroidism, but the effect may be transient (Lovejoy et al. 1997). There may be an enhanced sensitivity to the antiproteolytic effect of insulin during mild hyperthyroidism, which would tend to ameliorate the catabolic effect of T_3 (Tauveron et al. 1995).

Several investigators have examined the effect of thyroid hormone administration during fasting or reduced-energy diets, because T_3 concentrations decline with energy deprivation and this effect may contribute the fall in metabolic rate. Generally, thyroid hormone administration increases N losses during fasting or low energy diets (Vignati et al. 1978; Gardner et al. 1979; Abraham et al. 1985; Wolman et al. 1985), although sometimes this effect has been marginal (Pasquali et al. 1984) or absent (Wilson and Lamberts 1981).

TABLE 6.4. Effect of experimental hyperthyroidism on whole-body protein turnover as assessed by leucine kinetics.

Authors	Dose of T_3	Effect on proteolysis	Effect on protein synthesis
Gelfand et al. (1987)	150 μg/d x 1 week	↑45%	↑37%
Tsalikian and Lim (1989)	0.8 μg/kg/d x 1 week	↑15%	not done
Martin et al. (1991)	100 μg/d x 1 week	↑12%	↑9%
Tauveron et al. (1995)	2 μg/kg/d T_4 x 6 weeks, 1 μg/kg/d T_3 last 2 weeks	↑22%	↑24%
Lovejoy et al. (1997)	50–75 μg/d x 9 weeks	↑7%*	↑6%*

*Effect not statistically significant.

There may be some resistance to T_3 during energy deprivation. Nair et al. (1989) did not observe an increase in whole-body protein turnover after one week of T_3 administration (60 μg/d) to fasting obese subjects. Pasquali et al. (1984) did not find any change in 3-methylhistidine excretion when T_3 (40 μg/d) was given to obese subjects on a low energy diet. Wolman et al. (1985) found only a slight increase (~10%) in whole-body protein turnover ([^{15}N]glycine end product method) when T_3 (60 μg/d) was given under similar conditions.

There generally is agreement that hyperthyroidism is a catabolic state that leads to muscle wasting, but studies of the effects of hyperthyroidism on protein turnover have not produced consistent results. Although experimental hyperthyroidism in normal subjects increases protein turnover, there is evidence that spontaneous hyperthyroidism is associated with slower protein turnover. Morrison et al. (1988) found that whole-body protein turnover per kg body weight was ~25% below normal in untreated thyrotoxic patients. This comparison is difficult to interpret because the heights and weights of the patients and control groups were quite different, body composition was not measured, and it is unclear whether the groups were matched by gender. However, the fact that protein turnover increased substantially after treatment indicates that hyperthyroidism can inhibit protein turnover. There is generally weight loss with thyrotoxicosis, and the negative energy balance might contribute to the reduced protein turnover. Leg balances of tyrosine, phenylalanine, and other essential amino acids were markedly more negative during hyperthyroidism, suggesting net protein loss from muscle. However, net leg balance of 3-methylhistidine was not affected by thyroid status. There is some evidence that excretion of 3-methylhistidine in urine is increased in hyperthyroidism (Rodier et al. 1984; Alderberth et al. 1987), and other evidence that it is not (Hagg and Adibi 1985). In contrast to the Morrison et al. (1988) study, Hagg and Adibi (1985) found that treatment of hyperthyroidism reduced whole-body protein turnover. When muscle protein synthesis and degradation were measured in vitro, tissue from hyperthyroid patients had above-normal protein degradation but a normal rate of protein synthesis (Hasselgren et al. 1984a). In rat muscle, proteasomal proteolysis is enhanced by hyperthyroidism (Tawa, Odessey, and Goldberg 1997).

Markers of collagen turnover are increased in patients with elevated thyroid hormone levels and reduced in those with low thyroid hormone levels (Nagasaka et al. 1997; Persani et al. 1997). A reduction in the dose of T_4 reduces collagen turnover in subjects on exogenous T_4 therapy (Guo, Weetman, and Eastell 1997). The increase in collagen degradation appears to be greater than the increase in collagen formation in thyrotoxic patients. Bone specific markers suggest that at least some of the effects of thyroid hormones on collagen metabolism occur in bone, but other tissues cannot be excluded.

Do variations in thyroid hormone levels within the normal range influence protein metabolism? We were unable to find any significant relation between whole-body protein synthesis or breakdown and serum levels (free or total) of

T_4 or T_3 in 47 healthy young adults (Welle and Nair 1990b). In contrast, Motil et al. (1994b) found a significant correlation between whole-body protein synthesis and T_3 levels in 12 lactating women, and correlations between 3-methylhistidine excretion and both T_4 and T_3 levels.

Glucagon

Glucagon secretion by pancreatic alpha cells is regulated by glucose and amino acid concentrations. Low glucose levels and high levels of some amino acids are stimulatory. Glucagon secretion also is elevated during periods of stress. The primary action of glucagon is to enhance hepatic glucose production by increasing glycogenolysis and gluconeogenesis. When a high-protein, low-carbohydrate meal is consumed, the stimulation of glucagon secretion by the amino acids from the protein helps to prevent a drop in glucose. Because glucagon's effects are antagonized by insulin, the glucagon:insulin ratio is an important determinant of carbohydrate metabolism. Less is understood about the role of glucagon in protein metabolism. One of the difficulties in doing research in this area is that administration of glucagon raises glucose concentrations, which in turn raises insulin concentrations, which antagonizes the effect of glucagon. Thus, much of the research on the effect of glucagon has used somatostatin to suppress insulin secretion. Somatostatin also inhibits growth hormone secretion, which might have some influence on the results of these studies.

When glucagon is infused for 2 to 5 days in normal subjects who have been fed with only glucose, there is an increase in urea excretion and a decline in amino acid levels, but no change in 3-methylhistidine excretion (Fitzpatrick et al. 1977; Wolfe et al. 1979). Thus myofibrillar proteins may not contribute to the increase in urea production. After a 5 day glucagon infusion, the forearm balance of α-amino N actually became more positive (Wolfe et al. 1979), but an anabolic effect has not been reproduced in other studies when glucagon has been given for only a few hours.

The acute effects of glucagon (infusions of 5 hr or less) appear to be highly dependent on the hormonal and metabolic milieu. There is general agreement that glucagon lowers levels of glucogenic amino acids under most conditions. Nair et al. (1987b) infused somatostatin to prevent the rise in insulin associated with glucagon infusion, but this procedure reduced insulin levels below normal. Under these conditions, glucagon increased whole-body proteolysis (leucine R_a) by ~15% and leucine oxidation by ~90%, and reduced whole-body protein synthesis by ~15%. Similar effects were observed on forearm leucine kinetics in a similar study, even though there was a small increase in insulin levels during the glucagon + somatostatin infusion (Pacy et al. 1990). In contrast, other studies in which normal insulin levels were maintained during glucagon infusion have indicated that glucagon has no effect on (Couet et al. 1990; Tessari et al. 1996a; Battezzati et al. 1998), or even suppresses (Hartl et al. 1990) whole-body proteolysis. Even when proteolysis is unchanged or reduced, glucagon increases oxidation of leucine and phenylalanine (Hartl et al. 1990; Tessari et al. 1996a).Total urea production was diminished by glucagon in the study of Hartl et al. (1990), even though leucine oxidation was markedly increased. Whole-body protein synthesis (non-oxidized leucine or phenylalanine R_d) is reduced ~5% to 15% by glucagon when insulin is maintained at normal levels (Hartl et al. 1990; Tessari et al. 1996a). Glucagon acutely stimulates the synthesis of fibrinogen (Tessari et al. 1997).

When amino acid are delivered by intravenous infusion, glucagon decreases amino acid concentrations by increasing uptake of amino acids from the circulation and intracellular degradation of the amino acids, with increased gluconeogenesis and ureagenesis (Boden et al. 1990; Charlton, Adey, and Nair 1996). Glucagon does not affect the suppression of whole-body proteolysis induced by hyperaminoacidemia, but it does inhibit the stimulation of whole-body protein synthesis (Charlton, Adey, and Nair 1996).

In rats, glucagon levels in the pathophysiologic range can inhibit protein synthesis in skeletal muscle, while modestly stimulating hepatic protein synthesis and not affecting protein synthesis in the heart (Preedy and Garlick 1985, 1988). Very high glucagon levels can increase hepatic proteolysis and RNA degradation in rats, an effect associated with an increase in the volume of autophagic vacuoles (Bleiberg-Daniel et al. 1994). Glucagon also prevents the antiproteolytic effect of insulin in perfused rat liver (Vom Dahl et al. 1991). This effect may be related in part to effects on hepatocyte volume or potassium uptake.

Cytokines

Cytokines are hormone-like polypeptides secreted by lymphocytes, macrophages, fibroblasts, endothelial cells, and other cells. Cytokine production is induced by various stressors and pathogenic conditions, such as infection, inflammation, cancer, and trauma. Under such conditions, muscle proteolysis is usually increased. Animal studies have implicated several cytokines as potentially important mediators of protein catabolism in muscle, including tumor necrosis factor-α (TNFα, also known as cachectin), interleukin-1 (IL-1), and IL-6 (Cooney, Kimball and Vary 1997). Administration of TNFα to rats can increase muscle proteolysis (Flores et al. 1989; Goodman 1991), but inhibits hepatic proteolysis (Flores et al. 1989; Ling, Schwartz, and Bistrian 1997). TNFα generally does not stimulate muscle proteolysis in vitro, suggesting that TNFα must induce one or more other factors to stimulate proteolysis (Kettelhut and Goldberg 1988; Goodman 1991; Fang et al. 1997), although there is evidence that it can directly increase expression of mRNAs encoding ubiquitin and a proteasomal subunit (Llovera et al. 1997). TNFα has little effect on muscle protein synthesis in rats (Charters and Grimble

1989; Ling, Schwartz, and Bistrian 1997). It can increase hepatic protein synthesis acutely (Charters and Grimble 1989), although this effect was not observed when it was given for several days (Ling, Schwartz, and Bistrian 1997). In dogs, a brief infusion of TNFα did not stimulate whole-body proteolysis, but increased leucine oxidation and reduced whole-body protein synthesis (Sakurai, Zhang, and Wolfe 1994). A 6 h infusion in dogs increased N excretion without altering efflux of amino acids from muscle, apparently by increasing intestinal N losses (Evans, Jacobs, and Wilmore 1993). IL-1 administration does not appear to affect muscle proteolysis in rats (Goldberg et al. 1988; Flores et al. 1989; Goodman 1991; Fang et al. 1997), although an IL-1 receptor antagonist can inhibit proteolysis and promote protein synthesis in muscle of septic rats (Fang et al. 1997; Cooney, Kimball, and Vary 1997). TNFα and IL-1 may have a synergistic effect on muscle catabolism (Flores et al. 1989). IL-6 administration or overexpression in rodents increases muscle proteolysis, but the effect may be indirect (Tsujinaka et al. 1996; Goodman 1994). IL-6 antibodies inhibit muscle atrophy in tumor-bearing mice and transgenic mice that overexpress IL-6 (Tsujinaka et al. 1996; Fujita et al. 1996). Not all cytokines have a catabolic effect on muscle. IL-15 is a cytokine highly expressed in muscle that promotes muscle fiber hypertrophy and myosin heavy chain accumulation in cultured myotubes (Quinn, Haugk, and Grabstein 1995).

There is not much information about the role of cytokines in regulating protein metabolism in humans. Conditions that are associated with high cytokine levels are characterized by negative N balance and muscle catabolism, but other catabolic hormones (cortisol, glucagon) also are typically elevated. A single dose of TNFα in cancer patients slightly increased whole-body protein turnover, but it is unclear whether this response merely reflects the elevated cortisol levels, fever, and general hypermetabolism induced by TNFα (Starnes et al. 1988). Forearm efflux of amino acids was markedly increased after TNFα administration, but arterial plasma amino acid levels declined. In cancer patients whose limbs were perfused for treatment of irresectable sarcomas or melanoma metastases, efflux of tyrosine, phenylalanine, and total amino acids was not affected by TNFα, but the study was limited by its brief duration and the lack of an adequate control period with stable amino acid efflux (De Blaauw et al. 1997). In burn patients, IL-6 levels (but not TNFα levels) correlated with phenylalaninemia and the 3-methylhistidine/creatinine ratio in urine (De Bandt et al. 1994). In addition to direct effects of cytokines on protein metabolism, the anorexia induced by cytokines can contribute to loss of muscle mass.

The "acute phase" proteins are serum proteins, synthesized by the liver, that have a role in extracellular proteolysis, blood clotting, fibrinolysis, immunity, and other responses to tissue injury, cancer, trauma, and infection (Kushner 1982). The acute phase response involves increased levels of some proteins (e.g. α_1-antichymotrypsin, C-reactive protein, α_1-acid glycoprotein, haptoglobin, fibrinogen, ceruloplasmin, amyloid A, ferritin), which appears to be mediated by increased production of these proteins. Levels of other proteins (e.g. albumin, transferrin) decline, although it is unclear whether reduced synthesis or increased proteolysis is responsible. The acute phase response of human hepatocytes may be mediated primarily by IL-6 and IL-8 (Wigmore et al. 1997; Wigmore, Fearon, and Ross 1997).

Prostaglandins

Prostaglandins are derivatives of C_{20} polyunsaturated fatty acids. Their production is ubiquitous in human tissues, and their effects are autocrine or paracrine rather than endocrine. In vitro studies and in vivo studies in animals suggest that prostaglandins have an important role in muscle protein synthesis and degradation (Palmer 1990). Prostaglandin $F_{2\alpha}$ stimulates protein synthesis without affecting protein degradation, whereas prostaglandin E_2 increases proteolysis without affecting protein synthesis (Palmer 1990; Rooyackers and Nair 1997).

There is very little information about the role of prostaglandins on protein metabolism in humans. Infusion of prostaglandin E_1 was found to mimic the acute phase protein response described in the preceding section, an effect that may be mediated by increased cytokine production (Whicher et al. 1984). Direct intraarterial infusion of prostaglandin E_1 into the forearm changed net amino acid balance from negative to positive (Stiegler et al. 1989). Indomethacin, which inhibits prostaglandin production, had no effect on whole-body protein breakdown or synthesis according to the [^{15}N]glycine end product method (McNurlan et al. 1987). Prostaglandin synthesis inhibitors also failed to alter protein turnover or N balance in elderly arthritis patients (Gann et al. 1988). In cancer patients with high cortisol and cytokine levels, there was no statistically significant difference in whole-body or hepatic (nonsecreted) protein synthesis between those receiving ibuprofen, which inhbits prostaglandin production, and those not receiving ibuprofen (Preston et al. 1995). However, there were some trends suggestive of reduced protein turnover and reduced N excretion with ibuprofen treatment. Other studies have indicated improved postsurgical N balance with ibuprofen treatment (Palmer 1990). These studies of prostaglandin synthesis inhibitors are limited by the nonspecific nature of the reduction in prostaglandin production, especially since the animal studies indicate anatagonistic roles of different prostaglandins. Further information on the effects of increasing or reducing the production of specific prostaglandins is needed before their role in protein metabolism is understood.

7 Regulation by Metabolic Substrates

Metabolic substrates are used to produce energy (mainly glucose, fatty acids, amino acids, and their byproducts) and to synthesize larger molecules (e.g., amino acids used for protein synthesis). As already discussed in the context of nutritional influences, provision of metabolic substrates in the diet generally results in protein anabolism. However, because food intake changes the levels of many metabolic substrates and hormones, it is not obvious from the response to meals how each metabolic substrate influences protein metabolism. This chapter defines the role of individual metabolic substrates.

Amino Acids

Amino acids have received the most attention with regard to substrate regulation of protein metabolism. Because they are the building blocks of protein, their potential importance is obvious. They also share with glucose, fatty acids, and other substrates the potential to serve as fuel for production of energy (ATP) by mitochondria, and this role should not be overlooked.

Effect of Amino Acids on Whole-Body Protein Metabolism

A high concentration of amino acids is associated with a marked increase in amino acid oxidation, but nevertheless is anabolic (Giordano, Castellino, and DeFronzo 1996). Amino acids enhance protein anabolism both by inhibiting proteolysis and stimulating protein synthesis. When a mixture of amino acids is infused intravenously, whole-body proteolysis declines, based on measurements of the R_a of leucine or other essential amino acids (Figures 7.1 and 7.2). In such studies, the amino acid mixtures do not contain every amino acid and do not influence the concentrations of every amino acid to the same extent. Generally, essential amino acids are included, and glutamine is excluded because of its instability in solution at neutral pH. Figure 7.1 illustrates the dose-response curve for the suppression of whole-body proteolysis as a function of the increase in total amino acid levels induced by infusing a

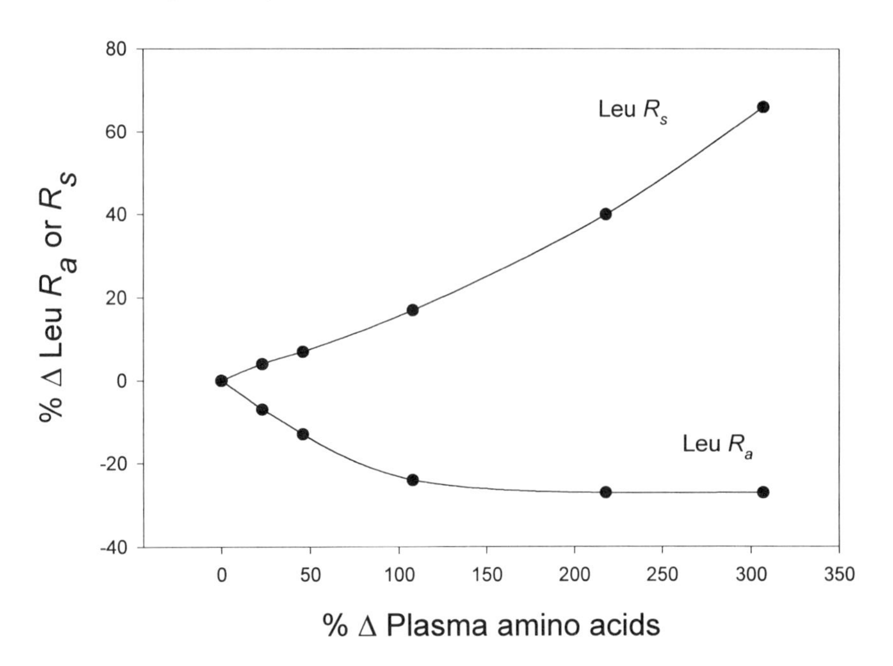

FIGURE 7.1. Percent change in whole-body protein synthesis (nonoxidized leucine R_d=R_s) and protein breakdown (leucine R_a) as a function of the percent increase in total amino acid concentration during infusion of a mixture of 15 amino acids. Data of Giordano, Castellino, and DeFronzo (1996).

mixture of 15 amino acids in normal young adults. Figure 7.2 illustrates the percent suppression of whole-body proteolysis as a function of increasing leucine concentrations during infusion of a mixture of amino acids in several studies. In these studies, there were minimal changes in insulin levels during amino acid administration. Although concentrations of amino acids other than leucine also increased, leucine was chosen as the index of hyperaminoacidemia because data on many of the other amino acids were not published in some of the papers. The use of leucine as the index of hyperaminoacidemia in Figure 7.2 should not be taken to indicate that leucine alone is responsible for the effect of hyperaminoacidemia.

Generally, a rise in amino acids comparable to what is observed after a typical meal reduces whole-body proteolysis ~10-20%. The use of amino acid R_a to determine the effect of hyperaminoacidemia probably leads to an underestimation of the suppression of proteolysis. Before amino acid administration, the plasma tracer enrichment is higher than the intracellular tracer enrichment, which causes an underestimation of R_a (see Chapter 3 for a discussion of this problem). Use of plasma KIC enrichment rather than plasma leucine enrichment ameliorates this problem to some extent when labeled

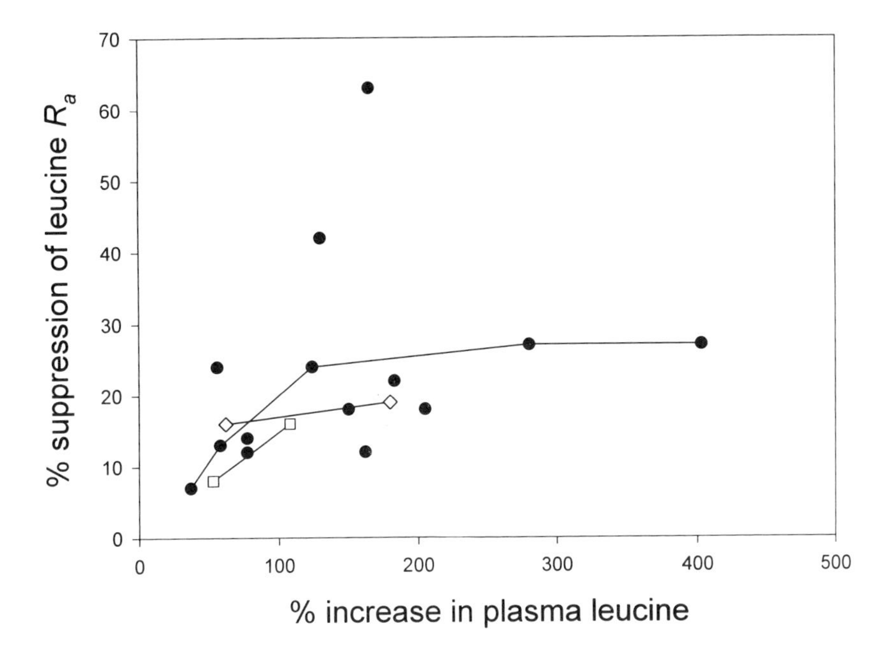

FIGURE 7.2. Suppression of whole-body proteolysis (leucine R_a) as a function of hyperaminoacidemia during infusion of amino acid mixtures in several studies of healthy subjects. Levels of most amino acids increased in these studies, but only the increase in leucine concentrations was plotted because it was measured in all studies. These are mean values of several subjects in each study. The individual circles represent data from studies of adults (Castellino et al. 1987; Tessari et al. 1987; Gelfand et al. 1988; Bennet et al. 1989; Fukagawa et al. 1989; Bennet et al. 1990b; Luzi, Castellino, and DeFronzo 1996). The circles connected by lines represent adults studied at several levels of hyperaminoacidemia in the same study (Giordano, Castellino, and DeFronzo 1996). The open symbols connected by lines represent data from 3-day-old (diamonds) and 3-week-old (squares) infants studied at two levels of hyperaminoacidemia (Poindexter et al. 1997).

leucine is used as a tracer, but does not prevent the problem entirely. When amino acid levels are high, a greater fraction of the free amino acid pool comes from the infused amino acids rather than from endogenous proteolysis, so that plasma and intracellular enrichments converge. Because R_a is underestimated less during amino acid administration than it is before the infusion, the suppression of R_a is underestimated.

Amino acid administration increases whole-body protein synthesis, as reflected by the nonoxidized portion of leucine R_d (Figures 7.1 and 7.3). An exception to this general finding was the failure of amino acid administration to stimulate nonoxidized leucine R_d in infants (Poindexter et al. 1997). In that study, the nonhydroxylated portion of phenylalanine R_d , another index of protein synthesis, was increased. It is interesting that both leucine and phenyalanine R_a indicated reduced proteolysis during hyperaminoacidemia

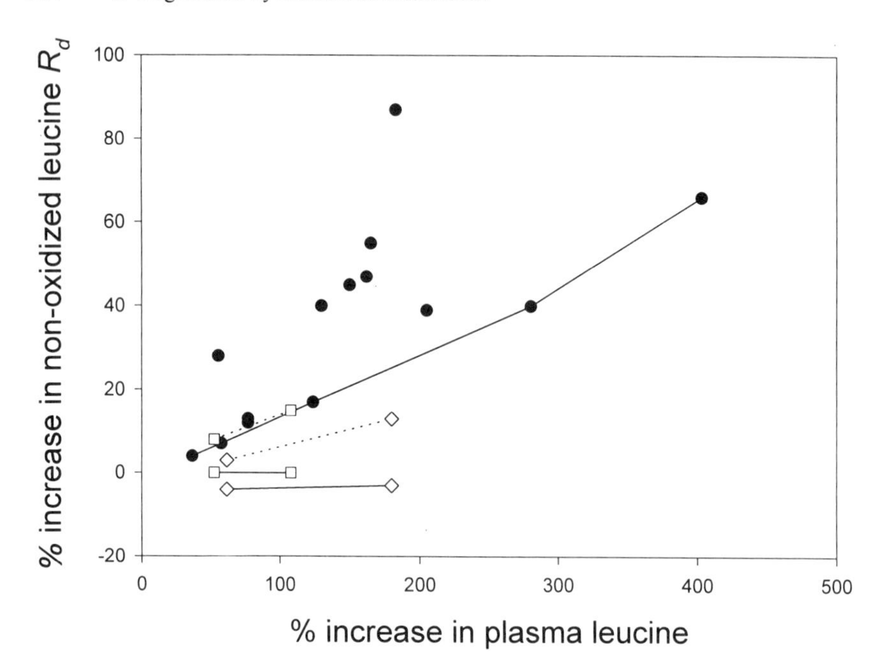

FIGURE 7.3. Stimulation of whole-body protein synthesis (nonoxidized portion of leucine R_d) as a function of hyperaminoacidemia during infusion of amino acid mixtures in several studies of healthy subjects. Levels of most amino acids increased in these studies, but only the increase in leucine concentrations was plotted because it was measured in all studies. These are mean values of several subjects in each study. The individual circles represent data from studies of adults (Castellino et al. 1987; Tessari et al. 1987; Gelfand et al. 1988; Bennet et al. 1989; Fukagawa et al. 1989; Bennet et al. 1990b; Luzi, Castellino, and DeFronzo 1996). The circles connected by lines represent adults studied at several levels of hyperaminoacidemia in the same study (Giordano, Castellino, and DeFronzo 1996). The open symbols connected by lines represent data from 3 day old (diamonds) and 3 week old (squares) infants studied at two levels of hyperaminoacidemia (Poindexter et al. 1997). The open symbols connected by dotted lines represent the protein synthesis in infants calculated from phenylalanine rather than leucine tracer data (Poindexter et al. 1997).

in infants, but that the different tracers gave discrepant results regarding the effect on protein synthesis. During amino acid infusion, much of the R_a of a traced amino acid is from the infusion rather than proteolysis. Thus, small errors in the estimation of proteolytic rate generally have little effect on the calculated rate of protein synthesis.

Effect of Amino Acids on Protein Metabolism in Muscle and Limbs

High amino acid levels promote a positive amino acid balance in the forearm or legs, by inhibiting proteolysis and stimulating protein synthesis (Table 7.1).

TABLE 7.1. Effect of infusing a mixture of amino acids on limb and muscle proteolysis and protein synthesis.

Reference	Δ% amino acid	Δ% limb R_a	Δ% limb R_d	Δ% FRS
Gelfand et al. (1988)	+165 leu	-42 leu, leg	+82 leu	
Bennet et al. (1989)	+91 phe			+40 leu[a], ant. tibialis
	+77 leu			+14 leu[b]
Bennet et al. (1990b)	+91 phe	-9 phe, leg	+32 phe	
	+77 leu	-17 leu	+25 leu	
Fryburg et al. (1995a)	+68 phe 3 hr	-26 phe, 3 hr, forearm	+34 phe, 3 hr	
	+93 phe 6 hr	-7 phe, 6hr	+40 phe, 6 hr	
Svanberg et al. (1996)	leu/phe	phe, forearm and leg	phe	
	+43/+96	-15	+9	
	+85/+165	-43	+44	
	+214/+354	-70	+45	
Biolo et al. (1997)	+200 leu	-11 phe, leg	+123 phe	+53 phe[b], v. lateralis
	+125 phe	+35 leu	+203 leu	
		-3 lys	+33 lys	
Volpi et al. (1998)	+161 phe	-35 phe, leg	+73 phe	+98 phe[b], v. lateralis
	+147 leu	-9 leu	+63 leu	
	+65 lys	+90 lys	+128 lys	

These are mean values. Not all changes were statistically significant.
[a]Using plasma KIC enrichment as index of leucyl-tRNA enrichment.
[b]Using tissue free amino acid enrichment as index of aminoacyl-tRNA enrichment.

However, many of the changes in proteolysis (i.e. limb R_a) shown in Table 7.1 were not statistically significant, which might be the result of either a small effect in muscle or to the inherent variability of the tracer balance method. However, the suppression of proteolysis in the limbs by amino acids could be underestimated by the tracer dilution method, as discussed above for whole-body studies. The study of Svanberg et al. (1996) indicated that limb proteolysis was suppressed by hyperaminoacidemia, but that myofibrillar proteolysis (3-methylhistidine efflux from the limbs) was not affected.

The stimulation of limb protein synthesis by amino acid administration seems to be more robust than the suppression of proteolysis. Based on leg amino acid uptake during hyperaminoacidemia, it was estimated that skeletal muscle uptake accounts for ~25% to 30% of total amino acid disposal (Gelfand et al. 1986). These limb studies probably reflect mainly effects on muscle, but some contribution of skin, fat, and bone cannot be excluded. In rabbits, skin protein turnover is increased by an amino acid infusion, but net protein balance in skin is unaffected (Zhang et al. 1998).

More direct studies of isotope incorporation into muscle proteins also indicate that hyperaminoacidemia stimulates muscle protein synthesis (Table 7.1). A problem with such studies, as with tracer balance studies in the limb, is the uncertainty about the tracer enrichment of the aminoacyl-tRNA. When Bennet et al. (1989) used plasma KIC enrichment as the index of leucyl-tRNA enrichment when [^{13}C]leucine was the tracer, they found that an amino acid infusion increased muscle protein synthesis by 40%. However, when the enrichment of free leucine in the tissue was used as the index, the increase was only 14% and was no longer statistically significant. Biolo et al. (1997) and Volpi et al. (1998) achieved somewhat higher amino acid levels, and found a 50% to 100% increase in muscle protein synthesis using the tissue free phenylalanine enrichment as the index of phenylalanyl-tRNA enrichment. A study of pigs, in which the aminoacyl-tRNA enrichment was directly measured, also indicated that amino acid administration stimulates muscle protein synthesis (Watt, Corbett, and Rennie 1992). In human abdominal muscle incubated in vitro, a very high amino concentration in the medium (10 times plasma level) increased the incorporation of leucine into proteins by ~50% (Lundholm and Schersten 1975).

Splanchnic Tissues

Splanchnic tissues account for ~70% of the whole-body amino acid disposal during an intravenous amino acid infusion (Gelfand et al. 1986). Much of this splanchnic disposal is for hepatic ureagenesis, with the carbons from the amino acids either being oxidized, used for gluconeogenesis, or released into the systemic circulation as ketoacids or other products. Some of an amino acid load is used for protein synthesis, but it is unclear how much hyperaminoacidemia affects splanchnic protein synthesis in humans. Leucine tracer balance indicated that leucine uptake by splanchnic tissues was

38% greater during amino acid infusion than it was in the postabsorptive state, although fractional extraction declined during hyperaminoacidemia (Gelfand et al. 1988). Leucine uptake into the liver is not a valid index of protein synthesis because the leucine can be removed through other pathways. In goats, hyperaminoacidemia increased hepatic protein synthesis by 30% according to the [^{3}H]valine flooding dose method, but did not affect muscle protein synthesis (Tauveron et al. 1994).

Leucine release from splanchnic tissues, an index of splanchnic proteolysis, declined 50% during an amino acid infusion that raised plasma concentrations 2 to 3 fold (Gelfand et al. 1988). This finding is consistent with the inhibitory effect of amino acids on proteolysis in perfused rat livers (Mortimore and Poso 1988). This inhibition is most pronounced as amino acid levels are increased from very low levels to twice the normal plasma levels, with very little further inhibition of proteolysis at very high amino acid levels.

Mechanism of Amino Acid Effects on Protein Metabolism

An adequate supply of all amino acids is necessary to support protein synthesis, so the lack of any one of the amino acids could halt protein synthesis. Each amino acid must be attached to a tRNA molecule by a specific aminoacyl-tRNA synthase before being incorporated into a protein. These synthases have K_m values that are far below the usual intracellular concentrations of amino acids in human tissues, even after many hours of food deprivation (Tischler, Desautels, and Goldberg 1982). Only at the very low amino acid levels near the K_m values would reduced availability of aminoacyl-tRNAs become rate-limiting for protein synthesis. In perfused rat liver, raising the amino acid concentration of the perfusate increases intracellular amino acid levels and protein synthesis without affecting the concentrations of aminoacyl-tRNAs (Flaim et al. 1982b). Thus, the stimulatory effect of hyperaminoacidemia on protein synthesis must be explained by some mechanism other than an increase in aminoacyl-tRNA concentrations. However, there is the possibility that not all aminoacyl-tRNA in the cell is available for protein synthesis and that somehow high amino acid levels increase the formation of aminoacyl-tRNAs that are channeled into the protein synthesizing pathway (Negrutskii and Deutscher 1991). Some amino acids can influence the initiation phase of protein synthesis (Li and Jefferson 1978; Flaim et al. 1982a; Kimball et al. 1989). The suppression of protein synthesis by essential amino acid deprivation may be mediated by increased phosphorylation of eIF-2α (Everson et al. 1989). In cultured hepatoma and myotube cells, increasing amino acid concentrations (particularly leucine) promote initiation by stimulating phosphorylation of PHAS-I and increasing the activity of p70 S6 kinase, which phosphorylates ribosomal protein S6 (Patti et al. 1998).

Although amino acid deprivation induces macroautophagy (increased lysosomal proteolysis) in perfused tissue and cultured cells (Lee and Marzella 1994), it is unclear whether or not this mechanism is important in mediating the effect of physiologic variations in amino acid levels on proteolysis. Hormonal responses to high amino acid levels, such as insulin and glucagon secretion, cannot explain the suppression of proteolysis. The increase in insulin levels during amino acid infusion is too small to account for the effect, and the increase in glucagon levels would be expected to increase rather than reduce proteolysis.

Effects of Individual Amino Acids on Protein Metabolism

Branched-Chain Amino Acids

There is evidence that branched chain amino acids (leucine, isoleucine, valine), particularly leucine, mediate much of the effect of hyperaminoacidemia on protein metabolism. In normal subjects, infusion of branched chain amino acids (Louard, Barrett, and Gelfand 1990; Louard, Barrett, and Gelfand 1995) reduces whole-body and forearm proteolysis by as much as 40%, as indicated by phenylalanine R_a. When labeled leucine is used as the tracer in such studies, whole-body and forearm leucine R_a may decrease, not change, or increase. The problem in using a leucine tracer during branched chain amino acid administration is that the high leucine level can cause a change in the ratio of intracellular to plasma tracer enrichment, thereby causing an underestimate of the suppression of proteolysis. The nonoxidized portion of whole-body leucine R_d, an index of protein synthesis, increases during branched chain amino acid infusion.

When only the branched chain amino acids are infused, levels of other amino acids decline because of the suppression of proteolysis. Thus, forearm protein synthesis (phenylalanine R_d) is unchanged or even inhibited by branched chain amino acid infusion, probably because of the reduced levels of other essential amino acids (Louard, Barrett, and Gelfand 1990; Louard, Barrett, and Gelfand 1995). Leucine disposal in the forearm increases markedly during branched chain amino acid infusion, but this is not an index of protein synthesis because the leucine can be deaminated or oxidized in muscle. In spite of the reduced protein synthesis in the limbs, the net effect of branched chain amino acid administration on limb and whole-body protein metabolism is an anabolic one in brief studies (Louard, Barrett and Gelfand 1990; Louard, Barrett and Gelfand 1995).

Chronically high levels of branched chain amino acids do not appear to influence whole-body proteolysis and protein synthesis. Persons with branched chain α-ketoacid deficiency (maple syrup urine disease) have elevated levels of branched chain amino acids because they cannot oxidize them. Children with this disorder, with plasma leucine levels ~3-fold higher than normal, have normal rates of whole-body protein breakdown and synthesis (Thompson et al. 1990). Thus, either the acute effect of branched chain amino acids that is observed in normal subjects disappears after chronic exposure to high levels, or the effect depends on some metabolite that is not produced by children with this disease.

Leucine

Several studies with isolated rat diaphragms implicated leucine as being of primary importance for the anabolic effect of branched chain amino acid mixtures in muscle, including both suppression of proteolysis and stimulation of protein synthesis (Buse and Reid 1975; Fulks, Li, and Goldberg 1975, Buse and Weigand 1977; Hedden and Buse 1982; Tischler, Desautels, and Goldberg 1982). However, leucine does not stimulate protein synthesis in rat muscle in vivo (McNurlan, Fern, and Garlick 1982). There have been several studies in human subjects to evaluate the effect of leucine on protein metabolism. Eriksson, Hagenfeldt, and Wahren (1981) demonstrated the specificity of leucine compared to other branched chain amino acids in lowering plasma amino acid levels. Sherwin (1978) found that leucine had a nitrogen sparing effect during prolonged fasting in obese humans. However, other research indicated that leucine did not spare nitrogen in fasting obese patients or postoperative patients, whereas KIC was effective (Mitch, Walser, and Sapir 1981; Sapir et al. 1983). Leucine administration can acutely suppress whole-body proteolysis, as indicated by reduced R_a of several essential amino acids, when leucine levels are increased by more than 200% over normal postabsorptive concentrations (Nair, Schwartz, and Welle 1992; Hoffer et al. 1997). When the increase in leucine concentrations is modest (<100%), there is no clear effect on whole-body proteolysis, although plasma levels of other amino acids decline and protein synthesis may be slightly increased (Schwenk and Haymond 1987).

There is evidence leucine has an antiproteolytic effect in muscle, but results have been inconsistent. Increasing the plasma leucine concentrations ~4-fold can promote net retention of tyrosine and phenylalanine in the forearm or leg, an effect that is mediated by reduced limb proteolysis rather than increased protein synthesis (Abumrad et al. 1982; Nair, Schwartz, and Welle 1992). This effect is not always statistically significant (Hagenfeldt, Eriksson, and Wahren 1980), and one study even found increased forearm N losses after a large oral dose of leucine in postabsorptive subjects (Aoki et al. 1981). In human muscle fiber bundles isolated from abdominal muscle, addition of only leucine (at 10 times plasma concentrations) to the incubation medium reduced proteolysis only 7.5% (Lundholm et al. 1981a). In the same tissue, a complete amino acid mixture reduced proteolysis 20%. Excretion of 3-methylhistidine was not affected by leucine in a study in which it spared N in fasting subjects, suggesting that reduced myofibrillar proteolyis is not required for leucine to have an anabolic effect (Sherwin 1978). However, in postoperative patients, addition of either leucine or its ketoacid to intravenous glucose suppressed 3-methylhistidine excretion, although only its ketoacid reduced N excretion (Sapir et al. 1983). In isolated rat muscle, a metabolite of leucine appears to be responsible for the inhibition of proteolysis, whereas leucine itself is responsible for the stimulation of protein synthesis (Tischler, Desautels, and Goldberg 1982).

Flooding doses of leucine have been used to measure the synthesis rate of muscle proteins. This procedure results in leucine concentrations several times higher than normal. The apparent rate of muscle protein synthesis is more than 50% faster during hyperleucinemia than it is when leucine levels are not elevated (Garlick et al. 1989; Smith et al. 1992). This effect might be caused by underestimation of protein synthesis when leucine levels are not high enough to minimize differences among the various indices of aminoacyl-tRNA labeling (Garlick et al. 1994). We did not find an effect of hyperleucinemia on muscle (leg) protein synthesis, using the phenylalanine tracer balance method (Nair, Schwartz, and Welle 1992). Because high concentrations of leucine inhibit proteolysis and thereby reduce the concentrations of other amino acids, the failure of leucine to promote muscle protein synthesis does not rule out the possibility that leucine has an important role in mediating the increase in protein synthesis when there is an increase in the level of all amino acids. However, it does indicate that leucine is not the only amino acid that influences protein synthesis.

The effect of leucine on proteolysis in tissues other than muscle has not been investigated in humans. In perfused rat liver, high leucine concentrations (2 times plasma level) suppress proteolysis more than high concentrations of any other amino acids (Mortimore et al. 1987; Mortimore and Poso 1988). Leucine at plasma concentrations is ineffective in suppressing proteolysis in the perfused liver, but is effective at half the normal plasma concentration. Leucine shares this multiphasic pattern of inhibition of proteolysis with several other amino acids (Mortimore et al. 1987; Mortimore and Poso 1988). The loss of an antiproteolytic effect of leucine at the normal plasma concentration is abolished by adding insulin to the perfusate, and the antiproteolytic effect at half the normal plasma amino acid concentration is abolished by adding glucagon (Mortimore et al. 1987).

Smith et al. (1994) reported that hyperleucinemia stimulates albumin synthesis in humans. They used plasma ketoisovaleric acid as the index of hepatic valyl-tRNA enrichment during [^{13}C]valine infusion. It is possible that high leucine levels altered the relation between this index and the true hepatic valyl-tRNA enrichment. This uncertainty about the precursor labeling makes it difficult to determine the extent to which leucine affects hepatic protein synthesis.

Isoleucine

A reduction of the plasma isoleucine concentration to ~10% of the normal level, induced by infusing insulin and amino acids other than isoleucine, reduced whole-body protein synthesis by 12% without influencing albumin or fibrinogen synthesis (Lecavalier, De Feo, and Haymond 1991). It is unlikely that this effect is specific for isoleucine. The effect of reducing other amino acids to a similar extent was not determined. Threonine levels were reduced by ~55% by the same method, without affecting whole-body protein metabolism. Whereas infusion of leucine at ~0.5 $\mu mol \times kg^{-1} \times min^{-1}$ caused a modest reduction in plasma amino acid levels, reflecting reduced proteolysis, an isomolar infusion of isoleucine did not have this effect (Schwenk and Haymond 1987).

Glutamine

Glutamine is the most abundant amino acid in the body. Plasma glutamine concentrations (~600 nmol/ml after overnight fasting) are higher than those of any other amino acid, and glutamine is very concentrated in muscle (>10 mmol/kg). Glutamine is an important substrate for energy production in the intestine, lymphocytes, and malignant cells, and therefore could modulate their protein metabolism by affecting availability of ATP (Higashiguchi et al. 1993; Wasa et al. 1996).

In healthy subjects, infusing enough glutamine to double the plasma glutamine concentration does not affect whole-body proteolysis, but slightly increases whole-body protein synthesis (Hankard, Haymond, and Darmaun 1996). With a threefold increase in plasma glutamine levels, there may be a slight decrease in whole-body proteolysis (Perriello et al. 1997). Glutamine infusion can improve N balance after surgery, more than would be expected from repletion of free glutamine in muscle (Hammarqvist et al. 1989). However, glutamine supplementation of a standard parenteral nutrition formula did not improve N balance or influence protein turnover in trauma patients (Long et al. 1995). A modest depletion of glutamine, induced by phenylbutyrate, did not influence whole-body proteolysis in normal subjects, but it increased leucine oxidation and reduced whole-body protein synthesis an average of 11% (Darmaun et al. 1998).

Animal studies suggest that glutamine stimulates protein synthesis and inhibits protein degradation in muscle (MacLennan, Brown, and Rennie 1987; Jepson et al. 1988; MacLennan et al. 1988; Wu and Thompson 1990; Zhou and Thompson 1997). In humans, there is evidence that glutamine attenuates the postsurgical decline in muscle protein synthesis, as reflected by total ribosome and polyribosome concentrations (Hammarqvist et al. 1989; Blomqvist et al. 1995). Tracer leucine incorporation also indicated that glutamine supplementation (as alanyl-glutamine) substantially increased muscle protein synthesis in postsurgical patients relative to isonitrogenous supplementation with alanine and glycine (Barua et al. 1992).

In human cancer cells grown in culture, glutamine regulates cell proliferation and protein synthesis (Wasa et al. 1996). These cells have an extremely high rate of glutamine uptake, much faster than that needed to support protein synthesis. The suppression of protein synthesis by glutamine deprivation may be related to reduced energy production because glutamine's role as the preferred metabolic substrate.

Other Amino Acids

Enteral infusion of a large amount of glycine, enough to raise plasma glycine levels 14-fold, reduced whole-body proteolysis by 11% in healthy volunteers (Hankard, Haymond and Darmaun 1996). Protein synthesis also declined, so that the nitrogen sparing effect was trivial. It is unclear whether the same amount of energy provided by a substrate other than glycine would have the same effect. There is no evidence that high phenylalanine concentrations affect the rate of protein synthesis in animals (Garlick et al. 1994). However, a large dose of phenylalanine appears to stimulate human muscle protein synthesis (Smith et al. 1998a). Infusing threonine at a rate equivalent to ~50% of its normal postabsorptive R_a does not influence whole-body protein metabolism in healthy subjects (Schwenk and Haymond 1987). A large dose of threonine increases muscle protein synthesis, whereas arginine, serine, and glycine do not (Smith et al. 1998a).

Glucose and Its Metabolites

The major difficulty in assessing the effect of glucose on protein metabolism is that it is the primary determinant of insulin secretion. Raising glucose levels in vivo normally causes hyperinsulinemia and the associated effects on protein metabolism (Chapter 6), and lowering glucose usually requires hyperinsulinemia. A number of studies have demonstrated a protein anabolic effect of glucose administration, but could not differentiate between the effect of glucose per se and that of insulin. Such research is not reviewed because it cannot provide any insights into the independent role of glucose or its metabolites (mainly lactate and pyruvate, but also a number of intracellular glycolytic intermediates). Hyperglycemia associated with diabetes cannot be used as a model to examine the role of glucose in protein metabolism, because diabetes also involves insulin deficiency or insulin resistance.

Only two studies have attempted to directly evaluate the independent effect of glucose on protein metabolism in humans (Flakoll, Hill, and Abumrad 1993; Heiling et al. 1993). Both studies involved the use of a leucine tracer to examine whole-body protein metabolism. In one study, there was no significant change in proteolysis associated with different levels of glycemia (from normal postabsorptive to ~3 times normal) at three different insulin levels within the physiologic range (Heiling et al. 1993). No measures of protein synthesis were made. In the other study, protein metabolism was examined during supraphysiologic hyperinsulinemia at two levels of glycemia—normal and twice normal (Flakoll, Hill, and Abumrad 1993). Leucine levels were maintained constant during hyperinsulinemia by a variable amino acid infusion. When glucose levels were normal, there was a very low rate of proteolysis (only 20% of normal postabsorptive) due to the very high insulin concentration. Proteolysis then increased to 55% of the postabsorptive rate when hyperglycemia was induced while the very high insulin level was maintained. Even though proteolysis was increased when the glucose level was elevated, hyperglycemia was not catabolic because protein synthesis increased as much as proteolysis. The investigators did not find any hormonal changes that would explain increased protein turnover during hyperglycemia, and concluded that the elevated glucose (or a metabolite) per se was responsible for the effect. A weakness of this study was the lack of a control group to demonstrate that protein turnover would not have rebounded even if hyperglycemia had not been induced. Thus, further evidence is needed before we can accept the conclusion that hyperglycemia increases protein turnover.

In healthy infants, a very modest increase in plasma glucose—from 75 to 90 mg/dl—did not affect whole-body proteolysis even though insulin levels increased slightly (Denne et al. 1995b). Proteolysis was similar in these infants during fasting, glucose infusion, and isocaloric lipid infusion. The differences in glucose and insulin concentrations among the various experimental conditions were too small to allow any conclusion about the importance of glucose and insulin in the regulation of proteolysis in infants.

Studies in isolated tissues avoid the problem of hormonal responses to altered glucose levels. The effect of glucose on protein metabolism has been examined in isolated rat diaphragms (Fulks, Li, and Goldberg 1975; Hedden and Buse 1982). Glucose slightly reduced proteolysis, whereas β-hydroxybutyrate and octanoate were ineffective. In the absence of insulin, glucose had little effect on protein synthesis in one study, but in another study was found to be more effective than pyruvate or lactate in supporting protein synthesis. Although lactate did not sustain ATP levels as well as glucose, the failure of pyruvate to support protein synthesis as well as glucose could not be explained by reduced ATP or phosphocreatine levels. In the presence of high insulin concentrations, glucose stimulated protein synthesis, an effect similar to the increase in whole-body protein synthesis when glucose levels are raised during hyperinsulinemia in humans (Flakoll, Hill, and Abumrad 1993).

In summary, it appears that most of the anabolic effect of glucose is related to the fact that it is a fuel for ATP production. This ATP is required to sustain protein metabolism. Any additional effect of glucose on protein metabolism probably derives from its stimulation of insulin secretion.

Triglycerides, Fatty Acids, and Ketone Bodies

Most of the fatty acids in the circulation are in triglycerides packaged in lipoprotein particles—mainly very low density lipoproteins, except after a fatty meal when chylomicrons carry most of the circulating lipid. These particles are too large to enter cells. Tissues extract the fatty acids from these triglycerides by lipoprotein hydrolysis at the capillary endothelium through the action of the enzyme lipoprotein lipase. Free fatty acids also are released from adipose tissue into the circulation after adipocyte triglycerides are hydrolyzed by the action of hormone sensitive lipase. Muscle and liver cells can store a small amount of triglyceride that can be hydrolyzed for energy production. Although some of the studies cited in this section examined the effect of administering fat as triglycerides, it is very likely that any effects on protein metabolism are mediated by the free fatty acids that have entered the cells rather than by the amount of circulating triglyceride per se. The effect of fatty acids on protein metabolism, like that of glucose, primarily relates to the fact that fatty acids are an important substrate for ATP production. Triglyceride hydrolysis also produces glycerol, but the energy content of the glycerol is a small fraction of that of the fatty acids. Unlike glucose, triglycerides and free fatty acids have little effect on insulin or other hormones that influence protein metabolism.

The effect of infusing fat on protein metabolism has been examined in several studies. Ferrannini et al. (1986) noted a modest reduction of amino acid levels during lipid+heparin infusion, suggesting either reduced proteolysis or enhanced amino acid disposal. The heparin was used to stimulate lipoprotein lipase activity, thereby markedly increasing free fatty acid concentrations. Under these conditions, the very high free fatty acid levels increased insulin somewhat, which might account for the reduced amino acid levels. However, fat infusion had the same effect during insulin infusion, suggesting that the fatty acids had an independent effect on amino acid levels. Later studies examined whether fat infusion influences whole-body proteolysis or protein synthesis, using tracer methods. A 3-fold increase in free fatty acid levels was associated with a 20% decline in leucine R_a (vs. 8% decline with saline infusion) in one study, indicating reduced proteolysis (Beaufrere et al. 1992). However, when plasma KIC enrichment rather than leucine enrichment was used to calculate R_a, the effect (15%) was no longer significantly different from the saline effect (11%). Whole-body protein synthesis (nonoxidized leucine R_d) declined in parallel with R_a. Leucine oxidation, an index of the rate of protein loss when leucine is not available from dietary sources, was reduced by fat infusion. A fat preparation containing a 50%/50% mixture of medium-chain and long-chain triglycerides was as effective in suppressing proteolysis as one containing only long-chain triglycerides, but was relatively ineffective in suppressing leucine oxidation. In another study, lipid+heparin infusion was not associated with any change in whole-body proteolysis, as assessed by phenylalanine R_a (Walker et al. 1993). There was a modest decline in forearm proteolysis during lipid+heparin infusion in this study, but there was a similar decline in forearm protein synthesis so that net forearm phenylalanine (protein) balance was not affected by elevated free fatty acid levels. In contrast, in another study lipid infusion (either long-chain or a mixture of long- and medium-chain triglycerides) increased phenylalanine and overall amino acid balance in the forearm (Wicklmayr et al. 1987). In newborns, lipid infusion did not affect whole-body proteolysis (Denne et al. 1995b). Thus, when fat is given to healthy subjects, without any amino acids or glucose, there is some evidence for a modest reduction in protein turnover and possibly a slight anabolic effect.

In obese patients on a starvation diet, consumption of 300 to 500 kcal daily of only fat did not alter the rate of N loss (Vazquez, Morse, and Adibi 1985). Addition of a small amount of fat (220 kcal/d) to carbohydrate (300 kcal/d) did not potentiate the N sparing effect of the carbohydrate. Dietary fat may have merely substituted for the oxidation of endogenous fat under such conditions. Fatty acids provide most of the energy needed to survive during starvation, and this energy is very important in restraining the use of endogenous proteins to provide energy. In dogs starved for 2 days, inhibition of lipolysis with nicotinic acid increased whole-body proteolysis and leucine oxidation, and inhibited incorporation of leucine into proteins (Tessari et al. 1986a). Although comparable studies have not been done in fasting humans, it would be very surprising if limiting the supply of fatty acids during starvation did not have a similar effect.

Substitution of fat calories for carbohydrate calories may lead to some N loss in healthy subjects on a normal or marginal protein intake (Munro 1964b; Richardson et al. 1979). The effect is most dramatic when the carbohydrate intake is extremely low. If carbohydrate intake is too low, there can be a significant reduction in insulin levels and a reduction in blood pH from high free fatty acid and ketone body production. The acidosis can contribute to N loss (see section on acidosis in Chapter 9), and reduced secretion of insulin certainly could be have a catabolic effect. Because very low carbohydrate diets stimulate gluconeogenesis, more of the amino acids from the diet or proteolysis are diverted for glucose production, leaving less available for reincorporation into proteins. However, under most conditions carbohydrate and fat calories appear to be equally effective in maintaining the protein stores of the body.

Several studies examined the relative effectiveness of glucose and lipids in sparing protein during parenteral feeding (Tessari et al. 1996a). Generally, glucose and fat are both effective in sparing protein when added to an amino acid infusion. The combination of adequate calories and an adequate supply of essential amino acids is important, whereas the source of the calories seems to have little effect on N balance or protein turnover of adults or infants being fed parenterally (Jeejeebhoy et al. 1976; Nordenstrom et al. 1983; Albert et al. 1986; DeChalain et al. 1992; Jones et al. 1995).

When the production of acetyl-CoA from fatty acid β-oxidation exceeds acetyl-coA oxidation in the citric acid cycle, the excess acetyl-CoA is converted to ketone bodies (β-hydroxybutyrate and acetoacetate) in the liver. This occurs primarily when the glucose supply is limited, because reduced glycolysis leads to reduced availability of oxaloacetate for the citric acid cycle. During starvation, there is a marked increase in ketone body concentrations after 2 to 3 days, which coincides with a decline in N excretion. Thus, there has been much interest in whether or not ketone bodies have a protein-sparing effect. Sherwin et al. (1975) reported that infusion of Na DL-β-hydroxybutyrate into obese patients who had been fasting for several weeks reduced N excretion ~30%. Pawan and Semple (1983) reported that ingestion of 18 g of Na DL-β-hydroxybutyrate daily, by obese subjects receiving no other food, reduced N excretion ~50% relative to N excretion with an equal amount of glucose. Excretion of 3-methylhistidine also was markedly inhibited by Na DL-β-hydroxybutyrate ingestion, suggesting that inhibition of myofibrillar proteolysis was involved in the N sparing effect. There is a modest alkalosis associated with Na DL-β-hydroxybutyrate administration, and an effect of increased pH rather than an effect of the ketone cannot be excluded. The endogenous ketone bodies are weak acids that would have the opposite effect on pH.

In isolated rat diaphragms, ketone bodies have no effect on protein degradation or synthesis (Fulks, Li and Goldberg 1975; Hedden and Buse 1982). In humans, infusion of Na β-hydroxybutyrate for up to 8 hours does not affect whole-body proteolysis (Miles et al. 1983; Nair et al. 1988; Beaufrere et al. 1992). In one study there was a 10% increase in whole-body and muscle protein synthesis and reduced leucine oxidation during Na-DL-β-hydroxybutyrate infusion (Nair et al. 1988). The mild alkalosis induced by infusion of the Na salt of the ketone could not explain the effect on protein synthesis, because there was no effect of an equivalent change in pH induced by $NaHCO_3$ infusion. Leucine oxidation is a reflection of protein balance in subjects who are not consuming any protein, and the reduced leucine oxidation translates into an N retention of over 2 g/d in a 70 kg subject. Studies using the physiological D isomer did not replicate the protein sparing effect (reduced leucine oxidation and increased nonoxidized leucine R_d) in normal postabsorptive subjects (Beaufrere et al. 1992) or in septic patients (Beylot et al. 1994). The fact that no studies have demonstrated a protein-sparing effect of only the D isomer makes it difficult to draw any firm conclusions about the role of ketones in the adaptation of protein metabolism to starvation or other ketogenic situations. To the extent that the brain or other tissues use the ketones rather than glucose as an energy source during starvation, ketogenesis could help to conserve amino acids for protein synthesis by reducing the rate of gluconeogenesis from amino acids.

Summary

Metabolic substrates are needed to provide the energy required to sustain protein turnover. In general, the particular substrate being used to provide cellular energy does not seem to influence protein metabolism very much in humans. Glucose is the major regulator of insulin secretion, and therefore can have an additional effect beyond its role in providing energy. Amino acid oxidation usually is much less important than glucose and fat oxidation in terms of providing energy, but amino acids obviously have the additional role of being the building blocks of proteins. Changes in amino acid levels within the normal physiologic range do not appear to significantly affect the aminoacyl-tRNA concentrations, so the role of amino acids in regulating protein synthesis is more than just a "mass action" effect. Although effects of amino acids on both protein synthesis and protein degradation have been demonstrated, the molecular basis of the effects is not understood.

8
Physical Activity

Skeletal muscle comprises about half of the protein mass of the human body. The mass and composition of muscle proteins can vary according to the demands put on the muscle. Muscle protein mass declines when a muscle is not used for an extended period and increases when a muscle is repeatedly overloaded. Not only does physical activity influence total muscle protein mass, but it also affects the concentrations of specific proteins. For example, regular aerobic activity can increase the amount of enzymes involved in producing energy from metabolic substrates. The pattern of use also can influence which isoforms of various contractile proteins are expressed. Thus, changes in protein breakdown and synthesis are important not only in altering muscle mass in response to muscle use, but also in remodeling the muscle for better performance at the type of activity for which it is being used.

Many of the studies cited in this chapter deal with measures of whole-body protein metabolism. Intuitively, you might think that all of the effects of physical activity on whole-body protein metabolism are related to effects on the specific muscles involved. Because some types of physical activity might involve only one or a few muscle groups, the impact on whole-body protein metabolism could be too slight to be detectable. For example, synthesis of all muscle proteins might account for ~30% of whole-body protein synthesis in a typical person. If that person starts doing armcurl exercises to strengthen his elbow flexor muscles, he would be stimulating less than 10% of his total muscle mass. Even a 100% increase in protein synthesis in the elbow flexors would therefore increase whole-body protein synthesis less than 3%. However, some types of physical activity might have indirect effects on protein metabolism in tissues other than the specific muscles involved in the activity. For example, intense aerobic activity has many systemic effects that might acutely influence protein metabolism in many tissues. Increased levels of hormones and substrates, changes in the pattern of blood flow and substrate oxidation, and increased body temperature could mediate such effects. With long term changes in physical activity, there can be alterations in body composition and sensitivity to insulin, and these might influence protein metabolism as discussed elsewhere in this book. We should not be too surprised if some studies show larger effects of physical activity on whole-body protein metabolism than what we would expect from the effects on the muscle itself.

Effect of Inactivity on Protein Metabolism

In most studies of protein turnover employing intravenous amino acid tracers, the subjects are restricted to bed and are very inactive during the study. Whole-body protein turnover during these brief periods of bed rest may be slower than it is during light physical activity. According to the [^{15}N]glycine end product method, whole-body protein synthesis and degradation are ~25% slower when subjects are resting supine during the study than when they are ambulatory (Bettany et al. 1996). N balance does not seem to be affected during such brief periods of inactivity. In other studies of the effect of inactivity, the "baseline" or "control" studies have been done while subjects are resting in bed, so any such acute influence of inactivity on protein turnover is not detected. However, studies with intravenous ^{13}C tracers during bed rest have not produced lower estimates of whole-body turnover than studies with [^{15}N]glycine in ambulatory subjects (Chapter 4), so this interesting result needs to be verified with more research before we can accept the fact that brief inactivity has such a marked effect on whole-body protein turnover.

Schonheyder, Heilskov, and Olesen (1954) studied three normal young adults during a period of normal activity and during immobilization of the lower body in a cast. In all cases, there was negative N balance after 5 days of immobilization even though the diet was unchanged. The peak rate of N loss occurred ~10 days after the onset of immobilization. Analysis of the decay of urinary ^{15}N excretion after a single dose of [^{15}N]glycine suggested that the N losses occurred because of reduced whole-body protein synthesis rather than increased proteolysis.

Shangraw et al. (1988) studied whole-body protein metabolism in normal volunteers placed on strict bed rest for a week. With magnetic resonance imaging, they documented slight decreases in the volumes of leg and back muscles even after this short period of inactivity. Compared with N balance before bed rest, N balance during bed rest was more negative by ~1g/d, suggesting a cumulative protein loss of ~40 g over the period of inactivity. There was not a change in whole-body proteolysis or protein synthesis according to leucine kinetics, although a later study by the same group indicated that whole-body protein synthesis does decline after a week of bed rest if protein intake is low (Stuart et al. 1990). The slight increase in 3-methylhistidine excretion was not statistically significant, prompting the authors to conclude that myofibrillar proteolysis was not affected by bed rest. However, the mean increase in 3-methylhistidine excretion during bed rest corresponded to an increase in myofibrillar protein degradation of as much as 70 g, or more than the protein loss according to N balance. Thus, this study did not rule out the possibility that increased muscle protein degradation has some role in the muscle atrophy associated with bed rest. Another bed rest study that lasted 17 days also indicated that 3-methylhistidine excretion did not change (Stein and Schluter 1997). Bed rest for 2 weeks reduced thigh muscle mass by 4% and quadriceps muscle protein synthesis by ~50% (Ferrando et al. 1996). There was also a 30% reduction in the average rate of leg proteolysis during bed rest, but the effect did not reach statistical significance. Whole-body protein synthesis declined 14% according to the [^{15}N]alanine-urea end product method, but whole-body proteolysis did not change. If all muscle responded to bed rest as much as the quadriceps, the decrease in muscle protein synthesis could explain all of the decrease in whole-body protein synthesis. Even though inactivity reduces total energy expenditure, which would tend to reduce protein oxidation, there was an increase in urea excretion during bed rest. Thus, more of the amino acids from the diet and from proteolysis were used for oxidation rather than protein synthesis when subjects were inactive. The muscle atrophy and reduction in muscle protein synthesis during bed rest can be prevented by resistance exercises (Ferrando et al. 1997b; Bamman et al. 1998).

Further evidence for reduced protein synthesis in inactive muscles comes from a study of men whose legs were immobilized in casts after tibial fracture (Gibson et al. 1987). Thigh muscle volume in the immobilized leg was ~10% less than in the other leg after 5 weeks of immobilization. The average fractional rate of muscle protein synthesis was 26% slower in the immobilized leg. The slowing of protein synthesis could not be explained by reduced RNA concentrations (mostly ribosomal), which tended to increase (per g protein) in the immobilized muscles. Calculations based on the approximate change in muscle mass and muscle protein synthesis indicated that muscle proteolysis probably did not increase in the immobilized leg, and might have decreased slightly. However, this calculation was based on the assumption that the fractional rate of protein synthesis determined at the end of the study was representative of the synthesis rate throughout the period of immobilization, and therefore should be considered an educated guess rather than a real measure of proteolysis.

A more moderate reduction in muscle activity than complete bed rest or immobilization may not reduce muscle protein synthesis. Gibson et al. (1989) studied patients with unilateral osteoarthritis of the knee. There was clear atrophy of the quadriceps muscle in the affected leg, suggesting that use of this muscle was diminished. Surprisingly, the atrophic muscle had a more rapid fractional rate of protein synthesis than the normal, contralateral muscle. The atrophic muscle had a much higher DNA concentration, reflecting the fact that muscle fibers were shrunken rather than lost. Protein synthesis per unit of DNA, which directs the protein synthetic process, was ~30% slower in the atrophic muscle. It is interesting that percutaneous electrical stimulation restored muscle mass, but did not influence the rate of protein synthesis in the muscle.

Bed rest leads not only to muscle atrophy, but also to reduced bone mass. Various markers of bone collagen degradation are increased during bed rest (Smith et al. 1998b).

Space Flight

Even though astronauts are quite active in space, the effort involved in limb movements and maintaining posture is greatly reduced because of the weightlessness. Consequently, muscle atrophy is a significant problem during long space flights, and measurable atrophy occurs within a week (LeBlanc et al. 1995). A series of studies by Stein et al. (Stein, Leskiw, and Schluter 1993, 1996; Stein and Schluter 1997) examined whether the loss of protein during weightlessness is related to increased proteolysis or reduced protein synthesis. During a 9 day flight of the space shuttle Columbia, there was negative N balance on the first day and on the last 3 days. Because protein synthesis, assessed by the [^{15}N]glycine-NH_3 end product method, was increased during the flight, it follows that whole-body proteolysis was increased even more. During a 14 day shuttle mission, a similar pattern was observed, but N balance recovered to preflight values after 10 days. The catabolic state on the first day may be mediated in part by reduced energy and protein intake. In addition to inactivity, there was an increase in cortisol production and a transient increase in production of the cytokine IL-6, and perhaps other stress hormones, that might have contributed to the catabolic effect of space flight. The increase in fibrinogen synthesis early in the flight and on the day of returning to earth is similar to the response to other types of stress. In spite of the evidence for increased whole-body proteolysis and increased stress, there was no change in 3-methylhistidine excretion during the shuttle missions, suggesting that myofibrillar proteolysis was not increased.

Studies during the Skylab missions also documented N losses and reduced lean body mass during weightlessness (Leonard, Leach, and Rambaut 1983; Stein, Leskiw, and Schluter 1996; Stein and Schluter 1997). Skylab was a space station that orbited the earth in the 1970s. Food intake during the Skylab missions generally was reduced compared to preflight intake, but N losses were recorded even when energy intake was not reduced. However, there was a very high level of exercise on Skylab missions, so that negative energy balance might have contributed to the loss of body proteins. Unlike the space shuttle missions, the Skylab missions were associated with increased 3-methylhistidine excretion, suggesting increased myofibrillar proteolysis. The negative energy balance and increased physical activity on Skylab probably explains this finding, since weightlessness (shuttle mission) and inactivity (bed rest) do not increase 3-methylhistidine excretion. The intake of 3-methylhistidine during the Skylab mission was not determined, so it is possible that the increased 3-methylhistidine excretion resulted from more intake rather than increased myofibrillar proteolysis.

There is a loss of bone mass during space flight. Measures of collagen degradation, some relatively specific for bone, indicate that bone collagen degradation was increased during the Skylab missions (Smith et al. 1998b).

Effect of Increased Physical Activity on Protein Metabolism

This discussion distinguishes between two types of activity—"aerobic" and "resistance." The term aerobic reflects the fact that this type of activity requires an increased supply of O_2 to the muscles involved in the activity. Aerobic exercise also is referred to as "endurance" exercise. Generally, such activity is performed over several minutes to hours at a low percentage of the maximal force that can be generated by the muscles. Although the relative force of each contraction is small, the contractions are repeated frequently enough that the muscle is operating at a high proportion of its maximal ability to generate energy. Repeated bouts of aerobic exercise generally have little effect on muscle bulk, but induce cardiovascular and metabolic adaptations that improve the efficiency of O_2 and nutrient delivery to the muscles. Resistance exercises are muscle contractions against a large resistance, generally requiring over 60% of the maximum force that the muscle group can generate. Only a few contractions can be performed before the muscle becomes temporarily exhausted. Resistance exercises are much more effective than aerobic exercises in promoting muscle hypertrophy and strength. Although there are certain activities that are considered purely aerobic (e.g., running on a level surface) and others that are considered purely resistive (e.g., weightlifting), the distinction is not always clear. For example, running up several flight of stairs carrying a heavy load has both an aerobic and a resistive component. Nevertheless, the distinction between these types of activity is a convenient way to organize the research in this area.

Effect of Aerobic Activity on Protein Metabolism

Acute Effect of Activity on Protein Oxidation

Aerobic activity can require a several-fold increase in substrate oxidation in the contracting muscles, which may be sustained for long periods. In addition to the stimulation of protein oxidation to provide energy in the contracting muscles, the drain on circulating glucose and fatty acids during prolonged exercise could indirectly influence protein oxidation in other tissues. While protein is quantitatively less important than glucose and fat in fueling muscle contractions, amino acids from proteolysis or dietary proteins do contribute to energy production during exercise. The key issue is whether or not amino acid oxidation during exercise or recovery from exercise is any different than it is during periods of inactivity. Several studies indicated that total N excretion, urea excretion, and isotopically-determined urea production were not stimulated by aerobic activity (Wolfe et al. 1984; Carraro et al. 1990a; Carraro, Kimbrough, and Wolfe 1993; El-Khoury et al. 1997;). However, others indicated that urea and N excretion did increase either during exercise (Calles-Escandon et al. 1984) or after exercise (Rennie et al. 1980; Lemon,

Dolny, and Yarasheski 1997). There are several problems in using urea or total N excretion to evaluate the effect of exercise on protein oxidation. Losses of urea in sweat may account for a significant proportion of the urea losses during exercise (Lemon and Mullin 1980; Calles-Escandon et al. 1984), although this factor does not always account for the failure to detect increased urea losses during exercise (El-Khoury et al. 1997). A lack of change in urea or N excretion does not rule out increased protein oxidation over short periods when the body pools of free N or urea may be changing. Prolonged exercise can markedly increase the size of the urea pool of the body, corresponding to a drop in the α-amino-N levels (Haralambie and Berg 1976). Reduced renal blood flow could delay the excretion of urea that is produced during or shortly after exercise, so that the increase in excretion might not occur until several hours after the exercise (Rennie et al. 1980), or even the day after exercise (Lemon, Dolny, and Yarasheski 1997). However, not all investigators have observed an increase in the urea pool of the body during exercise (Carraro, Kimbrough, and Wolfe 1993). The amount of glycogen available to fuel muscle contractions may be a critical determinant of how much urea is produced during exercise (Lemon and Mullin 1980). It is safe to conclude that even when exercise does increase total amino acid oxidation, the contribution of amino acids to total energy expenditure is small compared to the contributions of carbohydrate and fat.

Branched chain amino acids are oxidized in muscle. Oxidation of leucine, and probably that of isoleucine and valine, is increased during aerobic exercise even when there is no evidence for an increase in total amino acid oxidation (Hagg, Morse, and Adibi 1982; Knapik et al. 1991; Millward et al. 1994; El-Khoury et al. 1997). The portion of whole-body leucine R_a that is used for oxidation is proportional to exercise intensity, and may reach 50% at maximal energy expenditure (Figure 8.1). As discussed in Chapter 3, not all of the labeled CO_2 is recovered in breath during the infusion of carbon-labeled substrates, because of sequestration in bicarbonate pools with a slow turnover. During intense activity, when CO_2 production is much greater than it is at rest, the recovery is closer to 100%. In fact, the rate of $^{13}CO_2$ expiration can transiently exceed the rate of ^{13}C infusion (as bicarbonate) when exercise is initiated after tracer has been infused for several hours. This transient effect must reflect some depletion of the prelabeled bicarbonate stores at the outset of the exercise session. Moreover, a shift in substrate oxidation from fat to glycogen during exercise can increase $^{13}CO_2$ expiration due to a higher natural abundance of ^{13}C in glucose than in fat. Not all studies have accounted for increased recovery of labeled CO_2 or shifts in the substrate mixture, and therefore exaggerated the influence of exercise on leucine oxidation. Nevertheless, exercise increases leucine oxidation even when the data are adjusted for the increased recovery of labeled CO_2 (El-Khoury et al. 1997). Most of the N produced by deamination of branched chain amino acids in muscle is used to synthesize alanine and glutamine, which can then be used by the liver (and kidneys to a lesser extent) for gluconeogenesis. Immediately after exercise, leucine oxidation declines if no leucine is provided by food and may even fall

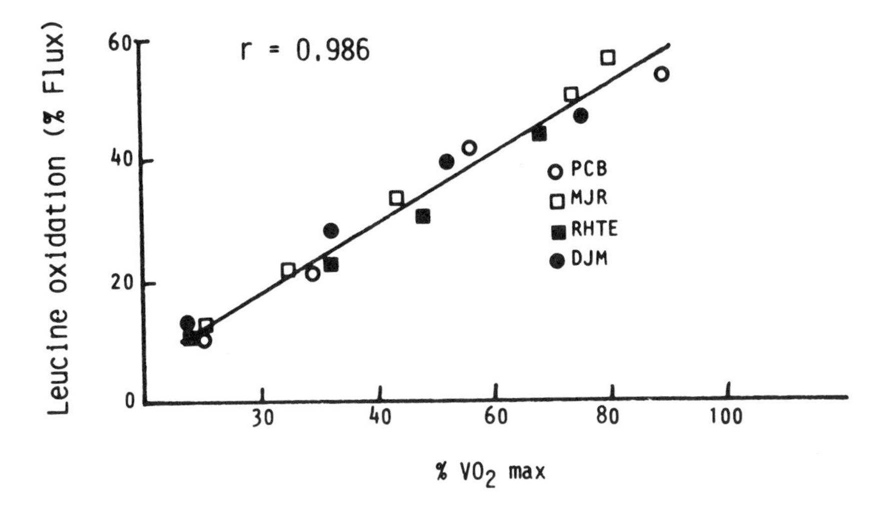

FIGURE 8.1. Leucine oxidation, as a fraction of leucine flux, in relation to exercise intensity in 4 subjects exercising on a bicycle ergometer after overnight fasting. From Millward et al. (1982), with permission.

slightly below the preexercise rate of leucine oxidation (Devlin et al. 1990; El-Khoury et al. 1997). In subjects fed protein-containing meals, leucine oxidation remains elevated during recovery from exercise (Forslund et al. 1998).

There is evidence that male athletes oxidize more protein and leucine than female athletes during exercise (Tarnopolsky et al. 1990; Phillips et al. 1993). The gender discrepancy is more than can be explained by the difference in body size or total energy expenditure during exercise. Females have a higher body fat content than males, and may rely on fat to fuel exercise more than males do.

There is an increase in NH_3 production by exercising muscle that apparently is unrelated to protein balance. Exercise induces deamination of adenosine monophosphate to inosine monophosphate, and the amount of NH_3 generated by this pathway is sufficient to increase plasma NH_3 levels 2-fold or more during very intense exercise (Babij, Matthews, and Rennie 1983).

Effect of Aerobic Training on Protein Oxidation and Protein Requirements

Studies done over only a few hours cannot elucidate whether or not the total daily protein or branched chain amino acid oxidation, and hence the minimum dietary protein or branched chain amino acid requirements are influenced by chronic changes in physical activity. Perhaps the most important thing to emphasize about this issue is that aerobic activity increases the total daily energy expenditure, so that energy intake must increase to maintain energy balance when physical activity is increased. If energy balance is not maintained, then the protein stores of the body generally will be depleted.

The question that has not been resolved to everyone's satisfaction is whether or not protein requirements are increased in the most active individuals even if they are consuming enough energy to maintain their body weight. While the increase in total protein oxidation does not appear to be out of proportion with the increase in total energy expenditure induced by increased physical activity, the fact that the essential branched chain amino acids may be selectively oxidized by active muscles raises the possibility that the protein concentration of the diet (or at least the branched chain amino acid concentration) should be increased in the most active individuals.

Butterfield and Calloway (1984) examined N balance as a function of the level of aerobic activity in 6 healthy young men over a period of 108 days. After an 18 day period on a generous protein allowance (1 $g \times kg^{-1} \times d^{-1}$), they were restricted to the minimum safe allowance suggested by the WHO (0.57 $g \times kg^{-1} \times d^{-1}$ of high quality proteins) for the remainder of the experiment. For the first 36 days on the lower protein intake, they walked on a treadmill for an hour daily and consumed a weight-maintening diet. By the end of this period, the average N balance was close to zero. Energy intake then was increased by 15% for 18 days while the exercise regimen stayed the same. N balance became positive, as expected. The amount of physical activity was then increased for the next 18 days so that there was no energy surfeit even though energy intake remained 15% above the original level. Once again, N balance was close to zero. Finally, energy intake was raised another 15% for 18 days while the more intense exercise regimen continued. N balance was even more positive during the energy surfeit when the exercise intensity was greater (Figure 8.2). This experiment suggested that increasing the amount of energy expended for physical activity does not increase the protein requirement as long as energy needs are met. It also indicated that the stimulation of protein accretion by an energy surfeit can be enhanced by exercise.

Butterfield (1987) reviewed the research done in this area before 1987, and concluded that no definite recommendations could be made about the protein requirements of those with a very high level of physical activity. The relevant studies often were terminated too soon, before N balance or body composition had stabilized, and did not control for the negative energy balance associated with increased activity. Although there is a transient N loss associated with increased activity, N balance may be restored within a few days when energy intake is adequate. Meredith et al. (1989) avoided the problem of acute adaptation by studying the protein requirements of men who had been doing aerobic exercises for 2 to 40 years. On average, N balance was achieved with a protein intake of 0.93 $g \times kg^{-1} \times d^{-1}$, slightly more than the U.S. recommended safe intake and >50% more than the WHO recommended intake (see Chapter 5). Unfortunately, no sedentary men were included in this study, so that it is impossible to determine whether these data demonstrate an increased protein requirement in more active individuals or merely technical differences between this study and other studies that defined protein requirements in more sedentary subjects (e.g., because of more

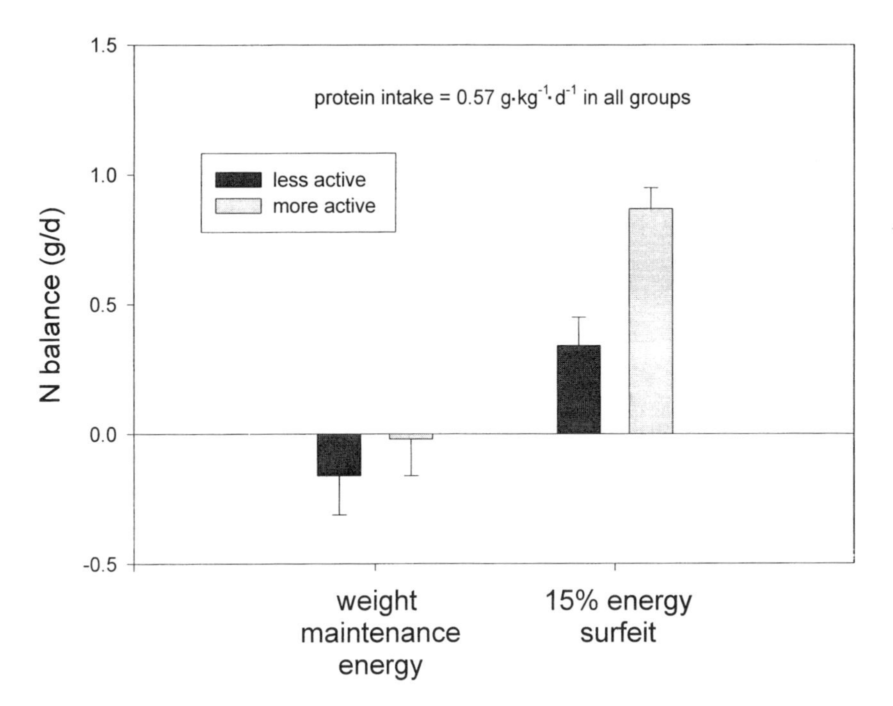

FIGURE 8.2. Mean ± SEM N balance as a function of activity level and energy intake in healthy men consuming 0.57 g×kg^{-1}×d^{-1} of protein daily. Increasing the level of physical activity did not increase the protein requirement when subjects were less active and were consuming a weight-maintaining diet. When energy intake exceeded energy expenditure by 15%, positive N balance was observed. N accretion was significantly greater during overfeeding when subjects were more active. Data of Butterfield and Calloway (1984).

thorough accounting for sweat losses or subtle differences in urine and stool collections or feeding protocols). The energy requirements of these men were ~25% to 50% greater than those of sedentary men, so the N balance of these active individuals would generally be achieved with a weight-maintaining diet even if the protein concentration (g protein/kcal energy) were no higher than that of sedentary persons.

Tarnopolsky et al. (1988) concluded that elite endurance athletes require 1.67 times more protein for maintenance of lean body mass than sedentary controls. Energy expenditure of these athletes was extremely high, over 4500 kcal/d. Thus, the projected protein requirement per kcal energy requirement was similar in athletes and controls. The conclusion was based on projections from studies done at two levels of protein intake that were higher than the minimum requirement, and there was no direct evidence that lean body mass or N balance could not be sustained with protein intakes similar to current recommended allowances. In a later study of athletes with similarly high energy requirements, N balance was slightly negative when protein intake was 0.8 to 1 g×kg^{-1}×d^{-1}, near the US recommended daily intake (Phillips et al.

1993). This study was done after 10 days of adaptation to the diet, which had a lower protein concentration that the athletes' habitual diets.

Lemon (1996) reviewed the research on protein requirements of active persons and concluded that those engaging in regular aerobic activity should consume 1.2–1.4 $g \times kg^{-1} \times d^{-1}$ of protein to ensure maintenance of lean body mass. In the United States, this level of protein intake generally would be achieved by weight-stable individuals with a high energy requirement, without any intentional effort to increase the protein concentration of the diet. Those whose diet has a low protein concentration might need to increase the protein concentration of the diet. However, the basis of this recommendation comes from studies of individuals who have generally been consuming a generous amount of protein. Because the protein requirement declines after a period of reduced protein intake, it seems very likely that active persons habitually consuming a diet with a below-average protein concentration could maintain muscle mass with less than 1.2 $g \times kg^{-1} \times d^{-1}$ of protein. As long as energy intake is adequate to maintain body weight, most persons probably do not need to make any special effort to increase their protein intake when increasing their physical activity. Increasing the protein concentration of the diet might help to minimize or prevent a reduction in muscle mass during exercise-induced weight loss, although the ideal protein intake in this situation cannot be specified based on our current knowledge.

Effect of Aerobic Activity on Protein Synthesis and Breakdown

Acutely, moderate intensity aerobic activity does not change, or slightly reduces, whole-body proteolysis according to the measures of the R_a of essential amino acids (Hagg, Morse, and Adibi 1982; Millward et al. 1982; Wolfe et al. 1984; Carraro et al. 1990a; Devlin et al. 1990; Knapik et al. 1991; Kanaley, Haymond and Jensen 1993; Tipton et al. 1996). However, the postprandial suppression of proteolysis may be reduced during recovery from an exercise session (Forslund et al. 1998). Because branched chain amino acids and possibly other amino acids are oxidized more rapidly during exercise, whole-body protein synthesis must be depressed during exercise. Although exercise does not appear to increase whole-body protein turnover acutely, there is some evidence that regular aerobic activity increases the basal rate of whole-body protein turnover (Garrel et al. 1988; Lamont, Patel, and Kalhan 1990).

Carraro et al. (1990b) found that the rate of muscle protein synthesis during 4 hours of treadmill exercise at moderate intensity was similar to the rate in resting muscle. Protein synthesis in the exercised muscle increased ~25% immediately after exercise, but whole-body protein synthesis did not change. Tipton et al. (1996) found a similar increase in the mean rate of protein synthesis in the posterior deltoid muscles after swimming, but the effect was not statistically significant. The stimulation of muscle protein synthesis after aerobic activity appears to be less than the stimulation induced by resistance exercises, which is discussed later in this chapter.

Several studies have examined the effect of physical activity on excretion of 3-methylhistidine, an index of the degradation of myofibrillar proteins. The results of studies examining the acute effect of an exercise bout have been inconsistent. Rennie et al. (1981) reported that during 225 min of treadmill exercise there was no change in plasma 3-methylhistidine excretion, reduced urinary 3-methylhistidine excretion, and reduced muscle 3-methylhistidine concentrations. Plasma concentrations and urinary excretion of 3-methylhistidine returned to preexercise levels shortly after exercise. They concluded that myofibrillar proteolysis is inhibited during aerobic exercise. In contrast, Dohm et al. (1982) found that plasma concentrations and urinary excretion of 3-methylhistidine increased immediately after exercise (running). Others have reported that 3-methylhistidine excretion is not affected by aerobic activity when data are expressed relative to creatinine excretion, even though the absolute amount of 3-methylhistidine excreted may be increased or decreased (Radha and Bessman 1983; Calles-Escandon et al. 1984). The acute effect of exercise on 3-methylhistidine excretion is difficult to interpret because changes in renal blood flow or other factors temporarily may alter the relation between myofibrillar proteolysis and 3-methylhistidine excretion. Studies comparing active and inactive subjects over periods of several days are probably more informative about the influence of activity on myofibrillar proteolysis. Dohm et al. (1985) performed a series of studies to address this problem. Subjects were placed on a meat-free diet to prevent variations because of dietary intake of 3-methylhistidine. They found that the acute effect of exercise was a reduction in 3-methylhistidine excretion (all data in this study are relative to creatinine) during exercise, followed by an increase for several hours. They also found that long-distance runners excreted an average of 19% more 3-methylhistidine than sedentary subjects, even though they did not exercise on the day of, and the day before, urine collection. Basketball players excreted 22% more 3-methylhistidine while in training than they did before the basketball season. In soldiers whose meat intake was not restricted, 3-methylhistidine excretion progressively increased during a week of increased activity, but increased 3-methylhistidine intake cannot be ruled out as the cause of increased excretion in this case. This series of studies indicates that, although there might be a transient inhibition of myofibrillar proteolysis during intense exercise, the total daily myofibrillar proteolysis is slightly greater in very active individuals than in sedentary persons. Myofibrillar synthesis must be similarly increased in these individuals because they maintain a normal muscle mass.

Eccentric exercise, in which the muscle lengthens during contraction (e.g., running downhill), causes a delayed increase in 3-methylhistidine excretion (Evans et al. 1986; Fielding et al. 1991). Eccentric contractions cause more muscle damage than concentric contractions, and cause muscle soreness. The increased proteolysis in muscles damaged by eccentric exercise may be part of an inflammatory response that includes mononuclear cell secretion of interleukin-1β and prostaglandin E_2 (Cannon et al. 1991).

Does exercise influence protein metabolism in tissues other than muscle, or in the muscles that are not used to perform the exercise? Devlin et al. (1990) found that whole-body protein synthesis increased slightly after intense bicycle exercise. They examined protein metabolism in the forearm, which was not involved in the exercise. In contrast to the whole-body effects, forearm proteolysis and protein synthesis were suppressed by the previous exercise. This reduced protein turnover may help to ensure that the amino acids used by the exercised muscle come from tissues other than nonexercised muscle. There is increased splanchnic efflux of branched chain amino acids after prolonged leg exercise, which is enough to supply most of the branched chain amino acid uptake by the exercising legs (Ahlborg et al. 1974). A recent study of dogs suggested that the gut rather than the liver is the source of increased splanchnic release of amino acids during exercise (Williams et al. 1996). Splanchnic uptake of gluconeogenic amino acids increases during exercise (Ahlborg et al. 1974). Several hours of walking does not affect albumin synthesis, but increases fibronectin and fibrinogen synthesis (Carraro et al. 1990a). Fibronectin synthesis increases during exercise, whereas fibrinogen synthesis increases only after the exercise is completed. There can be a slight increase in albumin synthesis during recovery from intense exercise, which may facilitate the expansion of plasma volume that follows intense exercise (Yang et al. 1998).

Stroud, Jackson, and Waterlow (1996) gave an interesting account of the effect of prolonged, strenuous exercise on whole-body protein metabolism. Two men walked 2300 km across Antartica, losing more than 20 kg of body weight in spite of extremely high energy intakes. They appeared to be near death at the end of the journey. No conclusions about physical activity per se can be made from this report because of the severe weight loss and extremely stressful conditions. In spite of their difficulties, they performed several determinations of whole-body protein turnover using the [^{15}N]glycine end product method. There was no consistent pattern of changes in protein turnover throughout the adventure or during recovery, compared with a determination before the expedition. They concluded that the maintenance of protein turnover, in spite of the weight loss and stress, indicated that protein turnover must have a vital physiological function.

Effect of Resistance Exercise on Protein Metabolism

Regular resistance exercise—muscle contractions against a greater load than that to which the muscle is accustomed—leads to increased muscle mass. The magnitude of the hypertrophy depends on the amount and frequency of overloading. Imaging methods generally indicate that the cross-sectional area of overloaded muscle groups increases about 5% to 25% within a few weeks with typical resistance exercise programs. Even greater muscle hypertrophy is evident in the most ardent body builders. The process of hypertrophy must result from a shift in the balance between protein synthesis and degradation.

Effect of Resistance Exercise on Protein Oxidation and Protein Requirements

Although the force of each muscle contraction is much greater during resistance exercise than during aerobic exercise, resistance exercise involves many fewer contractions so that there should be little need to oxidize very much protein to fuel the activity. However, many body builders are convinced that they must consume large amounts of protein to increase and maintain their muscle mass and often buy protein and amino acid supplements to increase their amino acid intake to more than three times the recommended allowance. It is easy to calculate that there is only a small protein requirement for supplying the amino acids needed to increase muscle mass. A 70 kg man trying to increase his muscle mass by a kg every month would need only 0.1 $g \times kg^{-1} \times d^{-1}$ of meat protein to meet the additional amino acid requirement. Although there eventually would be some increase in amino acid oxidation because of the elevated muscle mass, there is little theoretical support for the notion that strength athletes could benefit from consuming 2 to 3 times more protein than the amount recommended for the general population. As discussed below, there is some empirical evidence that high protein intakes facilitate N retention in those engaged in strength training, but there is insufficient supporting evidence (e.g., demonstration that muscle mass is increased by high protein intake) to support a recommendation for elevated protein requirements during resistance training.

Tarnopolsky et al. (1988) reported that body builders maintained N balance about as well as sedentary subjects when protein intake was 1 $g \times kg^{-1} \times d^{-1}$, slightly above the U.S. recommended allowance. When protein intake was raised to 2.7 $g \times kg^{-1} \times d^{-1}$ in the body builders, the average N balance was markedly positive, ~12 g/d. This level of positive N balance corresponds to an increase in lean body mass of ~11 kg in a month, so the N balance data are quite suspect. Moreover, sedentary subjects also had considerable N retention when their protein intake was raised to 1.9 $g \times kg^{-1} \times d^{-1}$, and the slope of the regression line relating N balance to protein intake was similar in body builders and sedentary subjects. Thus, this study merely shows that high protein intakes tend to produce erroneously high N retention data, as discussed in Chapter 5. At a high protein intake, the body builders excreted less urea per g of protein intake than sedentary subjects. At normal protein intakes, sedentary subjects and body builders excreted about the same amount of urea. In a later report, these investigators recommended that strength athletes should consume 1.76 $g \times kg^{-1} \times d^{-1}$ of protein, about twice the current recommended intake for the general population (Tarnopolsky et al. 1992). This recommendation was based on a regression of N balance as a function of protein intake over the range of 0.9 to 2.4 $g \times kg^{-1} \times d^{-1}$. At 0.9 $g \times kg^{-1} \times d^{-1}$ protein intake, a level that maintained protein mass in sedentary subjects, most strength athletes were in negative N balance. However, the athletes did not excrete more urea or oxidize more leucine than the sedentary subjects on the low protein diet. About half of the N deficit in the body builders was attributed to increased losses in sweat, which were estimated rather than directly measured. There was no significant decline in lean body mass over a 13 day period on the lower protein diet, as determined by underwater weighing, although the method might not have enough sensitivity to detect small changes.

Walberg et al. (1988) examined the effect of protein intake on N balance in weightlifters consuming hypocaloric diets. Subjects lost about 4 kg of body weight in a week. They appeared to be in negative N balance when protein intake was 0.8 $g \times kg^{-1} \times d^{-1}$, as expected. Surprisingly, they appeared to be in positive N balance when protein intake was 1.6 $g \times kg^{-1} \times d^{-1}$. Underwater weighing did not reveal a significant effect of protein intake on lean body mass, although the average decline in subjects on a high protein diet (1.4 kg) was less than that of subjects on the lower protein diet (2.7 kg). With a hypocaloric diet, the values for lean body mass by densitometry are complicated by the glycogen depletion (which also leads to intracellular water depletion), so that lean body mass changes may not reflect changes in protein mass.

Fern et al. (1991) reported that when young men participated in a resistance training program while consuming their usual amount of protein (1.3 $g \times kg^{-1} \times d^{-1}$), they gained 1.5 kg of body weight, nearly all of it lean tissue according to underwater weighing. Another group followed the same program, except they supplemented their usual intake with 2 $g \times kg^{-1} \times d^{-1}$ of protein. The supplemented subjects gained an additional 1.3 kg of total weight. The investigators stated that the increase in lean body mass was greater in the supplemented group, but the mean values were not reported. N balance measurements near the end of the study indicated that the supplemented subjects should have been gaining ~20 g of protein per d more than the unsupplemented subjects. However, as noted in Chapter 5, high protein intakes produce high N retention values in most studies, even in those who are not exercising. No direct measures of muscle mass were made, so it is unclear whether or not the apparent protein accretion was limited to muscle.

Lemon et al. (1992) reported that during intensive strength training, subjects receiving a carbohydrate supplement and consuming 0.6 to 1.5 $g \times kg^{-1} \times d^{-1}$ of protein were in negative N balance. Those receiving an isocaloric protein supplement and consuming 2 to 3 $g \times kg^{-1} \times d^{-1}$ of protein were in markedly positive N balance. A few of them would have been gaining over 2.5 kg of lean body mass every week if these N balance data were accurate, which is an incredibly rapid rate of protein accretion. Strength and muscle mass gains were not different between the groups, indicating that N balance did not accurately reflect protein balance. An earlier study (Rasch and Pierson 1962) also had indicated that a protein supplement did not enhance the effect of resistance training on arm girth or strength.

Lemon (1996) cited a study by Meredith et al. (1992) as evidence that protein supplementation enhances the muscle hypertrophy associated with resistance training. Men who consumed a supplement containing 23 g protein daily had an average increase of 13% in the cross sectional area (by computed tomography) of the trained muscle group after 12 weeks, which was more than the 5% increase in men who followed the same exercise program but did not receive a dietary supplement. This study does not allow any conclusions about the specific effect of protein intake, because the supplement increased energy intake significantly, and had a protein concentration that was not much higher than that of a typical diet. The increased energy intake alone could have promoted the muscle growth, in the absence of any additional protein and possibly even in the absence of resistance training.

Campbell et al. (1994b, 1995) found that while resistance training increased the energy requirement of older subjects slightly, training effects on body composition, N balance, and strength were similar whether the subjects consumed 0.8 or 1.6 $g \times kg^{-1} \times d^{-1}$ of protein. Although N balance was more positive with the higher protein intake, the same was true before resistance training. Gains in lean body mass were small whether protein intake was low or high. The exercise program did not increase the rate of leucine oxidation.

Based on this research, the only safe conclusion is that more information is needed to settle the issue of whether increased protein intake is necessary to promote or maintain muscle hypertrophy in those engaged in resistance exercises. The optimal study would not rely on N balance, but would involve long-term assessment of muscle mass by an accurate imaging method.

Effect of Resistance Exercise on Protein Breakdown and Synthesis

Numerous studies published since 1992 have examined the effect of resistance exercise on muscle protein synthesis in human subjects (Table 8.1). The consensus is that resistance exercises significantly accelerate the rate of protein synthesis in muscle. The magnitude of the response has been variable and probably depends mainly on the intensity of the exercise stimulus in the specific muscle group being examined, as well as on the delay between the exercise session and the determination of protein synthesis. The response peaks sometime within the first day after an exercise session, and still may be detectable 2 days after exercise. Stimulation of muscle protein synthesis by resistance exercises is evident in both men and women, and in both young and old subjects. In fact, older subjects tend to have a more robust response to the exercise than young subjects (Yarasheski, Zachwieja, and Bier 1993; Hasten et al. 1997). A preliminary report suggests that actin and myosin heavy chain synthesis increase to a similar extent after exercise (Hasten et al. 1998).

There is some evidence that persons accustomed to resistance exercise have a smaller acute stimulation of muscle protein synthesis after an exercise session than those unaccustomed to this type of exercise. Phillips et al. (1998) recently reported that the acute response in untrained subjects was more than twice that of trained subjects, although net protein anabolism was not diminished because the increase in proteolysis also was blunted in trained subjects. This finding is consistent with our failure to detect a significant increase in myofibrillar protein synthesis in subjects who had been training for three

TABLE 8.1. Effect of resistance exercise on muscle protein synthesis in humans.

Reference	Number of subjects	Age (yr)	Training duration	Delay since last exercise	Exercise intensity[a]	Muscle studied	Feeding condition	% Increase in protein synthesis
Chesley et al. (1992)	6♂	20s	Chronic	4 hr	4 sets of 6-12 reps @ 80% 1RM (to failure), 3 different exercises	Biceps brachii	Fed	50%
	6♂	20s		24 hr				109%
Yarasheski et al. (1992)	9♂	21-34	12 wk	20 hr	4 sets of 4-8 reps @75–90% 1RM, 2 different exercises	Vastus lateralis	Fed	38%
Yarasheski, Zachwieja and Bier (1993)	2♂ 4♀	20s	2 wk	5 hr	2-3 sets of 8-10 reps at 60-75% 1RM, 2 different exercises	Vastus lateralis	Overnight fast	53%
	4♂ 2♀	60-73						153%
MacDougall et al. (1995	6♂	20s	Chronic	36 hr	4 sets of 6-12 reps @ 80% 1RM (to failure), 3 different exercises	Biceps brachii	Fed	14%
Biolo et al. (1995b)	5♂	20s	Untrained	3 hr	5 sets of 10 reps @ 12RM, leg press; 4 sets of 8 reps @ 10RM, 3 different exercises	Whole leg and vastus lateralis	Overnight fast	108% whole leg 140% vastus lateralis
Yarasheski et al. (1995)	10♂	64–75	16 wk	12 hr	4 sets of 5-10 reps @75-90% 1RM, 3 different exercises	Vastus lateralis	Fed	48%
Welle, Thornton and Statt (1995)	5♂ 4♀	22-31	13 wk	24 hr	3 sets of 8 reps @ 80% 3RM, knee extension	Vastus lateralis (myofibrillar proteins only)	Overnight fast	2%
	5♂ 4♀	62-72						10%

Tipton et al. (1996)	7♀	20s	Chronic	2 hr	3 sets of 6-10 reps @ 65-80% 1RM, various exercises	Posterior deltoid	Overnight fast	7% resistance exercise alone 82% resistance exercise and swimming
Biolo et al. (1997)	6♂	20s and 30s	Untrained	4 hr	5 sets of 10 reps @ 12RM, leg press; 4 sets of 8 reps @ 10RM, 3 different exercises	Whole leg and vastus lateralis	Overnight fast, i.v. amino acid infusion	69% whole leg 44% vastus lateralis
Hasten et al. (1997)	6♂ + ♀ 6♂ + ♀ 6♂ + ♀	23-32 63-66 78-85	2 wk	?	?	Quadriceps	?	53% 111% 220%
Phillips et al. (1997)	4♂ 4 ♀	20s	Untrained	3 hr 24 hr 48 hr	8 sets of 8 reps @ 80% 1RM, knee extension	Vastus lateralis	Overnight fast	112% 65% 34%
Welle and Thornton (1998)	9♂ 9♀	62-75	1 wk	23 hr	5 sets of 10 reps @ 80% 1RM, knee extension	Vastus lateralis	Fed	27%
Phillips et al. (1998)	6 6	? ?	Untrained chronic	?	?	Leg	?	119% 49%
Hasten et al. (1998)	4♂ + ♀	23-30	2 wk	?	?	Vastus lateralis	?	91% total proteins 61% actin 64% myosin heavy chain

[a] Only exercises affecting the muscle group being studied are included here. Rep refers to number of times an exercise was repeated in a set, and set refers to a series of contractions without a rest. RM refers to repetition maximum, or amount of weight that can be lifted with good form a specified number of times. For example, 1RM is amount of weight that can be lifted once with good form, and 3RM is amount that can be lifted 3 times with good form.

months (Welle, Thornton, and Statt 1995), and the observation that trained swimmers did not have an increase in muscle protein synthesis after resistance exercises unless they were combined with swimming (Tipton et al. 1996). In contrast, Chesley et al. (1992) observed fairly robust increases in muscle protein synthesis after exercise in men who were accustomed to resistance training.

It is unclear whether or not there is an effect of nutrition on the stimulation of muscle protein synthesis by resistance exercise. Biolo et al. (1997) observed a much greater increase in muscle protein synthesis when it was measured during amino acid infusion than when it was determined in postabsorptive subjects (Biolo et al. 1995b). However, the amino acid infusion itself stimulated muscle protein synthesis, so that the stimulation attributable to exercise actually was somewhat less when amino acids were infused (Table 8.1). We did not observe any more stimulation of myofibrillar synthesis by resistance exercise when subjects were fed high-protein meals than when they were fed low-protein meals (Welle and Thornton 1998), but did not compare fed to postabsorptive subjects. The results summarized in Table 8.1 certainly do not leave the impression that fed subjects have any more of an increase in muscle protein synthesis than postabsorptive subjects.

Yarasheski et al. studied the effect of growth hormone on the response to resistance training. The mean increase in muscle protein synthesis after resistance training in subjects treated with growth hormone was higher than in those not given the hormone (46% versus 38% in young men, 80% versus 48% in older men), but the difference was not statistically significant (Yarasheski et al. 1992; Yarasheski et al. 1995). Moreover, growth hormone did not alter the baseline rate of muscle protein synthesis in experienced weight lifters (Yarasheski et al. 1993).

Resistance exercises increase not only muscle protein synthesis, but also muscle protein breakdown. The hypertrophic effect of resistance training results from a greater stimulation of synthesis than of proteolysis. Acutely, proteolysis in an exercised limb increases by 30% to 50% (Biolo et al. 1995b; Biolo et al. 1997; Phillips et al. 1997). Daily excretion of 3-methylhistidine generally increases somewhat at the beginning of a resistance exercise program (Hickson and Hinkelmann 1985; Frontera et al. 1988; Pivarnik, Hickson, and Wolinsky 1989; Campbell et al. 1995), although sometimes no increase is detected (Yarasheski, Zachwieja, and Bier 1993; Welle, Thornton, and Statt 1995; Phillips et al. 1997).

Do these changes in muscle protein metabolism translate into detectable effects on whole-body protein metabolism? Generally, changes in whole-body protein metabolism have not been found after resistance exercise, either acutely or after several weeks of training (Tarnopolsky et al. 1991; Yarasheski et al. 1992; Yarasheski, Zachwieja, and Bier 1993; Yarasheski et al. 1995; Tipton et al. 1996; Phillips et al. 1997). Sometimes, a slight increase (10% or less) in whole-body protein turnover is detected after an exercise session (Biolo et al. 1995b) or after several weeks of training (Campbell et al. 1995;

Welle, Thornton, and Statt 1995). Fern et al. (1991) reported that the average whole-body protein turnover was ~20% more rapid after a month of strength training, but that the effect was not statistically significant. However, strength training plus a high protein diet increased whole-body protein turnover by more than 100%. This remarkable effect must be related more to the high protein diet than the exercise. Tarnopolsky et al. (1992) noted that strength athletes had higher protein turnover rates than sedentary persons and that the difference was magnified as protein intake was increased. With a very high protein intake, protein turnover of the strength athletes was ~50% more rapid than that of sedentary subjects with the same protein intake. The athletes consumed 25% to 40% more energy, per kg body weight, than the sedentary individuals in this study, so it is difficult to separate the influence of the resistance training from that of the higher energy intake.

9
Topics of Clinical Interest

Many studies of protein metabolism in humans have included patients with various conditions of clinical importance. A common feature of most of the conditions discussed in this chapter is a loss of body proteins, especially from skeletal muscle. The effect of these conditions on protein metabolism (or the role of altered protein metabolism in the pathophysiology of a disease) often is difficult to determine, because there are multiple factors influenced by the disease that may alter protein metabolism (e.g., nutrition, physical activity, cytokine levels, medications). Moreover, severity of diseases, trauma, metabolic abnormalities, or other conditions can be quite variable, and early effects may be quite different from later effects. It is important to keep these problems in mind when trying to understand the relation between clinical problems and protein metabolism.

Acidosis

Acidosis refers to any condition in which the alkali content of the body fluids in reduced in relation to the acid content. The pH of the body fluids typically is reduced during acidosis, but may be within an acceptable range if increased H^+ production is compensated by respiratory and renal mechanisms. Common causes of acidosis are uncontrolled diabetes and starvation, which lead to increased production of acidic ketone bodies, and hypoventilation in patients with lung disease. In normal subjects in whom acidosis was induced by NH_4Cl administration (pH fell from 7.42 to 7.35), there was a 28% increase in proteolysis (leucine R_a) and protein synthesis (non-oxidized leucine R_d) (Reaich et al. 1992). There also was an increase in whole-body protein synthesis, which may be secondary to increased intracellular amino acid levels induced by the increased proteolysis. Leucine oxidation rose slightly, indicating a net catabolic effect in these postabsorptive subjects. A similar degree of acidosis increased daily N excretion only slightly when N intake was constant, whereas more severe acidosis (pH 7.30) increased N excretion by 5 g/d and reduced albumin synthesis by 24% (Ballmer et al. 1995). The increased excretion of NH_4^+ during acidosis may contribute to the negative N balance. In subjects with metabolic acidosis induced by prolonged fasting, reduction of NH_4^+ excretion by administration of bicarbonate and potassium reduced N excretion by 36% (Hannaford et al. 1982). Elimination of the acidemia with this treatment did not alter 3-methylhistidine excretion. Alkali therapy also restores normal growth rates in children whose growth is stunted by idiopathic renal tubular acidosis (McSherry and Morris 1978). Correction of acidosis in patients with chronic renal failure markedly reduces the rate of whole-body protein turnover, and reduces postabsorptive leucine oxidation by 30% (Reaich et al. 1993).

In rats, NH_4Cl-induced acidosis selectively stimulates the ubiquitin-proteasome pathway of proteolysis in skeletal muscle by a mechanism that requires the presence of glucocorticoid (May, Kelly, and Mitch 1986; Mitch et al. 1994; Price et al. 1994b). Acidosis does not affect muscle protein synthesis in rats (May, Kelly, and Mitch 1986). The rat studies involved a more severe acidosis than the acidosis in the human studies cited above.

Alcoholism

The acute effects of alcohol consumption on protein metabolism in healthy subjects were reviewed in Chapter 5. There is little immediate effect of alcohol on whole-body protein metabolism, but large doses in animals suppress protein synthesis in a number of tissues. Chronic alcohol abuse often is associated with malnutrition, liver cirrhosis, and other medical problems that may have independent effects on protein metabolism (see section on cirrhosis later in this chapter). In general, alcoholism causes loss of body proteins, especially in muscle. When alcohol calories were completely replaced by energy from fat and carbohydrate in alcoholic patients who had been consuming 200 g ethanol daily, with no change in N intake, N balance improved significantly (Bunout et al. 1987). The reduced mass of muscle proteins in chronic alcoholics appears to be mediated by reduced protein synthesis rather than increased muscle proteolysis (Martin and Peters 1985; Pacy et al. 1991). Pacy et al. (1991) reported muscle protein synthesis was significantly slower in alcoholic subjects than in control subjects, even though no alcohol was consumed for several hours before the study. There was a trend toward reduced whole-body proteolysis and protein synthesis in the alcoholics, but the difference was not significant statistically. Leucine oxidation was reduced slightly in the alcoholic patients. In contrast, Hirsch et al. (1995) reported that cirrhotic alcoholics had faster rates of whole-body proteolysis and protein synthesis than nonalcoholic subjects, although the effect was not evident in those who had abstained from alcohol in the previous month. The reason for the discrepancy between these studies is unclear. Even though Hirsch et al. (1995) found increased protein turnover in nonabstaining alcoholics, net protein balance in the postabsorptive state, as reflected by leucine oxidation, was not significantly affected.

A number of studies have examined the effect of chronic ethanol consumption on protein metabolism in rats (Preedy et al. 1994; Bonner et al. 1995; Bonner et al. 1996; Preedy et al. 1996; Preedy et al. 1997). Suppression of skeletal muscle protein synthesis is evident and may be mediated by reduced mRNA translation, reduced mRNA concentrations (but not those encoding actin and myosin), and reduced total RNA (mostly ribosomal) concentrations. Although cardiac protein synthesis is acutely suppressed by alcohol, there is a compensatory increase with chronic ingestion. Brain protein and RNA content declines with chronic ethanol ingestion, but the fractional rate of protein synthesis is not reduced. Cathepsin B activity (a lysosomal protease) is increased in the brain after chronic ethanol consumption, but activities of some other proteases are not.

Arthritis

Patients with rheumatoid arthritis often have a reduced body cell mass and depletion of muscle proteins (Gibson et al. 1991; Roubenoff et al. 1994). Loss of muscle mass may be related to inactivity, and in some cases glucocorticoid therapy (see section on glucocorticoids later in this chapter). Increased cytokine levels (TNFα, IL-1β) also may contribute to protein wasting (Roubenoff et al. 1994; Rall et al. 1996).

There have been few studies of protein metabolism in arthritic patients. Gibson et al. (1991) reported that patients with rheumatoid arthritis had a normal fractional rate of protein synthesis in quadriceps muscle, if they were not taking the glucocorticoid prednisolone. Those taking prednisolone had a slow fractional rate of muscle protein synthesis. The normal value was determined from the protein synthesis in the unaffected leg of patients with osteoarthritis. Because the patients with rheumatoid arthritis had evidence of muscle protein wasting (low protein to DNA ratio), even a normal fractional synthesis rate reflects a diminished absolute synthesis rate. Whole-body proteolysis (leucine R_a) after overnight fasting tends to be accelerated in rheumatoid arthritis patients, per kg of body cell mass (Rall et al. 1996). The rate of whole-body protein synthesis is not increased, reflecting greater net postabsorptive protein losses in the patients. Glucocorticoid therapy could not explain the elevated proteolysis. Patients on methotrexate had a slower rate of proteolysis than those not receiving methotrexate. After several weeks of strength training, no difference in protein metabolism was observed between arthritic patients and normal subjects. The whole-body proteolytic rate correlated with levels of growth hormone and TNFα. Higher glucagon levels were associated with slower rates of protein synthesis. Thus, the hormonal milieu in rheumatoid arthritis patients appears to promote a protein catabolic state.

Cancer

The cachexia that is often associated with cancer includes substantial loss of body proteins. A primary factor is malnutrition. However, other factors must be considered in evaluating the relation between cancer and protein metabolism. These include increased levels of cytokines and other catabolic hormones, the effects of chemotherapy or surgical treatment, and the protein metabolism of the tumor itself. Because these factors vary among patients, the effect of cancer per se on protein metabolism is difficult to define.

Whole-Body Protein Metabolism

Many studies have compared whole-body protein metabolism of cancer patients with that of healthy subjects or patients with non-malignant conditions (Table 9.1). The general consensus is that cancer is associated with increased whole-body protein turnover, even though not all studies have confirmed this effect (not all of the increases in protein turnover in cancer patients shown in Table 9.1 were statistically significant, because of considerable heterogeneity among individuals). Whenever whole-body proteolysis was reported to be increased, whole-body protein synthesis also was increased. Protein turnover seems to be elevated, per kg body weight, more in cachectic patients than in those who have not lost a significant amount of weight. Most studies did not account for differences in body composition between patients and controls, merely expressing protein turnover in relation to total body weight. If most of the weight loss in cachectic patients is adipose tissue and muscle (Heymsfield and McManus 1985), which have a slower protein turnover per kg than the whole-body average, then protein turnover per kg body weight would tend to increase even without any other influence of cancer. Comparisons with noncancer patients with similar degrees of cachexia probably are more meaningful, although changes in body composition associated with cancer may be different from those associated with simple malnutrition. For example, visceral mass may be preserved during weight loss better in cancer patients than in those with anorexia nervosa (Heymsfield and McManus 1985). In spite of these problems, the magnitude of the increase in whole-body protein turnover in cancer patients in many of the studies cited in Table 9.1 is too great to attribute it exclusively to altered body composition.

Tumor Protein Metabolism

Does protein metabolism in the malignant cells contribute very much to whole-body protein metabolism? Because the mass of a tumor rarely exceeds 1% of body weight, a tumor would need to have a very high protein turnover rate to have a detectable influence on whole-body protein metabolism. The few studies that have compared protein synthesis rates in tumors with rates in healthy adja-

TABLE 9.1. Whole-body proteolysis in cancer patients.

Reference	Tracer	Type of cancer	Results
Waterhouse and Mason (1981)	[^{14}C]leucine	Heterogeneous	Proteolysis ↑ 72% vs. malnourished patients with no cancer, per kg body weight
Norton, Stein and Brennan (1981)	[^{15}N]glycine	Heterogeneous	Proteolysis ↑ 52% vs. malnourished patients with no cancer, due to large effect in 3/7 cancer patients, per kg body weight
Heber et al. (1982)	[^{14}C]lysine	Non-oat cell lung	Proteolysis ↑ 68% vs. normal, per kg body weight
Kien and Camitta (1983)	[^{15}N]glycine	Leukemia or lymphoma (children)	Proteolysis ↑ 77% vs. normal, per kg body weight (6/8 cancer patients febrile during study)
Glass, Fern and Garlick (1983)	[^{15}N]glycine	Rectal	No change in proteolysis 12 weeks after surgical removal of tumor
Jeevanandam et al. (1984)	[^{15}N]glycine	Heterogeneous	Proteolysis ↑ 32% vs. malnourished patients with no cancer, per kg body weight
Emery et al. (1984)	[^{13}C]leucine	Heterogeneous	Proteolysis same as normal (but "normal" proteolysis was twice as high as currently accepted values)
Eden et al. (1984)	[^{14}C]tyrosine	Heterogeneous	Proteolysis ↑ 40% vs. malnourished patients with no cancer, per kg body cell mass (↑ 21% per kg body weight)
Ward et al. (1985)	[^{15}N]glycine	Heterogeneous	Postsurgical proteolysis ↑ 35% in patients with metastases vs. surgical patients without cancer, per kg body weight; no increase in proteolysis in patients with localized tumors
Jeevanandam, Lowry and Brennan (1987)	[^{15}N]glycine	Heterogeneous	Proteolysis ↑ 15–36%, depending on feeding conditions, vs. patients with benign gastrointestinal diseases, per kg body weight
Inculet et al. (1987)	[^{13}C]leucine	Miscellaneous sarcoma (noncachectic)	Proteolysis ↑ 16% vs. normal, per kg body weight
Borzotta, Clague and Johnston (1987)	[^{14}C]leucine	Gastrointestinal	Proteolysis ↑ 12% vs. patients with benign disease, per kg body weight; proteolysis ↑ 52% in patients with advanced cancer (Stage 4) vs. patients with localized cancer (Stages 1–3)

TABLE 9.1. (*continued*).

Reference	Tracer	Type of cancer	Results
Fearon et al. (1988)	[^{15}N]glycine	Lung or colon	Proteolysis ↑ >50% in cancer patients vs. weight-stable patients with benign gastrointestinal diseases, per kg lean body mass or total body weight; increased proteolysis in patients with benign diseases who were losing weight; weight-losing cancer patients had only slight ↑ in proteolysis vs. weight-losing controls
Harrison et al. (1989b)	[^{13}C]leucine	Colon	Only 3/15 patients with elevated proteolysis, per kg lean body mass, vs. patients with benign gastrointestinal disease
O'Keefe et al. (1990)	[^{14}C]leucine	Hepatocellular	Proteolysis ↑ 98% vs. normal, per kg body weight (↑ 28% with hepatic metastases of colon cancer)
Melville et al. (1990)	[^{13}C]leucine	Lung (noncachectic)	Proteolysis ↑ 15% vs. patients without cancer, per kg lean body mass
Shaw et al. (1991)	[^{14}C]leucine	Heterogeneous	Proteolysis during surgery ↑ 80% in cachectic cancer patients vs. surgical patients with no cancer, per kg body weight; normal proteolysis in noncachectic patients
Heslin et al. (1992b)	[^{14}C]leucine	Heterogeneous	Proteolysis ↑ 11% in cachectic cancer patients vs. normal, per kg body weight; proteolysis ↑ only 5% in noncachectic cancer patients; suppression of proteolysis by insulin+glucose+amino acids not affected by cancer
Preston, Fearon and McMillan (1995)	[^{15}N]glycine	Colon with hepatic metastases	Proteolysis ↑ 75% vs. normal, per kg body weight
Daley et al. (1996)	[^{13}C]leucine; [$^{2}H_{5}$]phenyl-alaninen	Heterogeneous (children)	Proteolysis ↑ 40% vs. normal, per kg body weight (historical data for normal values)

cent tissue have sometimes indicated modestly increased synthesis in the malignant samples (Table 9.2), but not enough to produce a significant change in whole-body protein synthesis. It is conceivable that even the healthy tissue in cancer patients has an elevated protein turnover rate. Heys et al. (1992) compared protein synthesis of colorectal mucosa tumors to that of rectal mucosa from patients undergoing hemorrhoidectomy. Malignant tumors had fractional rates of

TABLE 9.2. Fractional rate of protein synthesis in tumors.

Reference	Tissue	Fractional rate in tumor (%/d)	Fractional rate in healthy tissue[a] (%/d)
Stein et al. (1978)	Colon (n = 7)	15	9
	Stomach (n = 6)	19	22
	Liver (n = 2)	20	20
	Esophagus (n = 1)	26	11
	Jejunum (n = 1)	20	31
Heys et al. (1991)	Rectum (n = 6)	23	
	Breast (n = 15)	10	
Heys et al. (1992)	Colorectal mucosa (n=6)	22	9[b]
Hartl et al. (1997)	Rectum (n = 5)	26	31
Gore et al. (1997)	Colon (n = 5)	45	36

[a]Except for Heys et al. (1992) the healthy tissue is from same cancer patients from whom cancerous tissues were obtained.
[b]Rectal mucosa of patients undergoing surgery for hemorrhoids (n = 5). Benign colorectal tumors had fractional rate of synthesis of 37%/d (n = 6). Rectal mucosa of patients with inflammatory bowel disease had fractional rate of synthesis of 25%/d.

synthesis ~2 times more rapid than healthy tissue, and benign tumors had even faster synthesis rates. Metastatic tumors may synthesize proteins at twice the rate of their primary tumors (Shaw et al. 1991). Nevertheless, even a doubling of protein turnover in tumors is unlikely to significantly influence whole-body protein synthesis unless the tumor mass is unusually large. Thus it appears that whole-body effects of cancer must be mediated primarily by systemic effects of the malignancy.

Muscle Protein Metabolism

Based on current knowledge, there is no clear answer to the question of whether increased proteolysis or reduced protein synthesis is the primary event in the muscle wasting associated with cancer cachexia. Muscle obtained from cancer patients has more rapid proteolysis in vitro than muscle from subjects without cancer (Lundholm et al. 1976), but in vitro measurements do not necessarily reflect in vivo proteolysis. Heber et al. (1982) reported that 3-methylhistidine excretion (per g creatinine) was increased ~50% in lung cancer patients relative to healthy subjects. Most of these patients had lost >10% of their body weight. However, Lundholm et al. (1982) found that 3-methylhistidine efflux from the leg tended to be less in cancer patients than in healthy subjects, whether or not they were cachectic. Phenylalanine and tyrosine efflux from the leg are not elevated in cachectic cancer patients compared with well-nourished healthy subjects or cachectic patients with no cancer, suggesting that muscle protein balance is normal after overnight fasting (Bennegard et al. 1984; Pisters and Pearlstone 1993). These data do not preclude a change in protein turnover, and obviously do not reflect metabolic changes that cause muscle mass to decline.

There is some evidence that cancer patients have a slower fractional rate of muscle protein synthesis than normal, but also evidence that they have normal or even accelerated muscle protein synthesis. One of the studies suggesting reduced muscle protein synthesis in cancer patients was based on in vitro measurements, and may not reflect the in vivo rate (Lundholm et al. 1976). The decline in protein synthesis in vitro in muscle of cancer patients is not caused by reduced RNA concentrations, and may be related to reduced initiation of protein synthesis (Lundholm et al. 1978). In the other study suggesting reduced protein synthesis in muscle of cancer patients, the control data were much higher than any other estimate of fractional rate of synthesis in human muscle, and controls were much younger than patients (Emery et al. 1984). McNurlan et al. (1994b) reported values for muscle protein synthesis in postabsorptive cancer patients (mean=1.78%/d) that were similar to those they had observed previously (Garlick et al. 1989; McNurlan et al. 1991; McNurlan et al. 1993) in healthy young subjects (1.8–2.2%/d). Individual variations and the advanced age of the cancer patients easily could account for the very slight decrement in the cancer patients. Shaw et al. (1991) found that the fractional rate of muscle protein synthesis during surgery was 94% more rapid in cachectic cancer patients than in noncachectic patients undergoing surgery for nonmalignant conditions. However, the absolute muscle protein synthesis rate must have been increased much less than this because the mass of muscle proteins that is turning over is significantly diminished in the cachectic patients. In another study, determination of phenylalanine disposal in the forearm did not suggest any abnormality of protein synthesis in muscle of cancer patients (Newman et al. 1992).

Animal studies indicate that even small tumors can induce muscle wasting via activation of the ATP-dependent proteasomal pathway (Temparis et al. 1994; Llovera et al. 1995) and inhibition of muscle protein synthesis (Smith and Tisdale 1993; Lorite, Cariuk, and Tisdale 1997). These effects may be mediated by cytokines (eg. IL-6), and a 24 KDa protein that may act by increasing prostaglandin E_2 in muscle (Fujita et al. 1996; Todorov et al. 1996; Lorite, Cariuk, and Tisdale 1997). Anorexia also can contribute to reduced muscle protein synthesis in tumor-bearing animals (Lundholm et al. 1981b).

Effect of Cancer on Hepatic Protein Metabolism

Shaw et al. (1991) found that hepatic protein synthesis was increased ~40% in cachectic cancer patients, but not in noncachectic patients. Albumin synthesis followed the same pattern as hepatic tissue synthesis. Starnes et al. (1987) measured protein synthesis of human hepatocytes in vitro. Those from noncachectic cancer patients had rates of protein synthesis 3-fold faster than those from patients with benign disease. In contrast to the in vivo data of Shaw et al., hepatocytes from cachectic patients had slower than normal protein synthesis, suggesting that malnutrition reversed the effect of cancer. In another study of in vitro protein synthesis, liver from patients with advanced cancer synthesized proteins 50% faster than liver from surgical patients without cancer (Lundholm et al. 1978). No details about extent of cachexia or medical problems of the control group were presented. O'Keefe et al. (1990) noted that in vivo protein synthesis in the liver of a single patient with hepatocellular cancer was extremely high. They also noted that fractional rates of synthesis of fibrinogen, albumin, and transferrin were twice the normal rates in most patients with hepatocellular carcinoma. Albumin levels tended to be subnormal, so an increased fractional rate does not necessarily indicate an increased absolute synthesis rate. Fearon et al. (1991) noted reduced synthesis of proteins retained within the liver in patients with hepatic metastases of colorectal cancer who had elevated IL-6 levels, although there was evidence of increased synthesis of acute phase proteins (increased C-reactive protein concentration). Heys et al. (1992) did not detect any abnormality in hepatic protein synthesis rate in patients with colorectal tumors. Fearon et al. (1998) reported that albumin synthesis was normal in hypoalbuminemic, cachectic cancer patients with elevated levels of acute phase proteins. Thus, inconsistent results preclude any generalizations about the effect of cancer on hepatic protein synthesis. Variability in nutritional status, cytokine levels, and type of cancer (especially the presence or absence of malignant cells within the liver) can cause a wide range of effects on hepatic protein metabolism.

Effect of Therapy on Protein Metabolism

The protein metabolism of cancer patients depends not only on the effect of cancer itself, but also on the effects of therapy. Surgery elicits a protein catabolic response, as discussed later in this chapter. In general, chemotherapy causes negative N balance, even when nutritional status is constant (Herrmann et al. 1981; LeBricon et al. 1995). The anorexia that often accompanies chemotherapy can further worsen the catabolic effect. Whole-body protein breakdown and synthesis may decline in response to some cytotoxic agents (Herrmann et al. 1981).

Cancer patients seem to have fairly normal responses to anabolic stimuli, such as insulin, amino acids, oral or enteral feeding, parenteral nutrition, and growth hormone (Melville et al. 1990; Shaw et al. 1991; Heslin et al. 1992b; Newman et al. 1992; Pisters and Pearlstone 1993; McNurlan et al. 1994b; Wolf et al. 1992b; Hochwald et al. 1997). Although hormonal and nutritional support can improve nitrogen balance in short-term studies, such therapies do not always lead to clinical improvement or detectable increases in lean body mass (Pisters and Pearlstone 1993).

Cirrhosis

Hepatic cirrhosis is a common complication of chronic alcohol abuse, but can also be related to other causes. The liver accounts for perhaps 15% to 20% of the whole-body protein turnover (see Chapter 4), and is the major organ involved in the secretion of plasma proteins. It also is the primary site of synthesis of many of the nonessential amino acids, and is essential for removal of amino N via urea production. The loss of functional liver mass in cirrhotic patients can be compensated to a certain extent by altered function of the surviving hepatocytes, but altered protein and amino acid metabolism is apparent even in stable patients. Malnutrition is common among cirrhotic patients, who have a greater protein requirement than healthy people to maintain their body protein mass (Swart et al. 1988; Charlton 1996; Kondrup, Nielsen, and Juul 1997). Often there is muscle wasting in the more advanced stages of the disease. Serum albumin levels are low. This is presumed to be the result of reduced capacity for albumin synthesis and hemodilution secondary to water retention. Tissue edema can result from reduced colloid osmotic pressure secondary to hypoalbuminemia. Concentrations of the branched-chain amino acids are usually reduced, which can lead to reduced oxidation of branched-chain amino acids (Millikan et al. 1985; Mullen et al. 1986). As discussed below, this effect is not caused by reduced proteolysis, but by increased clearance from plasma. In contrast, the levels of aromatic amino acids are elevated. The resulting increase in the ratio of aromatic to branched-chain amino acids has been postulated to cause cirrhotic encephalopathy (Charlton 1996).

Several investigators concluded that whole-body protein turnover is normal in cirrhotic patients (Millikan et al. 1985; Mullen et al. 1986; Shanbhogue et al. 1987; Petrides et al. 1991). All of these studies failed to account adequately for altered body composition in cirrhotic patients. McCullough, Mullen, and Kalhan (1992) demonstrated that body cell mass is reduced in cirrhotic patients, per kg body weight or per kg lean body mass, even when there is no obvious clinical evidence of edema or ascites. Because protein metabolism occurs intracellularly, the protein metabolism should be expressed per kg body cell mass. In cirrhotic patients whose whole-body protein turnover was normal when expressed per kg body weight, or even per kg lean body mass, protein turnover was elevated per kg body cell mass. Several other studies provide evidence for increased whole-body protein turnover in cirrhotic patients, even without accounting for the reduced body cell mass per kg body weight (Swart et al. 1988; McCullough et al. 1992; Zillikens et al. 1993; Tessari et al. 1994b; Hirsch et al. 1995). Phenylalanine R_a is increased more than leucine R_a, but there is no satisfactory explanation for this observation (Tessari et al. 1993). Leucine R_a is elevated in cirrhotic patients when it is calculated based on plasma KIC enrichment, whereas it is normal when based on plasma leucine enrichment (see Chapter 3 for explanation of the use of plasma KIC enrichment to calculate leucine R_a) (McCullough et al. 1992; Tessari et al. 1994b). This discrepancy could be related to altered intracellular leucine metabolism in cirrhotic patients. The studies that employed [^{15}N]glycine as a tracer (Swart et al. 1988; Zillikens et al. 1993) used the average of the results obtained from urea and NH_3 enrichments. This approach may be problematic in that NH_3 enrichments are generally higher than urea enrichments (see Chapter 3), and cirrhotic patients excrete more NH_3 relative to urea than healthy subjects (Mullen et al. 1986).

Consistent with the finding of a greater whole-body phenylalanine R_a/leucine R_a ratio in cirrhotic patients, Morrison et al. (1990) reported that leg efflux of phenylalanine and tyrosine is increased whereas efflux of branched-chain amino acids is reduced. Phenylalanine and tyrosine are not metabolized by the leg except for their incorporation into proteins and release from proteins. Thus, net protein loss in the leg after an overnight fast appears to be greater in cirrhotic patients. Reduced branched chain amino acid efflux might be explained by increased oxidation. Efflux of 3-methylhistidine was not different in cirrhotic and control subjects. There was a great deal of individual variability in all of these efflux measurements. Daily excretion of 3-methylhistidine in urine is increased in patients with cirrhosis (Marchesini et al. 1981; Zoli et al. 1982). Although this increase could be related to more nonmuscle 3-methylhistidine production rather than to increased myofibrillar degradation in skeletal muscle (Charlton 1996), there is no direct evidence for this possibility.

Even though patients with cirrhosis are generally resistant to insulin with respect to glucose metabolism, they have a normal suppression of whole-body proteolysis in response to insulin administration (Petrides et al. 1991; Tessari et al. 1993). The changes in protein metabolism after meals are fairly normal in cirrhotic patients at the whole-body level, although splanchnic extraction of phenylalanine is reduced (Tessari et al. 1994b). In response to intravenous amino acids, patients with cirrhosis increase their protein synthesis less than normal, but they suppress their proteolysis more so that protein balance changes normally (Tessari et al. 1996d). Whole-body protein synthesis of patients with cirrhosis increases with increased food intake (Kondrup, Nielsen and Juul 1997), which is the normal response (Chapter 5).

You would expect that protein metabolism of the liver would be influenced by cirrhosis more than protein metabolism in any other tissue. Although determination of in vivo metabolism of proteins retained within the liver requires a liver biopsy, metabolism of proteins secreted by the liver can be evaluated readily. Studies of radiolabeled albumin decay curves generally indicate a prolonged half-life of serum albumin in cirrhotic patients (Wilkinson and Mendenhall 1963; Hasch, Jarnum, and Tygstrup 1967). Because the serum albumin pool is smaller (total body pool may be normal because of increased extravascular albumin), these data indicated that albumin synthesis is slower in cirrhotic patients. Determination of albumin synthesis with [^{14}C]carbonate [which is incorporated into the guanidine C of arginine and then into urea in the liver, so that urea enrichment reflects hepatic arginine enrichment (see Chapter 3)] also indicates that albumin synthesis is slower than normal in many cirrhotic patients, but also that it sometimes is faster than normal (Tavill, Craigie, and Rosenoer 1968; Rothschild et al. 1969). O'Keefe et al. (1981) found no evidence for reduced incorporation of tyrosine into albumin or globulin (mixture of proteins precipitated by ammonium sulfate) in stable cirrhotic patients, but found reduced albumin synthesis when there was fulminant hepatic failure. However, they used historical control data from another laboratory, and had no way of knowing whether the specific activity of plasma tyrosine reliably indicated that of intrahepatic tyrosine. Ballmer et al. (1993, 1996) used flooding doses of either [^{13}C]leucine or [$^{2}H_{5}$]phenylalanine to examine albumin synthesis in cirrhotic patients. Those with mild disease had normal albumin synthesis rates, whereas those with severe disease had diminished albumin synthesis. There was no clear relation between disease severity and fibrinogen synthesis.

In summary, secretion of plasma proteins is fairly well preserved in patients with mild to moderate cirrhosis, but albumin secretion is impaired with severe disease. Whole-body protein turnover tends to be more rapid in cirrhotic patients, especially when expressed per kg of body cell mass. It is unclear whether or not some of this increase in whole-body protein turnover is related to nutritional rehabilitation prior to the protein metabolism experiments, since cirrhotic patients often are malnourished before enrollment in research protocols. The response to insulin and meals appears to be fairly normal. Muscle protein degradation may be accelerated in cirrhosis, but definitive data are not available. A variety of hormonal changes could lead to alterations in protein metabolism in cirrhosis, including reduced concentrations of IGF-I, testosterone, cortisol, and thyroid hormones, and increased levels of growth hormone, glucagon, insulin, catecholamines, and estradiol (Charlton 1996).

Diabetes

The effect of insulin-dependent (type 1) diabetes mellitus (IDDM) on protein metabolism was reviewed in Chapter 6, in the context of the role of insulin in protein metabolism. Noninsulin-dependent diabetes (type 2, NIDDM) is much more common. This affliction results from insulin resistance rather than an absolute insulin deficiency, and is the focus of the present discussion. NIDDM is very often associated with obesity. The effect of obesity on protein metabolism is discussed later in this chapter.

In contrast to the rapid loss of lean body mass with untreated IDDM, there is no obvious loss of body proteins associated with NIDDM. Treatment of NIDDM with drugs or diet may slightly increase total body nitrogen, but not nearly as much as insulin therapy increases it (Walsh et al. 1976). Either the insulin resistance in NIDDM does not include the protein-sparing effects of insulin, or patients secrete enough insulin to compensate for any resistance to the effect of insulin on protein metabolism.

When subjects with NIDDM are compared with weight-matched subjects without diabetes, there generally is no difference in their postabsorptive whole-body protein turnover (Staten, Matthews, and Bier 1986; Welle and Nair 1990a; Biolo et al. 1992b; Luzi, Petrides, and DeFronzo 1993). In one study, postabsorptive proteolysis, per kg lean body mass, was elevated in patients with NIDDM (Denne et al. 1995a). Treatments that reduce the fasting blood glucose in patients with NIDDM do not reduce the fasting protein turnover rate (Staten, Matthews, and Bier 1986; Welle and Nair 1990a; Tessari et al. 1994a). Infusion of insulin, or increasing insulin and amino acid levels with meals, reduces whole-body proteolysis to a similar extent in NIDDM patients and nondiabetic subjects (Biolo et al. 1992b; Luzi, Petrides and DeFronzo, 1993; Denne et al. 1995a), suggesting that sensitivity to insulin's antiproteolytic effect is preserved in NIDDM. However, a study employing the [^{15}N]glycine end product method over several days indicated that whole-body protein breakdown and synthesis, per kg lean body mass, were ~25% more rapid in obese subjects with NIDDM than in nondiabetic obese subjects (Gougeon, Pencharz, and Sigal 1997). Moreover, the decline in protein turnover associated with a low-energy diet was diminished in patients with NIDDM, although insulin therapy normalized the rate of protein turnover. It is unclear whether or not methodological differences (use of [^{15}N]glycine rather than essential amino acid tracers) can explain the discrepancy between this study and the others that reported normal protein turnover in NIDDM.

Marchesini et al. (Marchesini et al. 1982) suggested that myofibrillar proteolysis is elevated in NIDDM patients in very poor control of their hyperglycemia, based on the observation that their 3-methylhistidine excretion was ~30% greater than that of nondiabetic subjects. Treatment with glibenclamide reduced 3-methylhistidine excretion 12%. The elevation of 3-methylhistidine excretion in NIDDM was much less than that associated with poorly controlled IDDM. Denne et al. (1995a) found that the postabsorptive rate of release of phenylalanine from the leg, which reflects both myofibrillar and nonmyofibrillar proteolysis, was significantly less in NIDDM patients than in nondiabetic subjects. Insulin infusion significantly reduced leg phenylalanine release in nondiabetic subjects, but had little effect in NIDDM patients. These patients were hyperinsulinemic, and apparently this hyperinsulinemia was enough to maximally inhibit limb proteolysis. It is interesting that the same patients had an elevated whole-body phenylalanine R_a, and normal suppression of whole-body phenylalanine R_a during insulin infusion, suggesting that there might be significant differences between viscera and limbs in the response to insulin in NIDDM.

Gestational diabetes is another common form of insulin resistance. Protein metabolism in women with gestational diabetes is discussed in the section on pregnancy later in this chapter.

Glucocorticoid Therapy

Synthetic glucocorticoids are used as anti-inflammatory agents or to suppress autoimmunity in a variety of conditions. In general, the effects of cortisol that were discussed in Chapter 6 also apply to the synthetic glucocorticoids. Commonly used ones are prednisone, prednisolone, dexamethasone, and betamethasone. The major effect relevant to the topic of protein metabolism is the muscle wasting that often accompanies treatment with synthetic glucocorticoids. Fluorinated glucocorticoids are more likely than nonfluorinated ones to cause this problem, but any glucocorticoid can induce muscle atrophy (Kaminski and Ruff 1994). Negative nitrogen balance and muscle wasting can occur even when there is clinical improvement and increased food intake associated with glucocorticoid therapy (Roubenoff et al. 1990).

In sufficiently high doses, glucocorticoids increase whole-body proteolysis. Daily doses of prednisone or methylprednisolone in excess of 0.4 $mg \times kg^{-1} \times d^{-1}$, given for a few days to healthy subjects, increase proteolysis ~20% (Garrel et al. 1988; Beaufrere et al. 1989; Horber and Haymond 1990; Berneis et al. 1997;). Dexamethasone has a similar effect at a lower dose (Louard et al. 1994). Leucine oxidation and nitrogen excretion increase in parallel with the increase in proteolysis. There may be some increase in whole-body protein synthesis, but less than the increase in proteolysis, so that the net effect is catabolic. 3-Methylhistidine excretion and forearm proteolysis do not increase, suggesting that myofibrillar proteolysis might not contribute to the whole-body effect (Garrel et al. 1988; Louard et al. 1994). The hyperinsulinemia that accompanies glucocorticoid administration may counteract any effect of the steroid on muscle proteolysis, since insulin's antiproteolytic effect in muscle does not appear to be affected (Louard et al. 1994). However, 3-methylhistidine excretion may be increased in patients with myopathy induced by chronic glucocorticoid treatment (Khaleeli et al. 1983). In animals, glucocorticoid administration increases 3-methylhistidine excretion or efflux of 3-methylhistidine and tyrosine from limbs, but the effect may be transient (Kayali, Young, and Goodman 1987; Bowes et al. 1996; Auclair et al. 1997). Prednisone treatment of boys with Duchenne muscular dystrophy reduces 3-methylhistidine excretion relative to muscle mass (creatinine excretion), but these children have muscle hypertrophy rather than muscle atrophy in response to prednisone (Rifai et al. 1995).

Animal studies consistently demonstrate that glucocorticoids inhibit protein synthesis in muscle (Rannels et al. 1978; Seene and Alev 1985; Kayali, Young, and Goodman 1987; Hickson, Czerwinski, and Wegrzyn 1995). There is very little information about the effect of glucocorticoids on protein synthesis in human muscle. Pacy and Halliday (1989) presented the case of a patient with acute steroid myopathy, whose fractional rate of muscle protein synthesis was 17% less than the mean value of healthy subjects. However, his protein synthesis rate was within the normal range, being less than one standard deviation below the mean value. More convincing evidence for reduced muscle protein synthesis in glucocorticoid-treated patients was presented by Gibson et al. (1991). In patients who had taken an average of 8 mg/d of prednisolone for several years for arthritis, the fractional rate of protein synthesis was 38% slower than it was in arthritis patients not taking any glucocorticoid. In boys with muscular dystrophy who were treated with prednisone (0.75 $mg \times kg^{-1} \times d^{-1}$) for 6 to 8 weeks, we did not observe any consistent change in muscle protein synthesis (Rifai et al. 1995). Two of 6 patients had a large decrease in muscle protein synthesis after treatment, one had a large increase, and the others had slightly increased muscle protein synthesis. Glucocorticoid treatment increases muscle mass in these patients, in contrast to its catabolic effect under most conditions.

In addition to muscle wasting, chronic glucocorticoid treatment can cause bone thinning. Markers of collagen degradation and synthesis generally suggest that glucocorticoids inhibit collagen synthesis, but do not accelerate collagen breakdown (Garrel et al. 1988; Pearce et al. 1998; Prummel et al. 1991; Co et al. 1993; Wolthers et al. 1997;).

Anabolic hormones, including growth hormone, IGF-I, and testosterone, can mitigate the protein catabolic effect of glucocorticoids (Horber and Haymond 1990; Bennet and Haymond 1992; Mauras and Beaufrere 1995; Oehri et al. 1996; Reid et al. 1996; Berneis et al. 1997). The muscle wasting can be reversed by exercise (Horber et al. 1985). Thus, it appears that glucocorticoids do not prevent normal responses to anabolic stimuli.

Heart Disease

Patients with chronic heart failure often exhibit muscle atrophy, sometimes referred to as cardiac cachexia (Mancini et al. 1992; Toth et al. 1997). Reduced food intake, inactivity, and hormonal changes (especially the increase in TNF-α and cortisol) may be important determinants of the loss of muscle protein mass (Anker et al. 1997). These patients typically are taking several medications, but their effects on protein metabolism is unknown. Morrison, Gibson, and Rennie (1988) reported that cachectic heart failure patients had elevated myofibrillar proteolysis, as indicated by efflux of 3-methylhistidine from the legs. Net tyrosine efflux from the legs also was more negative in cardiac patients than in healthy subjects, but no tracer was used to determine if this effect was related to increased proteolysis or reduced protein synthesis. In spite of the increased myofibrillar proteolysis, whole-body proteolysis, protein synthesis, and leucine oxidation were markedly reduced (~40%) in these patients, per kg total body weight. Although expression of data in terms of body cell mass probably would have diminished the apparent difference in protein turnover between heart failure patients and healthy subjects, the difference appears to be too great to be explained entirely by altered body composition.

Myocardial protein metabolism has been measured in patients with coronary artery disease (Young et al. 1991; McNulty et al. 1995). Insulin and branched-chain amino acids are anabolic—insulin by inhibiting myocardial proteolysis and branched chain amino acids by promoting myocardial protein synthesis. No data are available regarding whether protein metabolism is the same in healthy hearts.

HIV Infection and AIDS

Weight loss, including loss of lean body mass, is a common feature of HIV infection, especially when it progresses to AIDS (Mulligan and Bloch 1998). A few studies have examined protein metabolism in HIV infected patients, but results have been too variable to allow any conclusions about whether reduced protein synthesis or increased proteolysis causes the loss of protein.

Stein et al. (1990) studied patients with AIDS using the [^{15}N]glycine method. Compared with normal men, these patients had reduced whole-body protein turnover, per kg total body weight, and slower fibrinogen synthesis. Data were not reported in relation to lean body mass, which clouds interpretation somewhat because usually there is some loss of lean body mass by the time HIV infection progresses to AIDS. However, body weights in these patients did not seem to be unusually low. Most were taking antiretroviral drugs, which have unknown effects on protein metabolism. Lieberman et al. (1994) studied AIDS patients, also taking antiretroviral drugs, who had lost 8% to 25% of their body weight, using both [^{15}N]glycine (over 24 h) and [^{13}C]leucine kinetics (postabsorptive). They did not study a healthy group, but noted that postabsorptive protein turnover seemed to be higher than values published for normal subjects, and that N flux data seemed to be normal. However, the protein synthesis values based on [^{15}N]glycine were even lower than the "low" values for AIDS patients reported by Stein et al. (1990). Macallan et al. (1995) found that postabsorptive proteolysis, per kg lean body mass, was 25% more rapid in patients with Stage IV HIV infection than in healthy subjects. Leucine oxidation was not affected, and protein synthesis also was increased with HIV infection. Subjects with Stage II disease had values intermediate between healthy subjects and those with Stage IV disease. When subjects were fed, there were no differences in protein metabolism among the groups. Selberg et al. (1995) noted that whole-body protein turnover in AIDS patients, per kg body weight, tended to be very high compared with values in the literature for normal subjects, but they did not have a control group in their study. Their subjects were being fed parenterally. Salbe et al. (1995) did not observe any abnormality of postabsorptive protein metabolism in asymptomatic HIV infected patients.

Muscle protein synthesis, albumin synthesis, and bone collagen formation are similar in healthy subjects and HIV infected patients, including those with AIDS and weight loss (McNurlan et al. 1997; McNurlan et al. 1998). Those with AIDS have more rapid myofibrillar degradation (~35% higher 3-methylhistidine/creatinine ratio) than healthy subjects or those who are HIV positive without symptoms.

HIV infected patients are responsive to anabolic stimuli. The acute changes in protein metabolism associated with meals are normal (Macallan et al. 1995), and increasing protein intake for a few days stimulates whole-body protein synthesis (Selberg et al. 1995). HIV infected patients also have a normal decline in protein turnover and oxidation when protein intake is reduced below the minimum requirement (Salbe et al. 1995). Patients with HIV infection have the expected anabolic response to growth hormone and IGF-I (Mulligan et al. 1993; Lieberman et al. 1994; McNurlan et al. 1998; Mulligan, Tai, and Schambelan 1998), although AIDS patients with weight loss may have an impaired response to growth hormone (McNurlan et al. 1997; McNurlan et al. 1998).

Infection

Research on the effect of infection on protein metabolism has dealt with severe infections, those requiring hospitalization. The reduced food intake and bed rest that usually accompany such infections have a protein catabolic effect, as discussed elsewhere in this book (see Chapters 5 and 8). With severe and prolonged infections, protein wasting becomes a major clinical problem. The response to sepsis includes a large increase in urea production and N excretion. Protein losses of 5% to 15% of the whole-body protein mass are commonly observed over the course of a septic episode (Cooney, Kimball. and Vary 1997).

Whole-body proteolysis and protein synthesis generally are increased in patients with infection. The increase in breakdown is greater, so protein mass declines. Long et al. (1977) studied 3 septic patients and 2 normal subjects with [^{15}N]alanine. The average protein turnover was more rapid in the patients, but there was some overlap with normal values. Tomkins et al. (1983), using [^{15}N]glycine, found that infected children had higher protein turnover rates than uninfected children recovering from malnutrition, but differences in nutritional status could not be ruled out as the explanation for the differences in protein metabolism. The fact that treatment of the infection reduced proteolysis in well-nourished children provided more convincing evidence that infection stimulates proteolysis. Shaw et al. (1987) found that lysine R_a was more rapid in severely septic patients than in healthy subjects. Glucose infusion and parenteral nutrition had little effect on proteolysis, but did reduce net protein losses. Jahoor et al. (1989) reported that proteolysis was ~50% more rapid than normal in septic patients, leucine oxidation was 100% more rapid, and incorporation of leucine into proteins was ~30% more rapid. Very high insulin levels suppressed protein turnover normally. While these data show that septic patients can respond to extremely high insulin levels, they do not rule out some degree of resistance to the antiproteolytic effect of lower insulin levels. Beylot et al. (1994) found that proteolysis and leucine oxidation were ~30% more rapid than normal in septic patients, whereas whole-body protein synthesis was increased to a lesser extent. Arnold et al. (1993) studied patients with multiple organ failure, of whom many were septic. Proteolysis was more than 100% more rapid than normal. Leucine oxidation was increased 75%, and leucine incorporation into proteins was increased ~50%. In contrast to these results, Manary et al. (1997) reported that children with edematous malnutrition and lower respiratory infections had reduced protein turnover, per kg body weight, relative to uninfected children recovering from malnutrition. However, it is unclear whether or not the body composition and recent nutritional status of these groups were similar enough to allow meaningful conclusions about the effect of infection.

There is general agreement that infection is associated with increased myofibrillar proteolysis, as reflected by 3-methylhistidine excretion (Wannemacher et al. 1975; Long et al. 1977; Long et al. 1981; Tomkins et al. 1983). Usually, 3-methylhistidine excretion is expressed in relation to creatinine excretion, to adjust for variability in muscle mass and completeness of urine collections. However, there is marked rise in creatinine excretion during fever, so the 3-methylhistidine/creatinine ratio is not a valid index of myofibrillar proteolysis in septic patients. Instead, uncorrected 3-methylhistidine excretion, or excretion per kg body weight, has been used to evaluate the effect of infection. The time course of the elevation of 3-methylhistidine excretion follows that of fever, and the magnitude of the increase in 3-methylhistidine excretion correlates with the magnitude of the febrile response. Hypercaloric parenteral nutrition can suppress the increase in 3-methylhistidine excretion (Leverve et al. 1984). Animal studies indicate that increased proteasomal proteolysis is primarily responsible for the increased muscle protein degradation in sepsis, although there may be a minor contribution of calpain-mediated and lysosomal proteolysis (Cooney, Kimball, and Vary 1997; Tawa, Odessey, and Goldberg 1997).

The effect of infection on muscle protein synthesis in vivo has not been measured in humans. In vitro, muscle from septic patients has a slight increase in protein synthesis, but much less than the marked increase in degradation (Clowes et al. 1983). Sjolin et al. (1990) did some indirect estimates of muscle protein synthesis, based on 3-methylhistidine, tyrosine, and phenylalanine net effluxes from the leg. They concluded that muscle protein synthesis may be reduced in septic patients. Some animal studies also indicate that sepsis inhibits protein synthesis, by reducing mRNA translation efficiency via reduced peptide-chain initiation (Hasselgren et al. 1984b; Jepson et al. 1986; Cooney, Kimball, and Vary 1997). This effect is greater than the suppression caused by reduced food intake. However, animal studies do not always indicate that sepsis reduces muscle protein synthesis (Hasselgren et al. 1988).

The increased proteolysis associated with infection results in greater efflux of amino acids into the circulation. The amino acids are cleared more rapidly from the circulation in septic patients than in healthy subjects (Clowes et al. 1980; Clowes et al. 1985), and are used for oxidation, glucose production, and protein synthesis. Muscle protein synthesis is either suppressed or unchanged, so the increase in whole-body protein synthesis must be explained by increased protein synthesis in other tissues. In rats, sepsis inhibits renal protein synthesis, and does not alter splenic protein synthesis (Cooney et al. 1996). It may either increase or slightly inhibit intestinal protein synthesis (Von Allmen et al. 1992; Cooney et al. 1996). Cells of the immune system certainly could contribute to the increased protein synthesis during infection, although the extent to which they account for the whole-body effect is uncertain. The liver may account for much of the increase in amino acid clearance and protein synthesis during infection (Wilmore et al. 1980; Hasselgren et al. 1988; Vary and Kimball 1992; Cooney et al. 1996). The "acute phase protein" response to infection and other catabolic stimuli involves increased hepatic production of several proteins (see section on cytokines in Chapter 6), and animal studies indicate that infection increases hepatic protein synthesis (Hasselgren et al. 1984b; Jepson et al. 1986; Sax et al. 1988; Vary and Kimball 1992). However, in malnourished children with infections, the increase in levels of some of the acute phase proteins appears to be explained by reduced degradation rather than increased production (Morlese, Forrester, and Jahoor 1998).

The alterations in protein metabolism associated with infection may be mediated to a large extent by hormones and cytokines. Concentrations of cortisol, glucagon, and cytokines increase during infection. Many of the changes in protein metabolism observed in septic patients are similar to those observed with cortisol or cytokine administration (see Chapter 6). Inhibiting TNF production ameliorates the muscle protein wasting and the reduction in muscle protein synthesis in septic rats (Breuille et al. 1993; Jurasinski, Kilpatrick, and Vary 1995). An IL-1 receptor antagonist prevents the sepsis-induced inhibition of muscle protein synthesis in rats (Cooney et al. 1994). Blocking glucocorticoid receptors also can inhibit the proteolytic response to sepsis, but there is evidence that the reduction in protein synthesis does not require increased glucocorticoid activity (Cooney, Kimball, and Vary 1997). The in vitro rate of hepatic protein synthesis in patients with sepsis correlates with circulating levels of a peptide produced by cleavage of IL-1 (Clowes et al. 1985).

The fever associated with infection may promote a generalized increase in protein turnover and oxidation, simply because all metabolic reactions are temperature dependent. Baracos, Wilson, and Goldberg (1984) examined the effect of temperature on protein metabolism in isolated muscle and hepatocytes. Protein degradation in muscle increased 2-fold as temperature was increased from 33° C to 42° C. Muscle protein synthesis increased only slightly as temperature increased from 33° C to 39° C, then decreased when temperature was increased to 42° C. Thus, elevated temperature per se had a protein catabolic effect in muscle. In contrast to the >50% increase in muscle proteolysis in response to a temperature increase from 36° C to 39° C, there was little change in hepatocyte proteolysis.

Muscular Dystrophies

Myotonic dystrophy, the most common adult form of muscular dystrophy, is caused by the insertion of hundreds of additional CUG repeats in the 3′ untranslated region of the mRNA of a protein kinase. This mutation also affects many organ systems other than muscle. The mechanism whereby the CUG expansion causes impairment of muscle function is not known, but may involve abnormalities in the binding of CUG-binding protein to other transcripts containing CUG repeats (Philips, Timchenko and Cooper 1998). Myotonic dystrophy is associated with resistance to insulin's effects on glucose metabolism, but it is unclear whether or not there is resistance to insulin's protein anabolic effect. Halliday et al. (1988) found that the fractional rate of protein synthesis was ~30% slower than normal in quadriceps muscle of myotonic dystrophy patients (Halliday et al. 1988). Fractional rates of muscle protein synthesis in myotonic dystrophy patients studied in our laboratory (Thornton et al. 1993) are similar to values we observe in healthy young men. However, considering the muscle wasting in these patients, the total muscle protein synthesis must be reduced. On an absolute basis, patients with myotonic dystrophy have markedly reduced 3-methylhistidine excretion. When corrected for muscle mass, 3-methylhistidine excretion is fairly normal or only slightly elevated (Griggs and Rennie 1982; Halliday et al. 1985). Thus, reduced protein synthesis rather than increased protein degradation appears to be responsible for the muscle atrophy in myotonic dystrophy. Muscle protein synthesis is stimulated and lean body mass increased in these patients by administration of testosterone (Griggs et al. 1986) or growth hormone (Thornton et al. 1993).

Duchenne muscular dystrophy is the most common and devastating form of muscular dystrophy in children. It is an X-linked recessive disorder resulting from the lack of dystrophin, a large protein that links the actin filaments of the cytoskeleton with a transmembrane glycoprotein complex that binds laminin in the basement membrane (Ozawa et al. 1995). Whereas dystrophin is absent in Duchenne dystrophy, it is present in reduced amounts or altered form in Becker muscular dystrophy. Patients with Becker muscular dystrophy have slower disease progression and longer survival than those with Duchenne dystrophy. Although the absolute rate of 3-methylhistidine excretion is generally normal in patients with Duchenne dystrophy, it is markedly elevated in relation to muscle mass as reflected by creatinine excretion (Figure 9.1). Although nonskeletal muscle actin degradation could account for a larger fraction of 3-methylhistidine excretion in dystrophic patients, this problem is unlikely to account for much of the large discrepancy in 3-methylhistidine/creatinine ratios between normal and dystrophic subjects. The high rate of myofibrillar degradation provides indirect evidence for increased myofibrillar synthesis. If there were no increase in protein synthesis, muscle mass would decline much faster than what is observed in these patients. Because normal young children cannot be asked to submit to muscle biopsies and long tracer infusions, the normalcy of muscle protein synthesis early in the course of the disease has not been directly measured by isotope incorporation. In 5 to 8 year old boys with Duchenne dystrophy, the fractional rate of muscle protein synthesis was about twice the rate typically observed in healthy young men (Rifai et al. 1995). Whether this difference primarily reflects a fall in protein

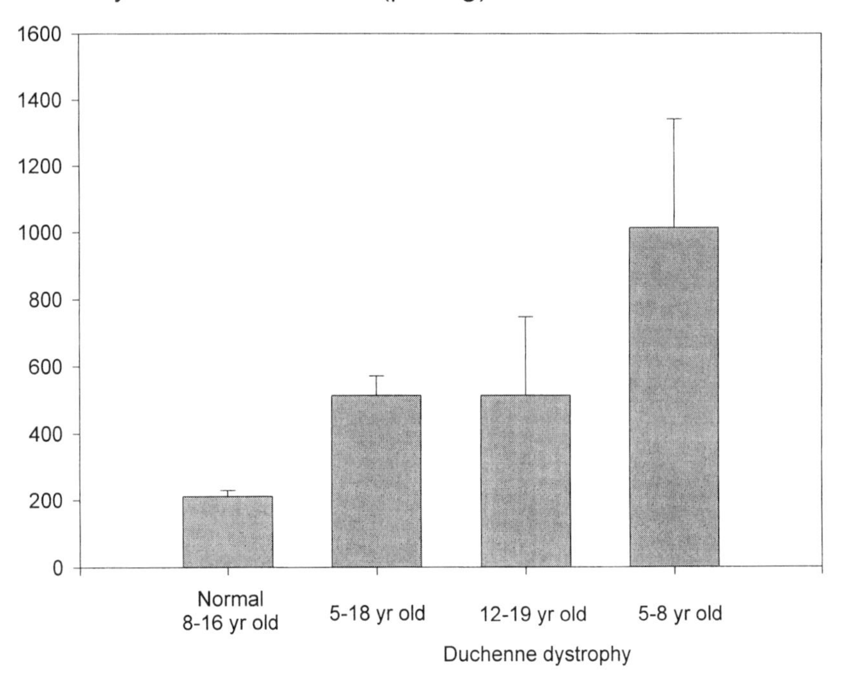

FIGURE 9.1. Evidence for increased myofibrillar protein degradation in children with Duchenne muscular dystrophy, as reflected by the mean ratio of 3-methylhistidine excretion to creatinine excretion (McKeran, Halliday, and Purkiss 1977; Ballard, Tomas, and Stern 1979; Rifai et al. 1995). Error bar represents one standard deviation.

synthesis with maturation is unclear. In young adult patients with Becker dystrophy, there was a similar elevation of muscle protein synthesis. It is likely that the high rate of proteolysis leads to a higher rate of protein synthesis because of elevated intramuscular amino acid levels. As the disease progresses, muscle protein synthesis may decline. Halliday et al. (1988) reported that the average fractional rate of muscle protein synthesis was markedly reduced in a group of patients with miscellaneous dystrophies (2 Duchenne, 1 Becker, 2 limb girdle, 1 scapuloperoneal). Individual data were not reported. In early stages of the disease, rapid protein synthesis may reflect muscle regeneration. In the later stages of the disease, muscle may lose its regeneration potential. Moreover, patients with more advanced dystrophy are confined to a wheelchair and the inactivity can contribute to slow protein synthesis.

In vitro protein synthesis has been measured in muscle preparations from patients with various forms of muscular dystrophy (Ionasescu 1975; Tomkins et al. 1982). Both normal and abnormal results were found, but the relevance to in vivo protein metabolism is unclear. There is evidence that polyribosmes from patients with Duchenne and Becker dystrophies have a higher proportion of collagen mRNA than polyribosomes from normal muscle. The in vivo studies discussed above involved measurement of noncollagen synthesis only.

Obesity

The excess weight in obese persons consists not only of fat, but also of protein. In general, about 25% of the excess weight is fat-free. This extra fat-free mass consists not only of the stromal component of adipose tissue, but also increased mass of muscle and viscera (Forbes 1987a). The protein accretion during weight gain is a natural consequence of overeating, as discussed in Chapter 5.

There is evidence, from leucine kinetics in postabsorptive subjects, that the rate of whole-body protein turnover in obese persons is somewhat faster than it is in normal-weight persons (Figure 9.2) even after accounting for the elevated protein mass in the obese subjects (Nair et al. 1983; Welle and Nair 1990a; Jensen and Haymond 1991; Welle et al. 1992a). Although there is agreement that protein turnover per person is greater in obese individuals, some investigators have not observed a significant difference between lean and obese subjects when protein turnover is expressed per kg lean body mass (Nair et al. 1987a; Caballero and Wurtman 1991; Luzi, Castellino, and DeFronzo 1996; Solini et al. 1997). Obese women with ovarian hyperandrogenism were found to have increased protein

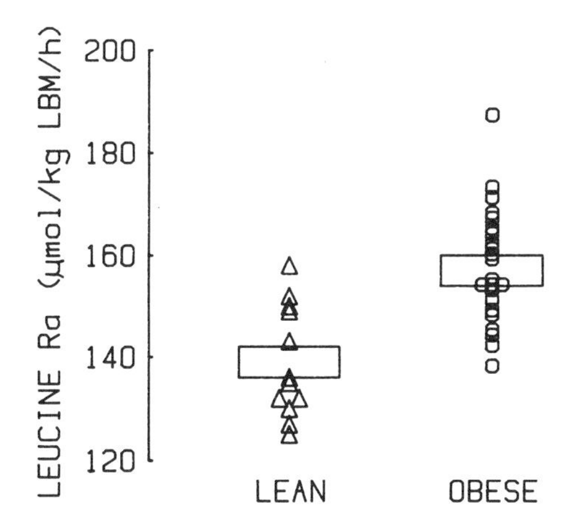

FIGURE 9.2. Increased whole-body proteolysis, per kg lean body mass (LBM), in obese women. LBM was determined from whole-body K. Each point represents an individual subject studied after an overnight fast. Rectangles indicate mean ± one standard error. Leucine R_a, determined by [^{13}C]leucine infusion, was the index of proteolysis. From Welle et al. Increased protein turnover in obese women, "Metabolism, vol 41, pp. 1028–1034, 1992, permission granted by W.B. Saunders Co.

turnover, per kg lean body mass, in a study in which obese women with normal androgen levels had normal protein turnover (Mauras et al. 1998b). The rate of whole-body protein turnover returns to normal when obese subjects lose the excess weight by dieting (Welle et al. 1994a).

One factor is that could increase protein turnover in obese subjects is the higher fractional rate of degradation of adipose tissue proteins (~8%/d in postabsorptive subjects) relative to the whole-body average (~2.5%/d) (see Chapter 4). Another potential factor is the insulin resistance that typically accompanies obesity. Whereas obesity certainly is associated with insulin resistance with respect to glucose metabolism, there is disagreement about whether it is associated with resistance to the antiproteolytic effect of insulin. My research (Welle et al. 1992a; Welle et al. 1994a) suggested that insulin resistance does not explain the faster whole-body proteolysis in obesity. The increase in whole-body leucine R_a in obese women was unrelated to indices of insulin resistance, and euglycemic hyperinsulinemia suppressed leucine R_a as much in obese women as in lean women. Other investigators also have failed to find any difference between lean and obese subjects in the suppression of whole-body proteolysis induced by insulin (Caballero and Wurtman 1991; Solini et al. 1997). However, there is evidence that the modest suppression of proteolysis induced by small increases in insulin levels is less in obese subjects, particularly those with upper-body obesity (Luzi, Castellino, and DeFronzo 1996; Jensen and Haymond 1991). Even if there is some insulin resistance with respect to antiproteolysis, it is unclear whether or not this can explain the elevated protein turnover in obesity, because there is compensatory hyperinsulinemia in obese persons that would tend to offset the insulin resistance.

The role of increased energy intake in the faster protein turnover in obese subjects is unclear. Although obese individuals must consume more energy than lean persons to maintain their elevated body weight, their energy requirement per kg lean body mass is not greater (Welle et al. 1992b).

Parenteral and Enteral Nutrition

A whole chapter easily could be devoted to the numerous studies that have investigated various aspects of protein metabolism in relation to parenteral or enteral nutrition in critically ill patients. However, little would be learned beyond what is known about nutritional influences on protein metabolism in general (Chapter 5) and the effects of individual metabolic substrates (Chapter 7). Nevertheless, a few general comments are in order because of the clinical importance of this topic.

The breakdown of body proteins (especially muscle proteins) in critically ill patients is an adaptive response that provides amino acids for synthesis of critical proteins to ensure survival (e.g., proteins needed for wound healing, tissue regeneration, replacement of lost blood, immune responses, etc.). Without nutritional support, the breakdown of endogenous proteins can provide the needed amino acids for a few days without significant impact on muscle protein mass, but prolonged recovery from critical illness produces substantial protein wasting. Obviously, provision of exogenous energy and amino acids helps to preserve the protein stores of the body in a patient who cannot eat normally. Energy in the form of either glucose or fat can spare protein to some extent, but even though they reduce amino acid oxidation they cannot eliminate it. Thus, amino acids must be provided to allow maintenance or accretion of body proteins. The amount of nonprotein energy provided is much more important than the source of energy in terms of preserving protein (Macfie, Smith, and Hill 1981; Nordenstrom et al. 1983; DeChalain et al. 1992; Jones et al. 1995), although sometimes differences in protein metabolism are noted when glucose and fat are compared (Bresson et al. 1991). While the qualitative effects of energy and amino acids on protein metabolism generally appear to be intact in critically ill patients, the amino acid and energy requirements for maintaining protein mass can be higher than normal because of elevated levels of catabolic hormones.

The protein sparing effect of nutritional support, as reflected by improved N balance, is not always correlated with better clinical outcome (Lin, Goncalves, and Lowry 1998; Silk and Green 1998). However, there is some evidence that mortality can be reduced by supplementing intravenous nutrition with glutamine or branched chain amino acids (Campbell 1998), which appear to be the most anabolic amino acids (see Chapter 7).

Pregnancy and Lactation

It is obvious that there must be protein accretion in normal pregnancy, not only for fetal growth but also for maternal supportive tissues. Figure 9.3 illustrates the time course of protein accumulation in a typical pregnancy, as well as the distribution of the protein at the time of birth. Most of the protein gain occurs toward the end of the pregnancy, and only about half of the protein gain is attributable to the fetus. The protein accretion is accomplished both by increased food intake and by reduced protein oxidation (when expressed as a fraction of the amount of protein consumed). Urea production rate declines in relation to protein intake, especially late in gestation (Olufemi, Whittaker, and Lind 1991; Forrester et al. 1994; McClelland, Persaud, and Jackson 1997; Kalhan et al. 1998). N salvage via intestinal hydrolysis of urea is enhanced in pregnant women, so less of the urea that is produced is excreted.

Studies of whole-body protein metabolism in normal pregnancy have not produced consistent results regarding either protein synthesis or breakdown, as summarized in Table 9.3. One of the problems associated with comparing whole-body protein metabolism in pregnant and nonpregnant women is the difference in body composition. Many of the studies simply expressed data per kg total body weight. Because much of the increase in body weight during pregnancy is

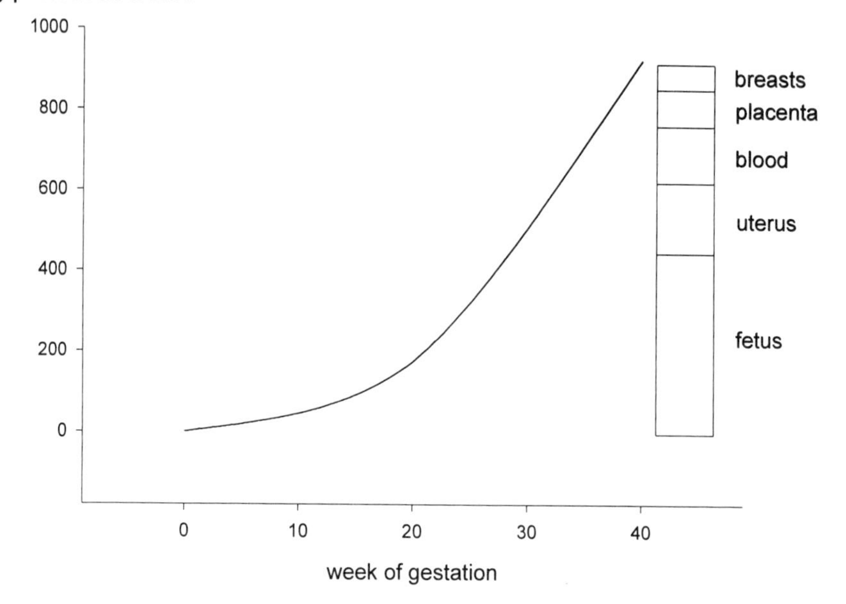

FIGURE 9.3. Time course of maternal/fetal protein accretion during a typical pregnancy, and distribution of the added protein at the end of pregnancy. Data provided by Robinson and Pendergast (1996).

fat and extracellular fluid rather than active fetal or maternal cell mass, the relative rate of protein turnover per kg cell mass in pregnant women is underestimated when data are expressed per kg body weight. Even when data are expressed per kg fat free mass, edema can create a similar problem, although use of fat free mass is an improvement over use of total body weight. Thompson and Halliday (1992) used potassium as an index of body cell mass to adjust their protein metabolism results, but used estimated rather than measured data for changes in whole-body potassium in pregnant women.

There is very little information on protein metabolism in individual tissues in human pregnancy. Fitch and King (1987) noted that 3-methylhistidine excretion was increased in pregnant women, even though their whole-body proteolysis (per kg) was not increased. Olufemi et al. (1991) reported that pregnant women have elevated rates of albumin synthesis. Infants have much higher rates of protein turnover than adults (Chapter 4), so fetal protein turnover certainly must contribute to any increase in whole-body protein turnover in pregnant women. Ovine fetuses retain only a quarter to a half of the protein that they synthesize (Meier et al. 1981). Fractional turnover rates of uterine and placental tissues have not been reported.

Pregnancy can induce a state of insulin resistance known as gestational diabetes. In normal pregnancy, there is some degree of insulin resistance with

TABLE 9.3. Comparisons of whole-body protein metabolism in pregnant and nonpregnant women.

Reference	Tracer	Fed or postabsorptive	Results
Studies showing reduced proteolysis and/or synthesis			
Denne, Patel and Kalhan (1991)	[^{13}C]leu	Postabsorptive	reduced proteolysis and synthesis per kg total or fat free mass in second half of pregnancy vs. nonpregnant women
Robinson et al. (1992)	[^{13}C]leu	Postabsorptive	Reduced proteolysis and synthesis per kg total or lean body mass in second trimester vs. nonpregnant women
Studies showing no change in protein turnover			
Fitch and King (1987)	[^{15}N]gly	Fed	Proteolysis and synthesis similar to nonpregnant women, per kg body weight, in last trimester
Olufemi et al. (1991)	[^{15}N]gly	Postabsorptive	Proteolysis and synthesis similar to nonpregnant women, per kg body weight, in last trimester
Cooper et al. (1992)	[^{13}C]leu and [$^{2}H_{5}$]phe	Postabsorptive	Proteolysis and synthesis similar to nonpregnant women, per kg body weight, in last trimester
Kalhan et al. (1998)	[^{13}C]leu	Postabsorptive and fed	Proteolysis and synthesis similar to nonpregnant women, per kg body weight, in all trimesters
Studies showing increased proteolysis and/or synthesis			
de Benoist et al. (1985)	[^{15}N]gly	Fed	Protein synthesis increased vs. nonpregnant women, per kg body weight, throughout pregnancy; proteolysis increased initially, then declines late in pregnancy
Thompson and Halliday (1992)	[^{13}C]leu	Postabsorptive	Increased proteolysis in third trimester, per kg body weight, vs. nonpregnant women; increased protein synthesis and proteolysis in second and third trimesters, per kg estimated fat free mass
Olufemi et al. (1991)	[^{13}C]leu	Postabsorptive	Increased proteolysis in third trimester, per kg body weight, vs. nonpregnant women

respect to stimulation of glucose uptake, but not with respect to inhibition of proteolysis (Cooper et al. 1992). Women with gestational diabetes generally have rates of leucine and phenylalanine appearance similar to those of non-diabetic pregnant women, even though they have higher insulin levels (Zimmer et al. 1996). There also is evidence for increased whole-body protein turnover in women with gestational diabetes relative to women with normal pregnancies (Fitch and King 1987; Robinson et al. 1992). The fact that proteolysis is increased or normal, in spite of hyperinsulinemia, suggests that there may be resistance to insulin's antiproteolytic effect in women with gestational diabetes. However, there is not enough evidence to rule out some other hormonal factor that can account for both insulin resistance and increased protein turnover.

A lactating woman produces ~6 to 9 g of milk protein daily if she is well nourished (Motil et al. 1998). When protein intake is adequate, the lactating mother does not need to sacrifice her own protein stores to produce milk. However, the protein requirement is higher in lactating than in nonlactating women. Even with increased energy intake, lactating women can have a significantly negative N balance with a protein intake of 1 $g \times kg^{-1} \times d^{-1}$, well above the safe level for nonlactating women (Motil et al. 1990). Excretion of 3-methylhistidine is reduced in lactating women whose protein intake is inadequate to maintain N balance, indicating that myofibrillar proteolysis is not activated to provide amino acids (Motil et al. 1990). Whole-body proteolysis and synthesis are similar in lactating and nonlactating women (Motil et al. 1989), even though the increased energy intake in lactating women might be expected to stimulate protein turnover (Chapter 5). Whole-body proteolysis is suppressed in response to protein restriction in lactating women (Motil et al. 1996). Thus, there is no overall increase in maternal protein degradation to supply amino acids for milk production when protein intake is inadequate. Instead, there is reduced oxidation of the amino acids generated by degradation of dietary and endogenous proteins. The amount of milk protein produced is a small fraction of whole-body protein synthesis, so it is not too surprising that there is no detectable increase in whole-body protein synthesis in lactating women.

Renal Failure

Renal failure often is associated with loss of lean body mass. Anorexia is common, so malnutrition may be one of the major reasons for reduced protein mass. Another common complication of renal failure is acidosis. The catabolic effect of acidosis can be ameliorated by alkali therapy, as discussed in the section on acidosis in this chapter. A number of other factors have the potential to influence protein metabolism in patients with renal disease, including uremic toxins, insulin resistance, amino acid losses during dialysis, reduced muscular activity, and various endocrine abnormalities (Guarnieri et al. 1997). The hormonal abnormalities can include elevated levels of cytokines and cortisol, and reduced levels of growth hormone, IGF-I, and testosterone. These factors vary according to stage of the disease and efficacy of treatment. Acute renal failure usually occurs in the setting of some other catabolic condition, such as trauma or sepsis (Druml 1998). This section deals with the effect of chronic renal dysfunction on protein metabolism.

In nephrotic patients, who are at increased risk of eventual renal failure, there is a significant amount of protein lost in the urine (Maroni et al. 1997). However, with renal failure, the opposite problem—retention of polypeptides—may be a significant problem (Kaysen and Rathore 1996). Not only is there reduced excretion of intact small peptides, but also retention of peptides that are normally degraded in healthy kidneys. The retained peptides include hormones and may include some peptides that have toxic effects.

In patients with chronic renal failure who are not acidotic, whole-body protein turnover per kg of body weight is either normal or slow (Conley et al. 1980; Hou et al. 1986; Berkelhammer et al. 1987; Goodship et al. 1987; Goodship et al. 1990; Castellino et al. 1992). In children with renal failure who had a slow rate of whole-body proteolysis, hemodialysis had no immediate effect on the proteolytic rate (Conley et al. 1980). The increase in whole-body protein synthesis induced by intravenous amino acid infusion is impaired in patients with chronic renal failure (Castellino et al. 1992). Reduced protein turnover may be more a manifestation of malnutrition rather than renal failure per se. However, uremic serum appears to contain a dialyzable factor, which may be a small peptide, inhibitory to protein synthesis in in vitro assays (Delaporte, Gros, and Anagnostopoulos 1980; Cernacek, Spustova, and Dzurik 1982).

Hou et al. (1986) reported that protein turnover ([^{15}N]glycine method) was markedly reduced in renal failure patients when they were consuming a diet with 1.2 g×kg^{-1}×d^{-1} of protein, but then normalized when they were placed on a low protein diet for 10 days (0.6 g×kg^{-1}×d^{-1}). Ordinarily, a low protein diet would be expected to reduce protein turnover (see Chapter 5), but a low protein diet can improve renal function in these patients and therefore could have some metabolic benefits. A more modest reduction in protein intake (from 1.0 to 0.6 g×kg^{-1}×d^{-1}) did not influence whole-body protein turnover ([^{13}C]leucine tracer) in another group of patients with chronic renal failure, who exhibited a normal decline in leucine oxidation when protein intake was reduced (Goodship et al. 1990). These patients did not have a slow rate of protein turnover at the higher protein intake. Renal failure patients can maintain their body protein mass for prolonged periods on very low protein intakes, supplemented with essential amino acids or their ketoacids, by appropriately reducing proteolysis and protein oxidation (Tom et al. 1995).

Renal failure typically causes resistance to insulin's effects on glucose metabolism. However, the suppression of whole-body proteolysis induced by insulin is normal in patients with chronic renal failure (Castellino et al. 1992).

Moreover, the effect of insulin on arterial levels of amino acids, and on their exchanges in leg and splanchnic tissues, is normal in uremic patients (Alvestrand et al. 1988). Thus, the insulin resistance of uremia does not appear to extend to insulin's effects on protein metabolism.

Garibotto et al. (1994; 1996; 1997) have used [^{3}H]phenylalanine kinetics to examine forearm protein metabolism in patients with chronic renal failure. In acidotic patients, both proteolysis and protein synthesis are increased in the forearm relative to values observed in healthy subjects. In nonacidotic patients, forearm protein metabolism is normal. The rate of forearm proteolysis in renal failure patients correlates with the degree of acidosis and the plasma cortisol concentration, which also correlates with the degree of acidosis. In malnourished hemodialysis patients, the basal rate of forearm protein synthesis is lower than what is typically observed in healthy subjects. Treatment with growth hormone stimulates forearm protein synthesis in these patients, without altering the rate of proteolysis.

Nephrotic patients can lose a significant amount of albumin in their urine, resulting in hypoalbuminemia. Renal failure patients undergoing continuous peritoneal dialysis can lose comparable amounts of albumin in the dialysis fluid, but maintain their albumin at a normal concentration by increasing synthesis and reducing degradation (Kaysen and Schoenfeld 1984). Uremic patients on a low protein diet tend to have slower rates of albumin synthesis than normal subjects or renal failure patients on a normal protein diet, but there is much individual variation in this effect (Varcoe et al. 1975).

Tissue Injury (Trauma, Surgery, Burns)

Patients recovering from severe injuries and major surgery are quite catabolic, not only because of increased levels of catabolic hormones, but also because of reduced nutrient intake and bed rest. Infection is a common complication in such patients, which can worsen the protein catabolic response (see section on infection in this chapter). The response to major tissue injury is similar to the response to sepsis, to the extent that many authors lump these conditions together when considering their effects on protein metabolism.

With severe burns and trauma, the negative N balance is mediated by a marked increase in whole-body proteolysis, as reflected by various tracers including [^{15}N]glycine and [^{13}C]leucine (Bilmazes et al. 1978a; Kien et al. 1978a; Birkhahn et al. 1980; Jahoor et al. 1988; Arnold et al. 1993; Petersen, Holaday, and Jeevanandam 1994; Flakoll, Wentzel, and Hyman 1995; Mansoor et al. 1997). Whole-body protein synthesis also increases, but the increase in synthesis is less than the increase in breakdown so that there is a net loss of protein. During recovery from surgery, there may be no change in protein turnover (Kien et al. 1978b, Harrison et al. 1989a), a reduction in synthesis without a change in proteolysis (O'Keefe, Sender, and James 1974; Crane et al. 1977), or increases in both protein breakdown and synthesis (Harrison et al. 1989a; Carli et al. 1990; Carli, Webster, and Halliday 1997). In infants and children, protein metabolism is quite variable in the immediate postoperative period, with most children exhibiting no major change in protein turnover and a few exhibiting a marked increase in proteolysis (Powis et al. 1998). Obviously, surgeons attempt to minimize the amount of tissue injury, so it is not surprising that sometimes no increase in proteolysis is observed after surgery. The typical reduction in food intake after surgery reduces protein synthesis, which may be a more important mediator of reduced protein synthesis than surgical stress (Clague et al. 1983).

Whether or not there is a detectable change in whole-body protein metabolism, the response to tissue injury involves a reprioritization of protein metabolism. It is essential to produce the proteins required for wound healing and restoration of lost blood, whereas maintenance of muscle mass becomes a low priority. Degradation of muscle proteins increases to ensure an adequate supply of amino acids for synthesis of critical proteins. Trauma increases muscle concentrations of mRNAs encoding proteins involved in various proteolytic pathways (Mansoor et al. 1996). Urinary excretion of 3-methylhistidine is increased after burn injuries, multiple fractures, and surgery (Williamson et al. 1977; Bilmazes et al. 1978a; Long et al. 1981; Rennie et al. 1984; Carli et al. 1990). Rennie et al. (1984) noted that the increase in 3-methylhistidine excretion after surgery did not correlate with efflux of this amino acid from the leg and suggested that increased skeletal muscle proteolysis might not account for the increase in urinary excretion of 3-methylhistidine. However, the preoperative value for leg efflux of 3-methylhistidine was made only for a brief period, and if extrapolated to 24 h for the entire skeletal muscle mass it would exceed the measured 24 h urinary excretion by more than 2-fold. Thus, the leg efflux of 3-methylhistidine in this study cannot be considered as representative of myofibrillar proteolysis throughout the day.

There is a reduction in muscle protein synthesis after head injury (Mansoor et al. 1997) or surgery (Essen et al. 1993; Tjader et al. 1996; Carli and Halliday 1997). The reduction in muscle protein synthesis in response to surgery can be detected immediately after the procedure, and appears to be caused by the surgery rather than the general anesthesia (Essen et al. 1992b). A reduction in both total ribosomal RNA and in polyribosomal RNA has been observed in skeletal muscle after surgery (Wernerman et al. 1987; Hammarqvist et al. 1992; Petersson et al. 1994). The decline in muscle protein synthesis within the first few days after surgery does not appear to be explained by reduced nutrient intake, for it is not prevented by parenteral feeding (Wernerman et al. 1987; Essen et al. 1993; Petersson et al. 1994; Tjader et al. 1996). However, administration of glutamine or its precursor α-ketoglutarate can prevent the early decline in ribosome and polyribosome concentrations after surgery (Wernerman et al. 1987; Petersson et al. 1994). The gradual decline in the protein/DNA ratio in muscle during the first month after surgery is prevented by parenteral feeding, indicating that reduced ad libitum food intake is an important factor in the muscle wasting associated with recovery from surgery (Petersson et al. 1995). In children recovering from burns, increasing the protein intake above the usual requirement has little effect on the rate of muscle protein synthesis or whole-body proteolysis, although it does stimulate protein synthesis in skin (Patterson et al. 1997a). Very high insulin levels can stimulate muscle protein synthesis in patients recovering from burns (Sakurai et al. 1995), although it is not clear if this effect is from the hyperinsulinemia itself or from the extra calories that must be given (as glucose) to prevent hypoglycemia. Surprisingly, insulin also stimulated muscle proteolysis in these patients, but less than it stimulated synthesis, so the net effect of insulin was anabolic.

In rats, trauma and surgery stimulate hepatic protein synthesis (Stein et al. 1976; Hasselgren et al. 1984b; Preedy et al. 1988). It is very likely that hepatic protein synthesis increases after major tissue injury in humans as well, but the response is much more complicated than an overall increase in the synthesis of all proteins produced by the liver. The acute phase response to injury includes increased levels of several proteins produced by the liver, and reduced concentrations of others. A marked increase in the synthesis of fibrinogen occurs in patients with various types of injury (Thompson et al. 1989a; Mansoor et al. 1997). Even though albumin concentrations decline after injury, the fractional and absolute synthesis rate of albumin is increased several days after the injury (Mansoor et al. 1997). It is possible that albumin synthesis is initially depressed after injury, then increases after plasma levels decline. It also is possible that even the initial decline in albumin is mediated by increased degradation or transcapillary escape rather than reduced synthesis. Although plasma fibronectin levels often increase after injury (Petersen et al. 1997), Thompson et al. (1989a) did not observe an increase in fibronectin synthesis in trauma patients, and found only a slight increase in patients with burn injury. However, neither group had significantly elevated fibronectin concentrations. Synthesis rates of other proteins that increase (C-reactive protein, α-1 antichymotrypsin, α-1 acid glycoprotein, α-1 antitrypsin, haptoglobin) or decrease (prealbumin, transferrin, immunogloulins) after injury (Petersen et al. 1997) have not been measured. Total lymphocyte protein synthesis was reported to increase ~40% to 50% the day after surgery (Essen et al. 1996).

The effect of trauma (femur fracture) and surgical stress on protein synthesis in several tissues has been surveyed in rats (Stein et al. 1976; Preedy et al. 1988). Trauma increased synthesis modestly in liver, heart, kidney, and muscle (anterior tibialis). The slight increase in lung protein synthesis was not significant statistically, but the average response was similar to that observed in other tissues. Surgery increased protein synthesis in heart, diaphragm, liver, and spleen, but did not alter protein synthesis in most tissues (lung, skeletal muscle, stomach, small intestine, uterus, kidney, brain, tibia). Note that these studies did not demonstrate the reduction in muscle protein synthesis that appears to be characteristic of tissue injury in humans, so the relevance of these data to the human response to injury is unclear. Burn injury (30% of body surface area) does reduce muscle protein synthesis slightly in rats (Fang et al. 1995).

Tissue injury increases levels of various catabolic hormones, including cortisol, glucagon, TNFα, and IL-6. Several lines of evidence suggest that the hormonal response mediates the protein catabolism. Combined infusion of cortisol, glucagon, and epinephrine for a few hours causes a delayed reduction in muscle protein synthesis (McNurlan et al. 1996), a rapid decline in muscle ribosome and polyribosome concentrations (Hammarqvist et al. 1994), and a delayed increase in albumin synthesis (McNurlan et al. 1996). Infusion of these hormones suppresses lymphocyte protein synthesis, in contrast to the increase in lymphocyte protein synthesis associated with surgical stress (Essen et al. 1996). Infusion of the cytokine TNFa also mimics some of the protein catabolic responses to tissue injury, including increased whole-body proteolysis and increased efflux of amino acids from the limbs (Starnes et al. 1988). IL-6 levels correlate with the increase in phenylalanine concentrations and the increase in urinary 3-methylhistidine excretion after burn injury (De Bandt et al. 1994). In rats, the increase in muscle proteolysis after burn injury is abolished by a glucocorticoid receptor antagonist (Fang et al. 1995). In addition to the increase in levels of catabolic hormones after tissue injury, there is also a reduction in levels of the anabolic hormone IGF-I (Petersen et al. 1997). Increasing the growth hormone and IGF-I levels after surgery or trauma, by exogenous growth hormone administration, can attenuate the catabolic response by reducing protein oxidation and promoting protein synthesis (Hammarqvist et al. 1992; Petersen, Holaday, and Jeevanandam 1994; Carli, Webster, and Halliday 1997). Growth hormone administration has only minor effects on the acute phase protein response of trauma patients (Petersen et al. 1997).

The pain associated with tissue injury may be one of the mediators of the catabolic response, probably via its effect on catabolic hormones (Carli and Halliday 1997). Continuous nociceptive blockade with an epidural anesthetic inhibits the post-surgical decline in muscle protein synthesis and the increase in whole-body proteolysis.

References

Abraham, R.R., Densem, J.W., Davies, P., Davie, M.W.J., and Wynn, V. 1985. The effects of triiodothyronine on energy expenditure, nitrogen balance and rates of weight and fat loss in obese patients during prolonged caloric restriction. *Int. J. Obesity* 9:433–442.

Abumrad, N.N., Robinson, R.P., Gooch, B.R., and Lacy, W.W. 1982. The effect of leucine infusion on substrate flux across the human forearm. *J. Surg. Res.* 32:453–463.

Ackermann, P.G., Toro, G., Kountz, W.B., and Kheim, T. 1954. The effect of sex hormone administration on the calcium and nitrogen balance in elderly women. *J. Gerontol.* 9:450–455.

Ahlborg, G., Felig, P., Hagenfeldt, L., Hendler, R., and Wahren, J. 1974. Substrate turnover during prolonged exercise in man. Splanchnic and leg metabolism of glucose, free fatty acids, and amino acids. *J. Clin. Invest.* 53:1080–1090.

Albert, J.D., Legaspi, A., Horowitz, G.D., Tracey, K.J., Brennan, M.F., and Lowry, S.F. 1986. Extremity amino acid metabolism during starvation and intravenous refeeding in humans. *Am. J. Physiol.* 251:E604–E610.

Alderberth, A., Angeras, U., Jagenburg, R., Lindstedt, G., Stenstrom, G., and Hasselgren, P.-O. 1987. Urinary excretion of 3-methylhistidine and creatinine and plasma concentrations of amino acids in hyperthyroid patients following preoperative treatment with antithyroid drug or β-blocking agent: results from a prospective, randomized study. *Metabolism* 36:637–642.

Aloia, J.F., Vaswani, A., Ma, R., and Flaster, E. 1996. Aging in women—the four-compartment model of body composition. *Metabolism* 45:43–48.

Alvestrand, A., DeFronzo, R.A., Smith, D., and Wahren, J. 1988. Influence of hyperinsulinemia on intracellular amino acid levels and amino acid exchange across splanchnic and leg tissues in uraemia. *Clin. Sci.* 74:155–163.

Amatruda, J.M., Biddle, T.L., Patton, M.L., and Lockwood, D.H. 1983. Vigorous supplementation of a hypocaloric diet prevents cardiac arrhythmias and mineral depletion. *Am. J. Med.* 74:1016–1022.

Ang, B.C., Halliday, D., and Powell-Tuck, J. 1995. Whole-body protein turnover in response to hyperinsulinemia in humans postabsorptively with [^{15}N]glycine as tracer. *Am. J. Clin. Nutr.* 61:1062–1066.

Anker, S.D., Chua, T.P., Ponikowski, P., Harrington, D., Swan, J.W., Kox, W.J. et al. 1997. Hormonal changes and catabolic/anabolic imbalance in chronic heart failure and their importance for cardiac cachexia. *Circulation* 96:526–534.

Antonetti, D.A., Kimball, S.R., Horetsky, R.L., and Jefferson, L.S. 1993. Regulation of rDNA transcription by insulin in primary cultures of rat hepatocytes. *J. Biol. Chem.* 268:25277–25284.

Aoki, T.T., Brennan, M.F., Muller, W.A., Soeldner, J.S., Alpert, J.S., Saltz, S.B. et al. 11976. Amino acid levels across normal forearm muscle and splanchnic bed after a protein meal. *Am. J. Clin. Nutr.* 29:340–350.

Aoki, T.T., Brennan, M.F., Fitzpatrick, G.F., and Knight, D.C. 1981. Leucine meal increases glutamine and total nitrogen release from forearm muscle. *J. Clin. Invest.* 68:1522–1528.

Apfelbaum, M. 1976. The effects of very restrictive high protein diets. *Clin. Endocrinol. Metab.* 5:417–430.

Arends, J., Schafer, G., Schauder, P., Bircher, J., and Bier, D.M. 1995. Comparison of serine and hippurate as precursor equivalents during infusion of [^{15}N]glycine for measurement of fractional synthetic rates of apolipoprotein B of very-low-density lipoprotein. *Metabolism* 44:1253–1258.

Arfvidsson, B., Zachrisson, H., Moller-Loswick, A.-C., Hyltander, A., Sandstrom, R., and Lundholm, K. 1991. Effect of systemic hyperinsulinemia on amino acid flux across human legs in postabsorptive state. *Am. J. Physiol.* 260:E46–E52.

Argiles, J.M., and Lopez-Soriano, F.J. 1996. The ubiquitin-dependent proteolytic pathway in skeletal muscle: its role in pathological states. *Trends in Pharmacol. Sci.* 17:223–226.

Arnold, J., Campbell, I.T., Samuels, T.A., Devlin, J.C., Green, C.J., Hipkin, L.J., MacDonald, I.A., Scrimgeour, C.M., Smith, K., and Rennie, M.J. 1993. Increased whole body protein breakdown predominates over increased whole body protein synthesis in multiple organ failure. *Clin. Sci.* 84:655–661.

Arslanian, S., and Suprasongsin, C. 1997. Testosterone treatment in adolescents with delayed puberty: changes in body composition, protein, fat, and glucose metabolism. *J. Clin. Endocrinol. Metab.* 82:3213–3220.

Arslanian, S.A., and Kalhan, S.C. 1996. Protein turnover during puberty in normal children. *Am. J. Clin. Nutr.* 270:E79–E84.

Atchley, D.W., Loeb, R.F., Richards, D.W.J., Benedict, E.M., and Driscoll, M.E. 1933. On diabetic acidosis. A detailed study of electrolyte balances following the withdrawal and reestablishment of insulin therapy. *J. Clin. Invest.* 12:297–326.

Attaix, D., Aurousseau, E., Manghebati, A., and Arnal, M. 1988. Contribution of liver, skin, and skeletal muscle to whole-body protein synthesis in the young lamb. *Br. J. Nutr.* 60:77–84.

Auclair, D., Garrel, D.R., Zerouala, A.C., and Ferland, L.H. 1997. Activation of the ubiquitin pathway in rat skeletal muscle by catabolic doses of glucocorticoids. *Am. J. Physiol.* 272:C1007–C1016.

Babij, P., Matthews, S.M., and Rennie, M.J. 1983. Changes in blood ammonia, lactate and amino acids in relation to workload during bicycle ergometer exercise in man. *Eur. J. Appl. Physiol.* 50:405–411.

Backeljauw, P.F., Underwood, L.E., and The GHIS Collaborative Group. 1996. Prolonged treatment with recombinant insulin-like growth factor-I in children with growth hormone insensitivity syndrome—a clinical research center study. *J. Clin. Endocrinol. Metab.* 81:3312–3317.

Baillie, A.G.S., and Garlick, P.J. 1991a. Responses of protein synthesis in different skeletal muscles to fasting and insulin in rats. *Am. J. Physiol.* 260:E891–E896.

Baillie, A.G.S., and Garlick, P.J. 1991b. Attenuated responses of muscle protein synthesis to fasting and insulin in adult female rats. *Am. J. Physiol.* 262:E1–E5.

Balagopal, P., Rooyackers, O.E., Adey, D.B., Ades, P.A., and Nair, K.S. 1997. Effects of aging on in vivo synthesis of skeletal muscle myosin heavy-chain and sarcoplasmic protein in humans. *Am. J. Physiol.* 273:E790–E800.

Balagopal, P., Ljungqvist, O.H., and Nair, K.S. 1997. Skeletal muscle myosin heavy-chain synthesis rate in healthy humans. *Am. J. Physiol.* 272:E45–E50.

Ballard, F.J., Tomas, F.M., and Stern, L.M. 1979. Increased turnover of muscle contractile proteins in Duchenne muscular dystrophy as assessed by 3-methylhistidine and creatinine excretion. *Clin. Sci.* 56:347–352.

Ballard, F.J., and Tomas, F.M. 1983. 3-Methylhistidine as a measure of skeletal muscle protein breakdown in human subjects: the case for its continued use. *Clin. Sci.* 65: 209–215.

Ballmer, P.E., McNurlan, M.A., Milne, E., Heys, S.D., Buchan, V., Calder, A.G. et al. 1990. Measurement of albumin synthesis in humans: a new approach employing stable isotopes. *Am. J. Physiol.* 259:E797–E803.

Ballmer, P.E., Walshe, D., McNurlan, M.A., Watson, H., Brunt, P.W., and Garlick, P.J. 1993. Albumin synthesis rates in cirrhosis: correlation with Child-Turcotte classification. *Hepatology* 18:292–297.

Ballmer, P.E., McNurlan, M.A., Hulter, H.N., Anderson, S.E., Garlick, P.J., and Krapf, R. 1995. Chronic metabolic acidosis decreases albumin synthesis and induces negative nitrogen balance in humans. *J. Clin. Invest.* 95:39–45.

Ballmer, P.E., Reichen, J., McNurlan, M.A., Sterchi, A.-B., Anderson, S.E., and Garlick, P.J. 1996. Albumin but not fibrinogen synthesis correlates with galactose elimination capacity in patients with cirrhosis of the liver. *Hepatology* 24:53–59.

Bamman, M.M., Clarke, M.S.F., Feeback, D.L., Talmadge, R.J., Stevens, B.R., Lieberman, S.A. et al. 1998. Impact of resistance exercise during bed rest on skeletal muscle sarcopenia and myosin isoform distribution. *J. Appl. Physiol.* 84:157–163.

Baracos, V., Wilson, E.J., and Goldberg, A.L. 1984. Effects of temperature on protein turnover in isolated rat skeletal muscle. *Am. J. Physiol.* 246:C125–C130.

Barinaga, M. 1998. Death by dozens of cuts. *Science* 280:32–34.

Bark, T.H., McNurlan, M.A., Lang, C.H., and Garlick, P.J. 1998. Increased protein synthesis after acute IGF-I or insulin infusion is localized to muscle in mice. *Am. J. Physiol.* 275:E118–E123.

Barle, H., Nyberg, B., Essen, P., Andersson, K., McNurlan, M.A., Wernerman, J. et al. 1997. The synthesis rates of total liver protein and plasma albumin determined simultaneously *in vivo* in humans. *Hepatology* 25:154–158.

Barrett, E.J., Revkin, J.H., Young, L.H., Zaret, B.L., Jacob, R., and Gelfand, R.A. 1987. An isotopic method for measurement of muscle protein synthesis and degradation *in vivo*. *Biochem. J.* 245:223–228.

Barrett, E.J., Jahn, L.A., Oliveras, D.M., and Fryburg, D.A. 1995. Chloroquine does not exert insulin-like actions on human forearm muscle metabolism. *Am. J. Physiol.* 268:E820–E824.

Barua, J.M., Wilson, E., Downie, S., Weryk, B., Cuschieri, A., and Rennie, M.J. 1992. The effect of alanyl-glutamine peptide supplementation on muscle protein synthesis in post-surgical patients receiving glutamine-free amino acids intravenously. *Proc. Nutr. Soc.* 51:104A(Abstract)

Bates, P.C., Grimble, G.K., Sparrow, M.P., and Millward, D.J. 1983. Myofibrillar protein turnover. Synthesis of protein-bound 3-methylhistidine, actin, myosin heavy chain and aldolase in rat skeletal muscle in the fed and starved states. *Biochem. J.* 214:593–605.

Battezzati, A., Simonson, D.C., Luzi, L., and Matthews, D.E. 1998. Glucagon increases glutamine uptake without affecting glutamine release in humans. *Metabolism* 47: 713–723.

Baumann, P.Q., Stirewalt, W.S., O'Rourke, B.D., Howard, D., and Nair, K.S. 1994. Precursor pools of protein synthesis: a stable isotope study in a swine model. *Am. J. Physiol.* 267:E203–E209.

Beaufrere, B., Horber, F.F., Schwenk, W.F., Marsh, H.M., Matthews, D., Gerich, J.E. et al. 1989. Glucocorticosteroids increase leucine oxidation and impair leucine balance in humans. *Am. J. Physiol.* 257:E712–E721.

Beaufrere, B., Chassard, D., Broussolle, C., Riou, J.P., and Beylot, M. 1992. Effects of D-β-hydroxybutyrate and long- and medium-chain triglycerides on leucine metabolism in humans. *Am. J. Physiol.* 262:E268–E274.

Beckett, P.R., Jahoor, F., and Copeland, K.C. 1997. The efficiency of dietary protein utilization is increased during puberty. *J. Clin. Endocrinol. Metab.* 82:2445–2449.

Bengtsson, B.-A., Eden, S., Lonn, L., Kvist, H., Stokland, A., Lindstedt, G. et al. 1993. Treatment of adults with growth hormone (GH) deficiency with recombinant human GH. *J. Clin. Endocrinol. Metab.* 76:309–317.

Bennegard, K., Lindmark, L., Eden, E., Svaninger, G., and Lundholm, K. 1984. Flux of amino acids across the leg in weight-losing cancer patients. *Cancer Res.* 44:386–393.

Bennet, W.M., Connacher, A.A., Scrimgeour, C.M., Smith, K., and Rennie, M.J. 1989. Increase in anterior tibialis muscle protein synthesis in healthy man during mixed amino acid infusion: studies of incorporation of [1-^{13}C]leucine. *Clin. Sci.* 76:447–454.

Bennet, W.M., Connacher, A.A., Scrimgeour, C.M., Jung, R.T., and Rennie, M.J. 1990a. Euglycemic hyperinsulinemia augments amino acid uptake by human leg tissues during hyperaminoacidemia. *Am. J. Physiol.* 259:E185–E194.

Bennet, W.M., Connacher, A.A., Scrimgeour, C.M., and Rennie, M.J. 1990b. The effect of amino acid infusion on leg protein turnover assessed by L-[^{15}N]phenylalanine and L-[1-^{13}C]leucine exchange. *Eur. J. Clin. Invest.* 20:41–50.

Bennet, W.M., Connacher, A.A., Smith, K., Jung, R.T., and Rennie, M.J. 1990c. Inability to stimulate skeletal muscle or whole body protein synthesis in Type 1 (insulin-dependent) diabetic patients by insulin-plus-glucose during amino acid infusion: studies of incorporation and turnover of tracer L-[1-^{13}C]leucine. *Diabetologia* 33:43–51.

Bennet, W.M., Connacher, A.A., Jung, R.T., Stehle, P., and Rennie, M.J. 1991. Effects of insulin and amino acids on leg protein turnover in IDDM patients. *Diabetes* 40:499–508.

Bennet, W.M., and Haymond, M.W. 1992. Growth hormone and lean tissue catabolism during long-term glucocorticoid treatment. *Clin. Endocrinol.* 36:161–164.

Bennet, W.M., O'Keefe, S.J.D., and Haymond, M.W. 1993. Comparison of precursor pools with leucine, α-ketoisocaproate, and phenylalanine tracers used to measure splanchnic protein synthesis in man. *Metabolism* 42:691–695.

Benyon, R.J., and Bond, J.S. 1986. Catabolism of intracellular protein: molecular aspects. *Am. J. Physiol.* 251:C141–C152.

Berg, T., Gjoen, T., and Bakke, O. 1995. Physiological functions of endosomal proteolysis. *Biochem. J.* 307:313–326.

Berkelhammer, C.H., Baker, J.P., Leiter, L.A., Uldall, P.R., Whittall, R., Slater, A. et al. 1987. Whole-body protein turnover in adult hemodialysis patients as measured by ^{13}C-leucine. *Am. J. Clin. Nutr.* 46:778–783.

Berneis, K., Ninnis, R., Girard, J., Frey, B.M., and Keller, U. 1997. Effects of insulin-like growth factor I combined with growth hormone on glucocorticoid-induced whole-body protein catabolism in man. *J. Clin. Endocrinol. Metab.* 82:2528–2534.

Berneis, K., Ninnis, R., and Keller, U. 1997. Ethanol exerts acute protein-sparing effects during postabsorptive but not during anabolic conditions in man. *Metabolism* 46:750–755.

Bettany, G.E.A., Ang, B.C., Georgiannos, S.N., Halliday, D., and Powell-Tuck, J. 1996. Bed rest decreases whole-body protein turnover in post-absorptive man. *Clin. Sci.* 90:73–75.

Beylot, M., Chassard, D., Chambrier, C., Guiraud, M., Odeon, M., Beaufrere, B. et al. 1994. Metabolic effects of a D-β-hydroxybutyrate infusion in septic patients: inhibition of lipolysis and glucose production but not leucine oxidation. *Critical Care Med.* 22:1091–1098.

Bhasin, S., Storer, T.W., Berman, N., Callegari, C., Clevenger, B., Phillips, J. et al. 1996. The effects of supraphysiologic doses of testosterone on muscle size and strength in normal men. *N. Engl. J. Med.* 335:1–7.

Bhasin, S., Storer, T.W., Berman, N., Yarasheski, K.E., Clevenger, B., Phillips, J. et al. 1997. Testosterone replacement increases fat-free mass and muscle size in hypogonadal men. *J. Clin. Endocrinol. Metab.* 82:407–413.

Bianda, T., Hussain, M.A., Glatz, Y., Bouillon, R., Froesch, E.R., and Schmid, C. 1997. Effects of short-term insulin-like growth factor-I or growth hormone treatment on bone turnover, renal phosphate reabsorption and 1,25 dihydroxyvitamin D_3 production in healthy man. *J. Intern. Med.* 241:143–150.

Bianda, T., Glatz, Y., Bouillon, R., Froesch, E.R., and Schmid, C. 1998. Effects of short-term insulin-like growth factor-I (IGF-I) or growth hormone (GH) treatment on bone metabolism and on production of 1,25-dihydroxycholecalciferol in GH-deficient adults. *J. Clin. Endocrinol. Metab.* 98:81–87.

Bilmazes, C., Kien, C.L., Rohrbaugh, D.K., Uauy, R., Burke, J.F., Munro, H.N. et al. 1978a. Quantitative contribution by skeletal muscle to elevated rates of whole-body protein breakdown in burned children as measured by N^{τ}-methylhistidine output. *Metabolism* 27:671–676.

Bilmazes, C., Uauy, R., Haverberg, L.N., Munro, H.N., and Young, V.R. 1978b. Muscle protein breakdown rates in humans based on N^{τ}-methylhistidine (3-methylhistidine) content of mixed proteins in skeletal muscle and urinary output of N^{τ}-methylhistidine. *Metabolism* 27:525–530.

Binnerts, A., Swart, G.R., Wilson, J.H.P., Hoogerbrugge, N., Pols, H.A.P., Birkenhager, J.C. et al. 1992. The effect of growth hormone administration in growth hormone deficient adults on bone, protein, carbohydrate and lipid homeostasis, as well as on body composition. *Clin. Endocrinol.* 37:79–87.

Biolo, G., Tessari, P., Inchiostro, S., Bruttomesso, D., Fongher, C., Sabadin, L. et al. 1992a. Leucine and phenylalanine kinetics during mixed meal ingestion: a multiple tracer approach. *Am. J. Physiol.* 262:E455–E463.

Biolo, G., Tessari, P., Inchiostro, S., Bruttomesso, D., Sabadin, L., Fongher, C. et al. 1992b. Fasting and postmeal phenylalanine metabolism in mild type 2 diabetes. *Am. J. Physiol.* 263:E877–E883.

Biolo, G., Gastaldelli, A., Zhang, X.-J., and Wolfe, R.R. 1994. Protein synthesis and breakdown in skin and muscle: a leg model of amino acid kinetics. *Am. J. Physiol.* 267:E467–E474.

Biolo, G., Fleming, R.Y.D., and Wolfe, R.R. 1995. Physiologic hyperinsulinemia stimulates protein synthesis and enhances transport of selected amino acids in human skeletal muscle. *J. Clin. Invest.* 95:811–819.

Biolo, G., Inchiostro, S., Tiengo, A., and Tessari, P. 1995a. Regulation of postprandial whole-body proteolysis in insulin-deprived IDDM. *Diabetes* 44:203–209.

Biolo, G., Maggi, S.P., Williams, B.D., Tipton, K.D., and Wolfe, R.R. 1995b. Increased rates of muscle protein turnover and amino acid transport after resistance exercise in humans. *Am. J. Physiol.* 268:E514–E520.

Biolo, G., Tipton, K.D., Klein, S., and Wolfe, R.R. 1997. An abundant supply of amino acids enhances the metabolic effect of exercise on muscle protein. *Am. J. Physiol.* 273:E122–E129.

Bird, J.L.E., and Tyler, J.A. 1994. Dexamethasone potentiates the stimulatory effect of insulin-like growth factor-I on collagen production in cultured human fibroblasts. *J. Endocrinol.* 142:571–579.

Birkhahn, R.H., Long, C.L., Fitkin, D., Geiger, J.W., and Blakemore, W.S. 1980. Effects of major skeletal trauma on whole body protein turnover in man measured by L-[1,^{14}C]-leucine. *Surgery* 88:294–300.

Bistrian, B.R., Sherman, M., and Young, V. 1981. The mechanisms of nitrogen sparing in fasting supplemented by protein and carbohydrate. *J. Clin. Endocrinol. Metab.* 53:874–878.

Bleiberg-Daniel, F., Lamri, Y., Feldmann, G., and Lardeux, B. 1994. Glucagon administration *in vivo* stimulates hepatic RNA and protein breakdown in fed and fasted rats. *Biochem. J.* 299:645–649.

Blomqvist, B.I., Hammarqvist, F., von der Decken, A., and Wernerman, J. 1995. Glutamine and α-ketoglutarate prevent the decrease in muscle free glutamine concentration and influence protein synthesis after total hip replacement. *Metabolism* 44:1215–1222.

Boden, G., Tappy, L., Jadali, F., Hoeldtke, R.D., Rezvani, I., and Owen, O.E. 1990. Role of glucagon in disposal of an amino acid load. *Am. J. Physiol.* 259:E225–E232.

Bohley, P., and Seglen, P.O. 1992. Proteases and proteolysis in the lysosome. *Experientia* 48:151–157.

Boirie, Y., Gachon, P., Corny, S., Fauquant, J., Maubois, J.-L., and Beaufrere, B. 1996. Acute postprandial changes in leucine metabolism as assessed with an intrinsically labeled milk protein. *Am. J. Physiol.* 271:E1083–E1091.

Boirie, Y., Gachon, P., and Beaufrere, B. 1997. Splanchnic and whole-body leucine kinetics in young and elderly men. *Am. J. Clin. Nutr.* 65:489–495.

Bollerslev, J., Moller, J., Thomas, S., Djoseland, O., and Christiansen, J.S. 1996. Dose-dependent effects of recombinant human growth hormone on biochemical markers of bone and collagen metabolism in adult growth hormone deficiency. *Eur. J. Endocrinol.* 135:666–671.

Bonadonna, R.C., Saccomani, M.P., Cobelli, C., and DeFronzo, R.A. 1993. Effect of insulin on system A amino acid transport in human skeletal muscle. *J. Clin. Invest.* 91:514–521.

Bonner, A.B., Swann, M.E., Marway, J.S., Heap, L.C., and Preedy, V.R. 1995. Lysosomal and nonlysosomal protease activities of the brain in response to ethanol feeding. *Alcohol* 12:505–509.

Bonner, A.B., Marway, J.S., Swann, M.E., and Preedy, V.R. 1996. Brain nucleic acid composition and fractional rates of protein synthesis in response to chronic ethanol feeding: comparison with skeletal muscle. *Alcohol* 13:581–587.

Borzotta, A.P., Clague, M.B., and Johnston, I.D.A. 1987. The effects of gastrointestinal malignancy on whole body protein metabolism. *J. Surg. Res.* 43:505–512.

Bouteloup-Demange, C., Boirie, Y., Dechelotte, P., Gachon, P., and Beaufrere, B. 1998. Gut mucosal protein synthesis in fed and fasted humans. *Am. J. Physiol.* 274:E541–E546.

Bowes, S.B., Benn, J.J., Scobie, I.N., Umpleby, A.M., Lowy, C., and Sonksen, P.H. 1993. Leucine metabolism in patients with Cushing's syndrome before and after successful treatment. *Clin. Endocrinol.* 39:591–598.

Bowes, S.B., Jackson, N.C., Papachristodoulou, D., Umpleby, A.M., and Sonksen, P.H. 1996. Effect of corticosterone on protein degradation in isolated rat soleus and extensor digitorum longus muscles. *J. Endocrinol.* 148:501–507.

Bowes, S.B., Umpleby, A.M., Cummings, M.H., Jackson, N.C., Carroll, P.V., Lowy, C. et al. 1997. The effect of recombinant human growth hormone on glucose and leucine metabolism in Cushing's syndrome. *J. Clin. Endocrinol. Metab.* 82:243–246.

Bresson, J.L., Bader, B., Rocchiccioli, F., Mariotti, A., Ricour, C., Sachs, C. et al. 1991. Protein-metabolism kinetics and energy-substrate utilization in infants fed parenteral solutions with different glucose-fat ratios. *Am. J. Clin. Nutr.* 54:370–376.

Breuille, D., Farge, M.C., Rose, F., Arnal, M., Attaix, D., and Obled, C. 1993. Pentoxifylline decreases body weight loss and muscle protein wasting characteristic of sepsis. *Am. J. Physiol.* 265:E660–E666.

Brillon, D.J., Zheng, B., Campbell, R.G., and Matthews, D.E. 1995. Effect of cortisol on energy expenditure and amino acid metabolism in humans. *Am. J. Physiol.* 268:E501–E513.

Brodsky, I.G., Balagopal, P., and Nair, K.S. 1996. Effects of testosterone replacement on muscle mass and muscle protein synthesis in hypogonadal men—a Clinincal Research Center study. *J. Clin. Endocrinol. Metab.* 81:3469–3475.

Bruce, A.C., McNurlan, M.A., McHardy, K.C., Broom, J., Buchanan, K.D., Calder, A.G. et al. 1990. Nutrient oxidation patterns and protein metabolism in lean and obese subjects. *Int. J. Obesity* 14:631–646.

Bunout, D., Petermann, M., Ugarte, G., Barrera, G., and Iturriaga, H. 1987. Nitrogen economy in alcoholic patients without liver disease. *Metabolism* 36:651–653.

Burman, P., Johansson, A.G., Siegbahn, A., Vessby, B., and Karlsson, F.A. 1997. Growth hormone (GH)-deficient men are more responsive to GH replacement therapy than women. *J. Clin. Endocrinol. Metab.* 82:550–555.

Buse, M.G., and Reid, S.S. 1975. Leucine. A possible regulator of protein turnover in muscle. *J. Clin. Invest.* 56:1250–1261.

Buse, M.G., and Weigand, D.A. 1977. Studies concerning the specificity of the effect of leucine on the turnover of proteins in muscles of control and diabetic rats. *Biochim. Biophys. Acta* 475:81–89.

Butterfield, G.E., and Calloway, D.H. 1984. Physical activity improves protein utilization in young men. *Br. J. Nutr.* 51:171–184.

Butterfield, G.E. 1987. Whole-body protein utilization in humans. *Med. Sci. Sports Exerc.* 19:S157–S165.

Butterfield, G.E., Thompson, J., Rennie, M.J., Marcus, R., Hintz, R.L., and Hoffman, A.R. 1997. Effect of rhGH and rhIGF-I treatment on protein utilization in elderly women. *Am. J. Physiol.* 272:E94–E99.

Caballero, B., and Wurtman, R.J. 1991. Differential effects of insulin resistance on leucine and glucose kinetics in obesity. *Metabolism* 40:51–58.

Calles-Escandon, J., Cunningham, J.J., Snyder, P., Jacob, R., Huszar, G., Loke, J. et al. 1984. Influence of exercise on urea, creatinine, and 3-methylhistidine excretion in normal human subjects. *Am. J. Physiol.* 246:E334–E338.

Campbell, I.T. 1998. Nutrition support in patients with multiple organ failure. *Curr. Opin. Nutr. Metab. Care* 1:211–216.

Campbell, W.W., Crim, M.C., Dallal, G.E., Young, V.R., and Evans, W.J. 1994a. Increased protein requirements in elderly people: new data and retrospective reassessments. *Am. J. Clin. Nutr.* 60:501–509.

Campbell, W.W., Crim, M.C., Young, V.R., and Evans, W.J. 1994b. Increased energy requirements and changes in body composition with resistance training in older adults. *Am. J. Clin. Nutr.* 60:167–175.

Campbell, W.W., Crim, M.C., Young, V.R., Joseph, L.J., and Evans, W.J. 1995. Effects of resistance training and dietary protein intake on protein metabolism in older adults. *Am. J. Physiol.* 268:E1143–E1153.

Cannon, J.G., Meydani, S.N., Fielding, R.A., Fiatarone, M.A., Meydani, M., Farhangmehr, M. et al. 1991. Acute phase response in exercise. II. Associations between vitamin E, cytokines, and muscle proteolysis. *Am. J. Physiol.* 260:R1235–R1240.

Caperna, T.J., Gavelek, D., and Vossoughi, J. 1994. Somatotropin alters collagen metabolism in growing pigs. *J. Nutr.* 124:770–778.

Carli, F., Webster, J., Ramachandra, V., Pearson, M., Read, M., Ford, G.C. et al. 1990. Aspects of protein metabolism after elective surgery in patients receiving constant nutritional support. *Clin. Sci.* 78:621–628.

Carli, F., and Halliday, D. 1997. Continuous epidural blockade arrests the postoperative decrease in muscle protein fractional synthetic rate in surgical patients. *Anesthesiology* 86:1033–1040.

Carli, F., Webster, J.D., and Halliday, D. 1997. Growth hormone modulates amino acid oxidation in the surgical patient: leucine kinetics during the fasted and fed state using moderate nitrogenous and caloric diet and recombinant human growth hormone. *Metabolism* 46:23–28.

Carlson, M.G., and Campbell, P.J. 1993. Intensive insulin therapy and weight gain in IDDM. *Diabetes* 42:1700–1707.

Carlson, M.G., Snead, W.L., and Campbell, P.J. 1994. Fuel and energy metabolism in fasting humans. *Am. J. Clin. Nutr.* 60:29–36.

Carraro, F., Hartl, W.H., Stuart, C.A., Layman, D.K., Jahoor, F., and Wolfe, R.R. 1990a. Whole body and plasma protein synthesis in exercise and recovery in human subjects. *Am. J. Physiol.* 258:E821–E831.

Carraro, F., Stuart, C.A., Hartl, W.H., Rosenblatt, J., and Wolfe, R.R. 1990b. Effect of exercise and recovery on muscle protein synthesis in human subjects. *Am. J. Physiol.* 259:E470–E476.

Carraro, F., Kimbrough, T.D., and Wolfe, R.R. 1993. Urea kinetics in humans at two levels of exercise intensity. *J. Appl. Physiol.* 75:1180–1185.

Carraro, F., Rosenblatt, J., and Wolfe, R.R. 1991. Isotopic determination of fibronectin synthesis in humans. *Metabolism* 40:553–561.

Castaneda, C., Dolnikowski, G.G., Dallal, G.E., Evans, W.J., and Crim, M.C. 1995. Protein turnover and energy metabolism of elderly women fed a low-protein diet. *Am. J. Clin. Nutr.* 62:40–48.

Castellino, P., Luzi, L., Simonson, D.C., Haymond, M., and DeFronzo, R.A. 1987. Effect of insulin and plasma amino acid concentrations on leucine metabolism in man. Role of substrate availability on estimates of whole body protein synthesis. *J. Clin. Invest.* 80:1784–1793.

Castellino, P., Luzi, L., Del Prato, S., and DeFronzo, R.A. 1990. Dissociation of the effects of epinephrine and insulin on glucose and protein metabolism. *Am. J. Physiol.* 258:E117–E125.

Castellino, P., Solini, A., Luzi, L., Barr, J.G., Smith, D.J., Petrides, A. et al. 1992. Glucose and amino acid metabolism in chronic renal failure: effect of insulin and amino acids. *Am. J. Physiol.* 262:F168–F176.

Cayol, M., Tauveron, I., Rambourdin, F., Prugnaud, J., Gachon, P., Thieblot, P. et al. 1995. Whole-body protein turnover and hepatic protein synthesis are increased by vaccination in man. *Clin. Sci.* 89:389–396.

Cayol, M., Boirie, Y., Prugnaud, J., Gachon, P., Beaufrere, B., and Obled, C. 1996. Precursor pool for hepatic protein synthesis in humans: effects of tracer route infusion and dietary proteins. *Am. J. Physiol.* 270:E980–E987.

Cayol, M., Boirie, Y., Rambourdin, F., Prugnaud, J., Gachon, P., Beaufrere, B. et al. 1997. Influence of protein intake on whole body and splanchnic leucine kinetics in humans. *Am. J. Physiol.* 272:E584–E591.

Cernacek, P., Spustova, V., and Dzurik, R. 1982. Inhibitors of protein synthesis in uremic serum and urine: partial purification and relationship to amino acid transport. *Biochem. Med.* 27:305–316.

Charlton, M.R. 1996. Protein metabolism and liver disease. *Balliere's Clin. Endocrinol. Metab.* 10:617–635.

Charlton, M.R., Adey, D.B., and Nair, K.S. 1996. Evidence for a catabolic role of glucagon during an amino acid load. *J. Clin. Invest.* 98:90–99.

Charlton, M.R., Balagopal, P., and Nair, K.S. 1997. Skeletal muscle myosin heavy chain synthesis in type 1 diabetes. *Diabetes* 46:1336–1340.

Charters, Y., and Grimble, R.F. 1989. Effect of recombinant human tumor necrosis factor α on protein synthesis in liver, skeletal muscle and skin of rats. *Biochem. J.* 258: 493–497.

Cheng, A.H.R., Gomez, A., Bergan, J.G., Lee, T.-C., Monckeberg, F., and Chichester, C.O. 1978. Comparative nitrogen balance study between young and aged adults using three levels of protein intake from a combination wheat-soy-milk mixture. *Am. J. Clin. Nutr.* 31:12–22.

Cheng, K.N., Dworzak, F., Ford, G.C., Rennie, M.J., and Halliday, D. 1985. Direct determination of leucine metabolism and protein breakdown in humans using L-[1-^{13}C, ^{15}N]-leucine and the forearm model. *Eur. J. Clin. Invest.* 15:349–354.

Cheng, K.N., Pacy, P.J., Dworzak, F., Ford, G.C., and Halliday, D. 1987. Influence of fasting on leucine and muscle protein metabolism across the human forearm determined using L-[1-^{13}C,^{15}N]leucine as the tracer. *Clin. Sci.* 73:241–246.

Cherel, Y., Attaix, D., Rosolowska-Huszcz, D., Belkhou, R., Robin, J.-P., Arnal, M. et al. 1991. Whole-body and tissue protein synthesis during brief and prolonged fasting in the rat. *Clin. Sci.* 81:611–619.

Chesley, A., MacDougall, J.D., Tarnopolsky, M.A., Atkinson, S.A., and Smith, K. 1992. Changes in human muscle protein synthesis after resistance exercise. *J. Appl. Physiol.* 73:1383–1388.

Chinkes, D., Rosenblatt, J., and Wolfe, R.R. 1993. Assessment of the mathematical issues involved in measuring the fractional synthesis rate of protein using the flooding dose technique. *Clin. Sci.* 84:177–183.

Chinkes, D., Klein, S., Zhang, X.-J., and Wolfe, R.R. 1996. Infusion of labeled KIC is more accurate than labeled leucine to determine human muscle protein synthesis. *Am. J. Physiol.* 270:E67–E71.

Clague, M.B., Keir, M.J., Wright, P.D., and Johnston, I.D.A. 1983. The effects of nutrition and trauma on whole-body protein metabolism in man. *Clin. Sci.* 65:165–175.

Clemmons, D.R., Snyder, D.K., Williams, R., and Underwood, L.E. 1987. Growth hormone administration conserves lean body mass during dietary restriction in obese subjects. *J. Clin. Endocrinol. Metab.* 64:878–883.

Clemmons, D.R., Smith-Banks, A., and Underwood, L.E. 1992. Reversal of diet-induced catabolism by infusion of recombinant insulin-like growth factor-I in humans. *J. Clin. Endocrinol. Metab.* 75:234–238.

Clowes, G.H.A., Heideman, M., Lindberg, B., Randall, H.T., Hirsch, E.F., Cha, C.-J. et al. 1980. Effects of parenteral alimentation on amino acid metabolism in septic patients. *Surgery* 88:531–543.

Clowes, G.H.A., George, B.C., Villee, C.A., and Saravis, C.A. 1983. Muscle proteolysis induced by a circulating peptide in patients with sepsis or trauma. *N. Engl. J. Med.* 308:545–552.

Clowes, G.H.A., Hirsch, E., George, B.C., Bigatello, L.M., Mazuski, J.E., and Villee, C.A. 1985. Survival from sepsis. The significance of altered protein metabolism regulated by proteolysis inducing factor, the circulating cleavage product of interleukin-1. *Ann. Surg.* 202:446–456.

Co, E., Chari, G., McCulloch, K., and Vidyasagar, D. 1993. Dexamethasone treatment suppresses collagen synthesis in infants with bronchopulmonary dysplasia. *Pediatr. Pulmonol.* 16:36–40.

Cobelli, C., Saccomani, M.P., Tessari, P., Biolo, G., Luzi, L., and Matthews, D.E. 1991. Compartmental model of leucine kinetics in humans. *Am. J. Physiol.* 261:E539–E550.

Cohn, J.S., Wagner, D.A., Cohn, S.D., Millar, J.S., and Schaefer, E.J. 1990. Measurement of very low density and low density lipoprotein apolipoprotein (apo) B-100 and high density lipoprotein apo A-I production in human subjects using deuterated leucine. Effect of fasting and feeding. *J. Clin. Invest.* 85:804–811.

Conley, S.B., Rose, G.M., Robson, A.M., and Bier, D.M. 1980. Effects of dietary intake and hemodialysis on protein turnover in uremic children. *Kidney Int.* 17:837–846.

Coomes, M.W. 1997. Amino acid metabolism. In *Textbook of Biochemistry with Clinical Correlations,* ed. T.M. Devlin, pp. 445–486. New York: Wiley-Liss.

Cooney, R., Owens, E., Jurasinski, C., Gray, K., Vannice, J., and Vary, T. 1994. Interleukin-1 receptor antagonist prevents sepsis-induced inhibition of protein synthesis. *Am. J. Physiol.* 267:E636–E641.

Cooney, R.N., Owens, E., Slaymaker, D., and Vary, T.C. 1996. Prevention of skeletal muscle catabolism in sepsis does not impair visceral protein metabolism. *Am. J. Physiol.* 270:E621–E626.

Cooney, R.N., Kimball, S.R., and Vary, T.C. 1997. Regulation of skeletal muscle protein turnover during sepsis: mechanisms and mediators. *Shock* 7:1–16.

Cooper, B.G., Walker, M., Daley, S.E., Olufemi, O.S., Bartlett, K., Halliday, D. et al. 1992. Whole body protein metabolism in late pregnancy and the effects of insulin. *Diabetologia* 35(Suppl 2):A177(Abstract)

Copeland, K.C., and Nair, K.S. 1994. Acute growth hormone effects on amino acid and lipid metabolism. *J. Clin. Endocrinol. Metab.* 78:1040–1047.

Coppack, S.W., Persson, M., and Miles, J.M. 1996. Phenylalanine kinetics in human adipose tissue. *J. Clin. Invest.* 98:692–697.

Couet, C., Fukagawa, N.K., Matthews, D.E., Bier, D.M., and Young, V.R. 1990. Plasma amino acid kinetics during acute states of glucagon deficiency and excess in heallthy adults. *Am. J. Physiol.* 258:E78–E85.

Coux, O., Tanaka, K., and Goldberg, A.L. 1996. Structure and functions of the 20S and 26S proteasomes. *Annu. Rev. Biochem.* 65:801–847.

Crane, C.W., Picou, D., Smith, R., and Waterlow, J.C. 1977. Protein turnover in patients before and after elective orthopaedic operations. *Br. J. Surg.* 64:129–133.

Crispell, K.R., Parson, W., Hollifield, G., and Brent, S. 1956. A study of the rate of protein synthesis before and during the administration of L-triiodothyronine to patients with myxedema and healthy volunteers using N-15 glycine. *J. Clin. Invest.* 35:164–169.

Cryer, D.R., Matsushima, T., Marsh, J.B., Yudkoff, M., Coates, P.M., and Cortner, J.A. 1986. Direct measurement of apolipoprotein B synthesis in human very low density lipoprotein using stable isotopes and mass spectrometry. *J. Lipid Res.* 27:508–516.

Cuneo, R.C., Salomon, F., Wiles, C.M., Hesp, R., and Sonksen, P.H. 1991. Growth hormone treatment in growth hormone-deficient adults. I. Effects on muscle mass and strength. *J. Appl. Physiol.* 70:688–694.

D'Avis, P.Y., Frazier, C.R., Shapiro, J.R., and Fedarko, N.S. 1997. Age-related changes in effects of insulin-like growth factor I on human osteoblast-like cells. *Biochem. J.* 324:753–760.

Daley, S.E., Pearson, A.D.J., Craft, A.W., Kerhahan, J., Wyllie, R.A., Price, L. et al. 1996. Whole body protein metabolism in children with cancer. *Arch. Dis. Child.* 75:273–281.

Darmaun, D., Matthews, D.E., and Bier, D.M. 1988. Physiological hypercortisolemia increases proteolysis, glutamine, and alanine production. *Am. J. Physiol.* 255:E366–E373.

Darmaun, D., Welch, S., Rini, A., Sager, B.K., Altomare, A., and Haymond, M.W. 1998. Phenylbutyrate-induced glutamine depletion in humans: effect on leucine metabolism. *Am. J. Physiol.* 274:E801–E807.

Davies, H.J.A., Baird, I.M., Fowler, J., Mills, I.H., Baillie, J.E., Rattan, S. et al. 1989. Metabolic response to low- and very-low-calorie diets. *Am. J. Clin. Nutr.* 49:745–751.

Davis, E.O., and Jenner, P.J. 1995. Protein splicing—the lengths some proteins will go. *Antonie van Leeuwenhoek* 67:131–137.

Davis, S.R., Barry, T.N., and Hughson, G.A. 1981. Protein synthesis in tissues of growing lambs. *Br. J. Nutr.* 46:409–419.

De Bandt, J.P., Chollet-Martin, S., Hernvann, A., Lioret, N., Du Roure, L.D., Lim, S.-K. et al. 1994. Cytokine response to burn injury: relationship with protein metabolism. *J. Trauma* 36:624–628.

de Benoist, B., Jackson, A.A., Hall, J.S.E., and Persaud, C. 1985. Whole-body protein turnover in Jamaican women during normal pregnancy. *Hum. Nutr. Clin. Nutr.* 39C:167–179.

De Blaauw, I., Eggermont, A.M.M., Deutz, N.E.P., De Vries, M., Buurman, W.A., and Von Meyenfeldt, M.F. 1997. TNF-α has no direct *in vivo* metabolic effect on human muscle. *Int. J. Cancer* 71:148–154.

De Boer, H., Blok, G.-J., and Van der Veen, E.A. 1995. Clinical aspects of growth hormone deficiency in adults. *Endocrine Rev.* 16:63–86.

De Feo, P., Gaisano, M.G., and Haymond, M.W. 1991. Differential effects of insulin deficiency on albumin and fibrinogen synthesis in humans. *J. Clin. Invest.* 88:833–840.

De Feo, P., Horber, F.F., and Haymond, M.W. 1992. Meal stimulation of albumin synthesis: a significant contributor to whole body protein synthesis in humans. *Am. J. Physiol.* 263:E794–E799.

De Feo, P., Volpi, E., Lucidi, P., Cruciani, G., Monacchia, F., Reboldi, G. et al. 1995. Ethanol impairs post-prandial hepatic protein metabolism. *J. Clin. Invest.* 95:1472–1479.

De Feo, P. 1996. Hormonal regulation of human protein metabolism. *Eur. J. Clin. Invest.* 135:7–18.

DeChalain, T.M.B., Michell, W.L., O'Keefe, S.J., and Ogden, J.M. 1992. The effect of fuel source on amino acid metabolism in critically ill patients. *J. Surg. Res.* 52:167–176.

DeHaven, J., Sherwin, R.S., Hendler, R., and Felig, P. 1980. Nitrogen and sodium balance and sympathetic-nervous-system activity in obese subjects treated with a low-calorie protein or mixed diet. *N. Engl. J. Med.* 302:477–482.

Del Prato, S., DeFronzo, R.A., Castellino, P., Wahren, J., and Alvestrand, A. 1990. Regulation of amino acid metabolism by epinephrine. *Am. J. Physiol.* 258:E878–E887.

Delaporte, C., Gros, F., and Anagnostopoulos, T. 1980. Inhibitory effects of plasma dialysate on protein synthesis in vitro: influence of dialysis and transplantation. *Am. J. Clin. Nutr.* 33:1407–1410.

Denne, S.C., Liechty, E.A., Liu, Y.M., Brechtel, G., and Baron, A.D. 1991. Proteolysis in skeletal muscle and whole body in response to euglycemic hyperinsulinemia in normal adults. *Am. J. Physiol.* 261:E809–E814.

Denne, S.C., Patel, D., and Kalhan, S.C. 1991. Leucine kinetics and fuel utilization during a brief fast in human pregnancy. *Metabolism* 40:1249–1256.

Denne, S.C., Brechtel, G., Johnson, A., Liechty, E.A., and Baron, A.D. 1995a. Skeletal muscle proteolysis is reduced in noninsulin-dependent diabetes mellitus and is unaltered by euglycemic hyperinsulinemia or intensive insulin therapy. *J. Clin. Endocrinol. Metab.* 80:2371–2377.

Denne, S.C., Karn, C.A., Wang, J., and Liechty, E.A. 1995b. Effect of intravenous glucose and lipid on proteolysis and glucose production in normal newborns. *Am. J. Physiol.* 269:E361–E367.

Deriaz, O., Fournier, G., Tremblay, A., Despres, J.-P., and Bouchard, C. 1992. Lean-body-mass composition and resting energy expenditure before and after long-term overfeeding. *Am. J. Clin. Nutr.* 56:840–847.

Devlin, J.T., Brodsky, I., Scrimgeour, A., Fuller, S., and Bier, D.M. 1990. Amino acid metabolism after intense exercise. *Am. J. Physiol.* 258:E249–E255.

Dohm, G.L., Williams, R.T., Kasperek, G.J., and van Rij, A.M. 1982. Increased excretion of urea and N^{τ}-methylhistidine by rats and humans after a bout of exercise. *J. Appl. Physiol.* 52:27–33.

Dohm, G.L., Israel, R.G., Breedlove, R.L., Williams, R.T., and Askew, E.W. 1985. Biphasic changes in 3-methylhistidine excretion in humans after exercise. *Am. J. Physiol.* 248:E588–E592.

Druml, W. 1998. Protein metabolism in acute renal failure. *Miner. Electrolyte Metab.* 24:47–54.

Eaton, S.B., and Konner, M. 1985. Paleolithic nutrition. A consideration of its nature and current implications. *N. Engl. J. Med.* 312:283–285.

Ebeling, P.R., Jones, J.D., O'Fallon, W.M., Janes, C.H., and Riggs, B.L. 1993. Short-term effects of recombinant human insulin-like growth factor I on bone turnover in normal women. *J. Clin. Endocrinol. Metab.* 77:1384–1387.

Eden, E., Ekman, L., Bennegard, K., Lindmark, L., and Lundholm, K. 1984. Whole-body tyrosine flux in relation to energy expenditure in weight-losing cancer patients. *Metabolism* 33:1020–1027.

Eisemann, J.H., Hammond, A.C., and Rumsey, T.S. 1989. Tissue protein synthesis and nucleic acid concentrations in steers treated with somatotropin. *Br. J. Nutr.* 62:657–671.

El-Harake, W.A., Furman, M.A., Cook, B., Nair, K.S., Kukowski, J., and Brodsky, I.G. 1998. Measurement of dermal collagen synthesis rate in vivo in humans. *Am. J. Physiol.* 274:E586–E591.

El-Khoury, A.E., Forslund, A.H., Olsson, R., Branth, S., Sjodin, A., Andersson, A. et al. 1997. Moderate exercise at energy balance does not affect 24-h leucine oxidation or nitrogen retention in healthy men. *Am. J. Physiol.* 273:E394–E407.

Elahi, D., McAloon-Dyke, M., Fukagawa, N.K., Sclater, A.L., Wong, G.A., Shannon, R.P. et al. 1993. Effects of recombinant human IGF-I on glucose and leucine kinetics in men. *Am. J. Physiol.* 265:E831–E838.

Emery, P.W., Edwards, R.H.T., Rennie, M.J., Souhami, R.L., and Halliday, D. 1984. Protein synthesis in muscle measured in vivo in cachectic patients with cancer. *Br. Med. J.* 289:584–586.

Erickson, G.F. 1995. The ovary: basic principles and concepts. A. Physiology. In *Endocrinology and Metabolism,* ed. P. Felig, J.D. Baxter, and L.A. Frohman, pp. 973–1015. New York: McGraw-Hill.

Eriksson, S., Hagenfeldt, L., and Wahren, J. 1981. A comparison of the effects of intravenous infusion of individual branched-chain amino acids on blood amino acid levels in man. *Clin. Sci.* 60:95–100.

Essen, P., McNurlan, M.A., Wernerman, J., Milne, E., Vinnars, E., and Garlick, P.J. 1992a. Short-term starvation decreases skeletal muscle protein synthesis rate in man. *Clin. Physiol.* 12:287–299.

Essen, P., McNurlan, M.A., Wernerman, J., Vinnars, E., and Garlick, P.J. 1992b. Uncomplicated surgery, but not general anesthesia, decreases muscle protein synthesis. *Am. J. Physiol.* 262:E253–E260.

Essen, P., McNurlan, M.A., Sonnenfeld, T., Milne, E., Vinnars, E., Wernerman, J. et al. 1993. Muscle protein synthesis after operation: effects of intravenous nutrition. *Eur. J. Surg.* 159:195–200.

Essen, P., McNurlan, M.A., Thorell, A., Tjader, I., Caso, G., Anderson, S.E. et al. 1996. Determination of protein synthesis in lymphocytes *in vivo* after surgery. *Clin. Sci.* 91:99–106.

Evans, D.A., Jacobs, D.O., and Wilmore, D.W. 1993. Effects of tumour necrosis factor on protein metabolism. *Br. J. Surg.* 80:1019–1023.

Evans, W.J., Meredith, C.N., Cannon, J.G., Dinarello, C.A., Frontera, W.R., Hughes, V.A. et al. 1986. Metabolic changes following eccentric exercise in trained and untrained men. *J. Appl. Physiol.* 61:1864–1868.

Everson, W.V., Flaim, K.E., Susco, D.M., Kimball, S.R., and Jefferson, L.S. 1989. Effect of amino acid deprivation on initiation of protein synthesis in rat hepatocytes. *Am. J. Physiol.* 256:C18–C27.

Fang, C.-H., James, H.J., Ogle, C., Fischer, J.E., and Hasselgren, P.-O. 1995. Influence of burn injury on protein metabolism in different types of skeletal muscle and the role of glucocorticoids. *J. Am. Coll. Surg.* 180:33–42.

Fang, C.-H., Li, B.G., James, H., Fischer, J.E., and Hasselgren, P.-O. 1997. Cytokines block the effects of insulin-like growth factor-I (IGF-I) on glucose uptake and lactate production in skeletal muscle but do not influence IGF-I-induced changes in protein turnover. *Shock* 8:362–367.

FAO 1970. *Amino-acid content of foods and biological data on proteins.* Rome: Food and Agriculture Organization of the United Nations

Fearon, K.C.H., Hansell, D.T., Preston, T., Plumb, J.A., Davies, J., Shapiro, D. et al. 1988. Influence of whole body protein turnover rate on resting energy expenditure in patients with cancer. *Cancer Res.* 48:2590–2595.

Fearon, K.C.H., McMillan, D.C., Preston, T., Winstanley, F.P., Cruickshank, A.M., and Shenkin, A. 1991. Elevated circulating interleukin-6 is associated with an acute-phase response but reduced fixed hepatic protein synthesis in patients with cancer. *Ann. Surg.* 213:26–31.

Fearon, K.C.H., Falconer, J.S., Slater, C., McMillan, D.C., Ross, J.A., and Preston, T. 1998. Albumin synthesis rates are not decreased in hypoalbuminemic cachectic cancer patients with an ongoing acute-phase protein response. *Ann. Surg.* 227:249–254.

Felig, P., Wahren, J., and Raf, L. 1973. Evidence of inter-organ amino-acid transport by blood cells in humans. *Proc. Natl. Acad. Sci. USA* 70:1775–1779.

Fern, E.B., Garlick, P.J., McNurlan, M.A., and Waterlow, J.C. 1981. The excretion of isotope in urea and ammonia for estimating protein turnover in man with [^{15}N]glycine. *Clin. Sci.* 61:217–228.

Fern, E.B., Garlick, P.J., and Waterlow, J.C. 1985. Apparent compartmentation of body nitrogen in one human subject: its consequences in measuring the rate of whole-body protein synthesis with ^{15}N. *Clin. Sci.* 68:271–282.

Fern, E.B., Bielinski, R.N., and Schutz, Y. 1991. Effects of exaggerated amino acid and protein supply in man. *Experientia* 47:168–172.

Ferrando, A., Lane, H.W., Stuart, C.A., Davis-Street, J., and Wolfe, R.R. 1996. Prolonged bed rest decreases skeletal muscle and whole body protein synthesis. *Am. J. Physiol.* 270:E627–E633.

Ferrando, A., Tipton, K., Doyle, D., Phillips, S., Cortiella, J., and Wolfe, R.R. 1997a. Net protein synthesis and amino acid uptake with testosterone injection. *FASEB J.* 11:A437(Abstract)

Ferrando, A., Tipton, K.D., Bamman, M.M., and Wolfe, R.R. 1997b. Resistance exercise maintains skeletal muscle protein synthesis during bed rest. *J. Appl. Physiol.* 82:807–810.

Ferrannini, E., Barrett, E.J., Bevilacqua, S., Jacob, R., Walker, M., Sherwin, R.S. et al. 1986. Effect of free fatty acids on blood amino acid levels in humans. *Am. J. Physiol.* 250:E686–E694.

Fielding, R.A., Meredith, C.N., O'Reilly, K.P., Frontera, W.R., Cannon, J.G., and Evans, W.J. 1991. Enhanced protein breakdown after eccentric exercise in young and older men. *J. Appl. Physiol.* 71:674–679.

Fisler, J.S., Drenick, E.J., Blumfield, D.E., and Swenseid, M.E. 1982. Nitrogen economy during very low calorie reducing diets: quality and quantity of dietary protein. *Am. J. Clin. Nutr.* 35:471–486.

Fitch, W.L., and King, J.C. 1987. Protein turnover and 3-methylhistidine excretion in non-pregnant, pregnant, and gestational diabetic women. *Hum. Nutr. Clin. Nutr.* 41C:327–339.

Fitzpatrick, G.F., Meguid, M.M., Gitlitz, P.H., and Brennan, M. 1977. Glucagon infusion in normal man: effects on 3-methylhistidine excretion and plasma amino acids. *Metabolism* 26:477–485.

Flaim, K.E., Copenhaver, M.E., and Jefferson, L.S. 1980. Effects of diabetes on protein synthesis in fast- and slow-twitch rat skeletal muscle. *Am. J. Physiol.* 239:E88–E95.

Flaim, K.E., Liao, W.S.L., Peavy, D.E., Taylor, J.M., and Jefferson, L.S. 1982a. The role of amino acids in the regulation of protein synthesis in perfused rat liver. II. Effects of amino acid deficiency on peptide chain initiation, polysomal aggregation, and distribution of albumin mRNA. *J. Biol. Chem.* 257:2939–2946.

Flaim, K.E., Peavy, D.E., Everson, W.V., and Jefferson, L.S. 1982b. The role of amino acids in the regulation of protein synthesis in perfused rat liver. I. Reduction in rates of synthesis resulting from amino acid deprivation and recovery during flow-through perfusion. *J. Biol. Chem.* 257:2932–2938.

Flakoll, P.J., Kulayat, M., Frexes-Steed, M., Hourani, H., Brown, L.L., Hill, J.O., and Abumrad, N.N. 1989. Amino acids augment insulin's suppression of whole body proteolysis. *Am. J. Physiol.* 257:E839–E847.

Flakoll, P.J., Hill, J.O., and Abumrad, N.N. 1993. Acute hyperglycemia enhances proteolysis in normal man. *Am. J. Physiol.* 265:E715–E721.

Flakoll, P.J., Wentzel, L.S., and Hyman, S.A. 1995. Protein and glucose metabolism during isolated closed-head injury. *Am. J. Physiol.* 269:E636–E641.

Flores, E.A., Bistrian, B.R., Pomposelli, J.J., Dinarello, C.A., Balckburn, G.L., and Istfan, N.W. 1989. Infusion of tumor necrosis factor/cachectin promotes muscle catabolism in the rat. A synergistic effect with interleukin 1. *J. Clin. Invest.* 83:1614–1622.

Fong, Y., Rosenbaum, M., Tracey, K.J., Raman, G., Hesse, D.G., Matthews, D.E. et al. 1989. Recombinant growth hormone enhances muscle myosin heavy-chain mRNA accumulation and amino acid accrual in humans. *Proc. Natl. Acad. Sci. USA* 86:3371–3374.

Forbes, G.B. 1987a. *Human Body Composition.* New York: Springer-Verlag.

Forbes, G.B. 1987b. Lean body mass-body fat interrelationships in humans. *Nutr. Rev.* 45:225–231.

Forbes, G.B., Brown, M.R., Welle, S.L., and Underwood, L.E. 1989. Hormonal response to overfeeding. *Am. J. Clin. Nutr.* 49:608–611.

Forrester, T., Badaloo, A.V., Persaud, C., and Jackson, A.A. 1994. Urea production and salvage during pregnancy in normal Jamaican women. *Am. J. Clin. Nutr.* 60:341–346.

Forslund, A.H., Hambraeus, L., Olsson, R., El-Khoury, A.E., Yu, Y.-M., and Young, V.R. 1998. The 24-h whole body leucine and urea kinetics at normal and high protein intakes with exercise in healthy adults. *Am. J. Physiol.* 275:E310–E320.

Freifelder, D., and Malacinski, G.M. 1993. *Essentials of Molecular Biology.* Boston: Jones and Bartlett.

Frexes-Steed, M., Warner, M.L., Bulus, N., Flakoll, P., and Abumrad, N.N. 1990. Role of insulin and branched-chain amino acids in regulating protein metabolism during fasting. *Am. J. Physiol.* 258:E907–E917.

Frontera, W.R., Meredith, C.N., O'Reilly, K.P., and Evans, W.J. 1988. Strength conditioning in older men: skeletal muscle hypertrophy and improved function. *J. Appl. Physiol.* 64:1038–1044.

Fryburg, D.A., Barrett, E.J., Louard, R.J., and Gelfand, R.A. 1990. Effect of starvation on human muscle protein metabolism and its response to insulin. *Am. J. Physiol.* 259:E477–E482.

Fryburg, D.A., Gelfand, R.A., and Barrett, E.J. 1991. Growth hormone acutely stimulates forearm muscle protein synthesis in normal humans. *Am. J. Physiol.* 260:E499–E504.

Fryburg, D.A., and Barrett, E.J. 1993. growth hormone acutely stimulates skeletal muscle but not whole-body protein synthesis in humans. *Metabolism* 42:1223–1227.

Fryburg, D.A. 1994. Insulin-like growth factor I exerts growth hormone- and insulin-like actions on human muscle protein metabolism. *Am. J. Physiol.* 267:E331–E336.

Fryburg, D.A., Gelfand, R.A., Jahn, L.A., Oliveras, D., Sherwin, R.S., Sacca, L. et al. 1995a. Effects of epinephrine on human muscle glucose and protein metabolism. *Am. J. Physiol.* 268:E55–E59.

Fryburg, D.A., Jahn, L.A., Hill, S.A., Oliveras, D.M., and Barrett, E.J. 1995b. Insulin and insulin-like growth factor-I enhance human skeletal muscle protein anabolism during hyperaminoacidemia by different mechanisms. *J. Clin. Invest.* 96:1722–1729.

Fryburg, D.A., and Barrett, E.J. 1995. Insulin, growth hormone, and IGF-I regulation of protein metabolism. *Diabetes Rev.* 3:93–112.

Fryburg, D.A. 1996. N^G-monomethyl-L-arginine inhibits the blood flow but not the insulin-like response of forearm muscle to IGF-I. *J. Clin. Invest.* 97:1319–1328.

Fu, A.Z., and Nair, K.S. 1998. Age effect on albumin and fibrinogen synthesis in humans. *FASEB J.* 12:A968(Abstract)

Fujita, J., Tsujinaka, T., Yano, M., Ebisui, C., Saito, H., Katsume, A. et al. 1996. Anti-interleukin-6 receptor antibody prevents muscle atrophy in colon-26 adenocarcinoma-bearing mice with modulation of lysosomal and ATP-ubiquitin-dependent proteolytic pathways. *Int. J. Cancer* 68:637–643.

Fukagawa, N.K., Minaker, K.L., Rowe, J.W., Goodman, M.N., Matthews, D.E., Bier, D.M. et al. 1985. Insulin-mediated reduction of whole body protein breakdown. Dose-response effects on leucine metabolism in postabsorptive men. *J. Clin. Invest.* 76:2306–2311.

Fukagawa, N.K., Minaker, K.L., Young, V.R., Matthews, D.E., Bier, D.M., and Rowe, J.W. 1989. Leucine metabolism in aging humans: effect of insulin and substrate availability. *Am. J. Physiol.* 256:E288–E294.

Fulks, R.M., Li, J.B., and Goldberg, A.L. 1975. Effects of insulin, glucose, and amino acids on protein turnover in rat diaphragm. *J. Biol. Chem.* 250:290–298.

Fuller, M.F., McWilliam, R., Wang, T.C., and Giles, L.R. 1989. The optimum dietary amino acid pattern for growing pigs. 2. Requirements for maintenance and for tissue protein accretion. *Br. J. Nutr.* 62:255–267.

Furuno, K., and Goldberg, A.L. 1986. The activation of protein degradation in muscle by Ca^{2+} or muscle injury does not involve a lysosomal mechanism. *Biochem. J.* 237:859–864.

Gambacciani, M., Ciaponi, M., Cappagli, B., Piaggesi, L., De Simone, L., Orlandi, R. et al. 1997. Body weight, body fat distribution, and hormonal replacement therapy in early postmenopausal women. *J. Clin. Endocrinol. Metab.* 82:414–417.

Gann, M.E., McNurlan, M.A., McHardy, K.C., Milne, E., and Garlick, P.J. 1988. Non-steroidal anti-inflammatory agents and protein turnover in the elderly. *Proc. Nutr. Soc.* 47:133A(Abstract)

Garcia-Sanz, J.A., Mikulits, W., Livingstone, A., Lefkovits, I., and Mullner, E.W. 1998. Translational control: a general mechanism for gene regulation during T cell activation. *FASEB J.* 12:299–306.

Gardner, D.F., Kaplan, M.M., Stanley, C.A., and Utiger, R.D. 1979. Effect of tri-iodothyronine replacement on the metabolic and pituitary responses to starvation. *N. Engl. J. Med.* 300:579–584.

Garibotto, G., Russo, R., Sofia, A., Sala, M.R., Robaudo, C., Moscatelli, P. et al. 1994. Skeletal muscle protein synthesis and degradation in patients with chronic renal failure. *Kidney Int.* 45:1432–1439.

Garibotto, G., Russo, R., Sofia, A., Sala, M.R., Sabatino, C., Moscatelli, P. et al. 1996. Muscle protein turnover in chronic renal failure patients with metabolic acidosis or normal acid-base balance. *Miner. Electrolyte Metab.* 22:58–61.

Garibotto, G., Barreca, A., Russo, R., Sofia, A., Araghi, P., Cesarone, A. et al. 1997. Effects of recombinant human growth hormone on muscle protein turnover in malnourished hemodialysis patients. *J. Clin. Invest.* 99:97–105.

Garlick, P.J., Millward, D.J., James, W.P.T., and Waterlow, J.C. 1975. The effect of protein deprivation and starvation on the rate of protein synthesis in tissues of the rat. *Biochim. Biophys. Acta* 414:71–84.

Garlick, P.J., Burk, T.L., and Swick, R.W. 1976. Protein synthesis and RNA in tissues of the pig. *Am. J. Physiol.* 230:1108–1112.

Garlick, P.J., Clugston, G.A., and Waterlow, J.C. 1980. Influence of low-energy diets on whole-body protein turnover in obese subjects. *Am. J. Physiol.* 238:E235–E244.

Garlick, P.J., and Fern, E.B. 1985. Whole-body protein turnover: theoretical considerations. In *Substrate and Energy Metabolism in Man,* ed. J.S. Garrow and D. Halliday, pp. 7–15. London: John Libbey.

Garlick, P.J., Wernerman, J., McNurlan, M.A., Essen, P., Lobley, G.E., Milne, E. et al. 1989. Measurement of the rate of protein synthesis in muscle of postabsorptive young men by injection of a 'flooding dose' of [1-^{13}C]leucine. *Clin. Sci.* 77:329–336.

Garlick, P.J., McNurlan, M.A., and Ballmer, P.E. 1991. Influence of dietary protein intake on whole-body protein turnover in humans. *Diabetes Care* 14:1189–1198.

Garlick, P.J., Wernerman, J., McNurlan, M.A., and Heys, S.D. 1991. Organ specific measurements of protein turnover in man. *Proc. Nutr. Soc.* 50:217–225.

Garlick, P.J., McNurlan, M.A., Essen, P., and Wernerman, J. 1994. Measurement of tissue protein synthesis rates in vivo: a critical analysis of contrasting methods. *Am. J. Physiol.* 266:E287–E297.

Garrel, D.R., Delmas, P.D., Welsh, C., Arnaud, M.J., Hamilton, S.E., and Pugeat, M.M. 1988. Effects of moderate pysical training on prednisone-induced protein wasting: a study of whole-body and bone protein metabolism. *Metabolism* 37:257–262.

Garrel, D.R., Moussalli, R., De Oliveira, A., Lesiege, D., and Lariviere, F. 1995. RU 486 prevents the acute effects of cortisol on glucose and leucine metabolism. *J. Clin. Endocrinol. Metab.* 80:379–385.

Gelfand, R.A., Glickman, M.G., Jacob, R., Sherwin, R.S., and DeFronzo, R.A. 1986. Removal of infused amino acids by splanchnic and leg tissues in humans. *Am. J. Physiol.* 250:E407–E413.

Gelfand, R.A., Hutchinson-Williams, K.A., Bonde, A.A., Castellino, P., and Sherwin, R.S. 1987. Catabolic effects of thyroid hormone excess: the contribution of adrenergic activity to hypermetabolism and protein breakdown. *Metabolism* 36:562–569.

Gelfand, R.A., Glickman, M.G., Castellino, P., Louard, R.J., and DeFronzo, R.A. 1988. Measurement of L-[1-^{14}C]leucine kinetics in splanchnic and leg tissues in humans. Effect of amino acid infusion. *Diabetes* 37:1365–1372.

Gelfand, R.A., and Barrett, E.J. 1987. Effect of physiologic hyperinsulinemia on skeletal muscle protein synthesis and breakdown in man. *J. Clin. Invest.* 80:1–6.

Gelfand, R.A., and Sherwin, R.S. 1986. Nitrogen conservation in starvation revisited: protein sparing with intravenous fructose. *Metabolism* 35:37–44.

Gersovitz, M., Munro, H.N., Udall, J., and Young, V.R. 1980. Albumin synthesis in young and elderly subjects using a new stable isotope methodology: response to level of protein intake. *Metabolism* 29:1075–1086.

Ghiron, L.J., Thompson, J.L., Holloway, L., Hintz, R.L., Butterfield, G.E., Hoffman, A.R., and Marcus, R. 1995. Effects of recombinant insulin-like growth factor-I and growth hormone on bone turnover in elderly women. *J. Bone Min. Res.* 10:1844–1852.

Gibson, J.N.A., Halliday, D., Morrison, W.L., Stoward, P.J., Hornsby, G.A., Watt, P.W., Murdoch, G., and Rennie, M.J. 1987. Decrease in human quadriceps muscle protein turnover consequent upon leg immobilization. *Clin. Sci.* 72:503–509.

Gibson, J.N.A., Morrison, W.L., Scrimgeour, C.M., Smith, K., Stoward, P.J., and Rennie, M.J. 1989. Effects of therapeutic percutaneous electrical stimulation of atrophic human quadriceps on muscle composition, protein synthesis and contractile properties. *Eur. J. Clin. Invest.* 19:206–212.

Gibson, J.N.A., Poyser, N.L., Morrison, W.L., Scrimgeour, A., and Rennie, M.J. 1991. Muscle protein synthesis in patients with rheumatoid arthritis: effect of chronic corticosteroid therapy on prostaglandin $F_{2\alpha}$ availability. *Eur. J. Clin. Invest.* 21:406–412.

Gibson, N.R., Fereday, A., Cox, M., Halliday, D., Pacy, P.J., and Millward, D.J. 1996. Influences of dietary energy and protein on leucine kinetics during feeding in healthy adults. *Am. J. Physiol.* 270:E282–E291.

Giesecke, K., Magnusson, I., Ahlberg, M., Hagenfeldt, L., and Wahren, J. 1989. Protein and amino acid metabolism during early starvation as refelcted by excretion of urea and methylhistidines. *Metabolism* 38:1196–1200.

Giordano, M., Castellino, P., and DeFronzo, R.A. 1996. Differential responsiveness of protein synthesis and degradation to amino acid availability in humans. *Diabetes* 45:393–399.

Giustina, A., Bussi, A.R., Jacobello, C., and Wehrenberg, W.B. 1995. Effects of recombinant human growth hormone (GH) on bone and intermediary metabolism in patients receiving chronic glucocorticoid treatment with suppressed endogenous GH response to GH-releasing hormone. *J. Clin. Endocrinol. Metab.* 80:122–129.

Glass, R.E., Fern, E.B., and Garlick, P.J. 1983. Whole-body protein turnover before and after resection of colorectal tumours. *Clin. Sci.* 64:101–108.

Glitz, D. 1997. Protein synthesis: translation and posttranslational modifications. In *Textbook of Biochemistry with Clinical Correlations.* ed. T.M. Devlin, pp. 713–753. New York: Wiley-Liss

Goldberg, A.L., Kettelhut, I.C., Furuno, K., Fagan, J.M., and Baracos, V. 1988. Activation of protein breakdown and prostaglandin E_2 production in rat skeletal muscle in fever is signaled by a macrophage product distinct from interleukin 1 or other known monokines. *J. Clin. Invest.* 81:1378–1383.

Goldberg, A.L. 1995. Functions of the proteasome: the lysis at the end of the tunnel. *Science* 268:522–523.

Golden, M., Waterlow, J.C., and Picou, D. 1977. The relationship between dietary intake, weight change, nitrogen balance, and protein turnover in man. *Am. J. Clin. Nutr.* 30:1345–1348.

Golden, M.H.N., Waterlow, J.C., and Picou, D. 1977. Protein turnover, synthesis and breakdown before and after recovery from protein-energy malnutrition. *Clin. Sci. Mol. Med.* 53:473–477.

Goodman, M.N. 1991. Tumor necrosis factor induces skeletal muscle protein breakdown in rats. *Am. J. Physiol.* 260:E727–E730.

Goodman, M.N. 1994. Interleukin-6 induces skeletal muscle protein breakdown in rats. *Proc. Soc. Exp. Biol. Med.* 205:182–185.

Goodship, T.H.J., Lloyd, S., Clague, M.B., Bartlett, K., Ward, M.K., and Wilkinson, R. 1987. Whole body leucine turnover and nutritional status in continuous ambulatory peritoneal dialysis. *Clin. Sci.* 73:463–469.

Goodship, T.H.J., Mitch, W.E., Hoerr, R.A., Wagner, D.A., Steinman, T.I., and Young, V.R. 1990. Adaptation to low-protein diets in renal failure: leucine turnover and nitrogen balance. *J. Am. Soc. Nephrol.* 1:66–75.

Gore, D.C., Wolfe, K.A., Foxx-Orenstein, A., and Hibbert, J. 1997. Assessment of human colon cancer protein kinetics in vivo. *Surgery* 122:593–599.

Gougeon, R., Pencharz, P.B., and Sigal, R.J. 1997. Effect of glycemic control on the kinetics of whole-body protein metabolism in obese subjects with non-insulin-dependent diabetes mellitus during iso- and hypoenergetic feeding. *Am. J. Clin. Nutr.* 65:861–870.

Griggs, R.C., and Rennie, M.J. 1982. Muscle wasting in muscular dystrophy: decreased protein synthesis or increased degradation? *Ann. Neurol.* 13:125–132.

Griggs, R.C., Halliday, D., Kingston, W., and Moxley, R.T. III. 1986. Effect of testosterone on muscle protein synthesis in myotonic dystrophy. *Ann. Neurol.* 20:590–596.

Griggs, R.C., Kingston, W., Jozefowicz, R., Herr, B.E., Forbes, G.B., and Halliday, D. 1989. Effect of testosterone on muscle mass and muscle protein synthesis. *J. Appl. Physiol.* 66:498–503.

Grinspoon, S.K., Baum, H.B.A., Peterson, S., and Klibanski, A. 1995. Effects of rhIGF-I administration on bone turnover during short-term fasting. *J. Clin. Invest.* 96:900–906.

Grinspoon, S.K., Baum, H., Lee, K., Anderson, E., Herzog, D., and Klibanski, A. 1996. Effects of short-term recombinant human insulin-like growth factor I administration on bone turnover in osteopenic women with anorexia nervosa. *J. Clin. Endocrinol. Metab.* 81:3864–3870.

Guarnieri, G., Toigo, G., Situlin, R., Ciocchi, B., and Biolo, G. 1997. Modulation of protein kinetics in chronic renal failure. *Miner. Electrolyte Metab.* 23:214–217.

Guo, C.-Y., Weetman, A.P., and Eastell, R. 1997. Longitudinal changes of bone mineral density and bone turnover in postmenopausal women on thyroxine. *Clin. Endocrinol.* 46:301–307.

Hagenfeldt, L., Eriksson, S., and Wahren, J. 1980. Influence of leucine on arterial concentrations and regional exchange of amino acids in healthy subjects. *Clin. Sci.* 59:173–181.

Hagg, S., Morse, E.L., and Adibi, S.A. 1982. Effect of exercise on rates of oxidation, turnover, and plasma clearance of leucine in human subjects. *Am. J. Physiol.* 242:E407–E410.

Hagg, S.A., and Adibi, S.A. 1985. Leucine metabolism in thyrotoxicosis: plasma aminogram and 3-methylhistidine excretion before and after treatment. *Metabolism* 34:813–816.

Halliday, D., and McKeran, R.O. 1975. Measurement of muscle protein synthetic rate from serial muscle biopsies and total body protein turnover in man by continuous intravenous infusion of L-[α-^{15}N]lysine. *Clin. Sci. Mol. Med.* 49:581–590.

Halliday, D., Ford, G.C., Edwards, R.H.T., Rennie, M.J., and Griggs, R.C. 1985. In vivo estimation of muscle protein synthesis in myotonic dystrophy. *Ann. Neurol.* 17:65–69.

Halliday, D., Cheng, K.N., Dworzak, F., Gibson, J.N.A., and Rennie, M.J. 1988. Rate of protein synthesis in skeletal muscle of normal man and patients with muscular dystrophy: a reassessment. *Clin. Sci.* 74:237–240.

Hamadeh, M.J., and Hoffer, L.J. 1998. Tracer methods underestimate short-term variations in urea production in humans. *Am. J. Physiol.* 274:E547–E553.

Hammarqvist, F., Wernerman, J., Ali, R., von der Decken, A., and Vinnars, E. 1989. Addition of glutamine to total parenteral nutrition after elective abdominal surgery spares free glutamine in muscle, counteracts the fall in muscle protein synthesis, and improves nitrogen balance. *Ann. Surg.* 209:455–461.

Hammarqvist, F., Stromberg, C., von der Decken, A., Vinnars, E., and Wernerman, J. 1992. Biosynthetic human growth hormone preserves both muscle protein synthesis and the decrease in muscle-free glutamine, and improves whole-body nitrogen economy after operation. *Ann. Surg.* 216:184–191.

Hammarqvist, F., von der Decken, A., Vinnars, E., and Wernerman, J. 1994. Stress hormone and amino acid infusion in healthy volunteers: short-term effects on protein synthesis and amino acid metabolism in skeletal muscle. *Metabolism* 43:1158–1163.

Han, K.-K., and Martinage, A. 1993. Post-translational chemical modifications of proteins—III. Current developments in analytical procedures of identification and quantitation of post-translational chemically modified amino acid(s) and its derivatives. *Int. J. Biochem.* 25:957–970.

Hankard, R.G., Haymond, M.W., and Darmaun, D. 1996. Effect of glutamine on leucine metabolism in humans. *Am. J. Physiol.* 271:E748–E754.

Hannaford, M.C., Leiter, L.A., Josse, R.G., Goldstein, M.B., Marliss, E.B., and Halperin, M.L. 1982. Protein wasting due to acidosis of prolonged fasting. *Am. J. Physiol.* 243:E251–E256.

Haralambie, G., and Berg, A. 1976. Serum urea and amino nitrogen changes with exercise duration. *Eur. J. Appl. Physiol.* 36:39–48.

Harris, C.I. 1981. Reappraisal of the quantitative importance of non-skeletal-muscle source of N^t-methylhistidine in urine. *Biochem. J.* 194:1011–1014.

Harrison, R.A., Lewin, M.R., Halliday, D., and Clark, C.G. 1989a. Leucine kinetics in surgical patients I: a study of the effect of surgical 'stress'. *Br. J. Surg.* 76:505–508.

Harrison, R.A., Lewin, M.R., Halliday, D., and Clark, C.G. 1989b. Leucine kinetics in surgical patients II: a study of the effect of malignant disease and tumour burden. *Br. J. Surg.* 76:509–511.

Hart, G.W. 1997. Dynamic O-linked glycosylation of nuclear and cytoskeletal proteins. *Annu. Rev. Biochem.* 66:315–335.

Hartl, W.H., Miyoshi, H., Jahoor, F., Klein, S., Elahi, D., and Wolfe, R.R. 1990. Bradykinin attenuates glucagon-induced leucine oxidation in humans. *Am. J. Physiol.* 259:E239–E245.

Hartl, W.H., Demmelmair, H., Jauch, K.-W., Schmidt, H.-L., Koletzko, B., and Schildberg, F.W. 1997. Determination of protein synthesis in human rectal cancer in situ by continuous [1-^{13}C]leucine infusion. *Am. J. Physiol.* 272:E796–E802.

Hasch, E., Jarnum, S., and Tygstrup, N. 1967. Albumin synthesis rate as a measure of liver function in patients with cirrhosis. *Acta Med. Scand.* 182:83–92.

Hassager, C., and Christiansen, C. 1989. Estrogen/gestagen therapy changes soft tissue body composition in postmenopausal women. *Metabolism* 38:662–665.

Hassager, C., Jensen, L.T., Johansen, J.S., Riis, B.J., Melkko, J., Podenphant, J. et al. 1991. The carboxy-terminal propeptide of type I procollagen in serum as a marker of bone formation: the effect of nandrolone decanoate and female sex hormones. *Metabolism* 40:205–208.

Hasselgren, P.-O., Alderberth, A., Angeras, U., and Stenstrom, G. 1984a. Protein metabolism in skeletal muscle tissue from hyperthyroid patients after preoperative treatment with antithyroid drug or selective β-blocking agent. Results from a prospective, randomized study. *J. Clin. Endocrinol. Metab.* 59:835–839.

Hasselgren, P.-O., Jagenburg, R., Karlstrom, L., Pedersen, P., and Seeman, T. 1984b. Changes of protein metabolism in liver and skeletal muscle following trauma complicated by sepsis. *J. Trauma* 24:224–228.

Hasselgren, P.-O., Pedersen, P., Sax, H.C., Warner, B.W., and Fischer, J.E. 1988. Current concepts of protein turnover and amino acid transport in liver and skeletal muscle during sepsis. *Arch. Surg.* 123:992–999.

Hasten, D.L., Pak, J.Y., Crowley, J.R., and Yarasheski, K.E. 1997. Effects of resistance exercise on muscle protein synthesis in young, late middle-aged, and old men and women. *FASEB J.* 11:A291(Abstract)

Hasten, D.L., Pak, J.Y., Crowley, J.R., and Yarasheski, K.E. 1998. Short-term resistance exercise increases the fractional synthesis rate of mixed muscle protein, myosin heavy chain, and actin in young men and women. *FASEB J.* 12:A414(Abstract)

Hayes, S.A., and Dice, J.F. 1996. Roles of molecular chaperones in protein degradation. *J. Cell Biol.* 132:255–258.

Heber, D., Chlebowski, R.T., Ishibashi, D.E., Herrold, J.N., and Block, J.B. 1982. Abnormalities in glucose and protein metabolism in noncachectic lung cancer patients. *Cancer Res.* 42:4815–4819.

Hedden, M.P., and Buse, M.G. 1982. Effects of glucose, pyruvate, lactate, and amino acids on muscle protein synthesis. *Am. J. Physiol.* 242:E184–E192.

Heiling, V.J., Campbell, P.J., Gottesman, I.S., Tsalikian, E., Beaufrere, B., Gerich, J.E. et al. 1993. Differential effects of hyperglycemia and hyperinsulinemia on leucine rate of appearance in normal humans. *J. Clin. Endocrinol. Metab.* 76:203–206.

Heitzman, R.J. 1979. The efficacy and mechanism of action of anabolic agents as growth promoters in farm animals. *J. Steroid Biochem.* 11:927–930.

Hendler, R., and Bonde, A.A. 1990. Effects of sucrose on resting metabolic rate, nitrogen balance, leucine turnover and oxidation during weight loss with low calorie diets. *Int. J. Obesity* 14:927–938.

Henson, L.C., and Heber, D. 1983. Whole body protein breakdown rates and hormonal adaptation in fasted obese subjects. *J. Clin. Endocrinol. Metab.* 57:316–319.

Hentze, M.W. 1997. eIF4G: a multipurpose ribosome adapter? *Science* 275:500–501.

Hermus, A.R., Smals, A.G., Swinkels, L.M., Huysmans, D.A., Pieters, G.F., Sweep, C.F. et al. 1995. Bone mineral density and bone turnover before and after surgical cure of Cushing's syndrome. *J. Clin. Endocrinol. Metab.* 80:2859–2865.

Herrmann, V.M., Garnick, M.B., Moore, F.D., and Wilmore, D.W. 1981. Effect of cytotoxic agents on protein kinetics in patients with metastatic cancer. *Surgery* 90:381–386.

Hershey, J.W.B. 1991. Translational control in mammalian cells. *Ann. Rev. Biochem.* 60:717–755.

Heslin, M.J., Newman, E., Wolf, R.F., Pisters, P.W.T., and Brennan, M.F. 1992a. Effect of hyperinsulinemia on whole body and skeletal muscle leucine carbon kinetics in humans. *Am. J. Physiol.* 262:E911–E918.

Heslin, M.J., Newman, E., Wolf, R.F., Pisters, P.W.T., and Brennan, M.F. 1992b. Effect of systemic hyperinsulinemia in cancer patients. *Cancer Res.* 52:3845–3850.

Heymsfield, S.B., Smith, R., Aulet, M., Bensen, B., Lichtman, S., Wang, J. et al. 1990. Appendicular skeletal muscle mass: measurement by dual-photon absorptiometry. *Am. J. Clin. Nutr.* 52:214–218.

Heymsfield, S.B., and McManus, C.B. 1985. Tissue components of weight loss in cancer patients. A new method of study and preliminary observations. *Cancer* 55:238–249.

Heys, S.D., Park, K.G.M., McNurlan, M.A., Calder, A.G., Buchan, V., Blessing, K. et al. 1991. Measurement of tumour protein synthesis *in vivo* in human colorectal and breast cancer and its variability in separate biopsies from the same tumour. *Clin. Sci.* 80:587–593.

Heys, S.D., Park, K.G.M., McNurlan, M.A., Keenan, R.A., Miller, J.D.B., Eremin, O. et al. 1992. Protein synthesis rates in colon and liver: stimulation by gastrointestinal pathologies. *Gut* 33:976–981.

Hicke, L. 1997. Ubiquitin-dependent internalization and down-regulation of plasma membrane proteins. *FASEB J.* 11:1215–1226.

Hickson, J.F., and Hinkelmann, K. 1985. Exercise and protein intake effects on urinary 3-methylhistidine excretion. *Am. J. Clin. Nutr.* 41:246–253.

Hickson, R.C., Czerwinski, S.M., and Wegrzyn, L.E. 1995. Glutamine prevents downregulation of myosin heavy chain synthesis and muscle atrophy from glucocorticoids. *Am. J. Physiol.* 268:E730–E734.

Higashiguchi, T., Hasselgren, P.-O., Wagner, K., and Fischer, J.E. 1993. Effect of glutamine on protein synthesis in isolated intestinal epithelial cells. *J. Parent. Ent. Nutr.* 17:307–314.

Hillier, T.A., Fryburg, D.A., Jahn, L.A., and Barrett, E.J. 1998. Extreme hyperinsulinemia unmasks insulin's effect to stimulate protein synthesis in the human forearm. *Am. J. Physiol.* 274:E1067–E1074.

Hilt, W., and Wolf, D.H. 1996. Proteasomes: destruction as a programme. *Trends in Biochem. Sci.* 21:96–102.

Hirsch, S., de la Maza, M.P., Petermann, M., Iturriaga, H., Ugarte, G., and Bunout, D. 1995. Protein turnover in abstinent and non-abstinent patients with alcoholic cirrhosis. *J. Am. Coll. Nutr.* 14:99–104.

Hochwald, S.N., Harrison, L.E., Heslin, M.J., Burt, M.E., and Brennan, M.F. 1997. Early postoperative enteral feeding improves whole body protein kinetics in upper gastrointestinal cancer patients. *Am. J. Surg.* 174:325–330.

Hoerr, R.A., Matthews, D.E., Bier, D.M., and Young, V.R. 1993. Effects of protein restriction and acute refeeding on leucine and lysine kinetics in young men. *Am. J. Physiol.* 264:E567–E575.

Hoffer, L.J., Bistrian, B.R., Young, V.R., Blackburn, G.L., and Matthews, D.E. 1984a. Metabolic effects of very low calorie weight reduction diets. *J. Clin. Invest.* 73:750–758.

Hoffer, L.J., Bistrian, B.R., Young, V.R., Blackburn, G.L., and Wannemacher, R.W. 1984b. Metabolic effects of carbohydrate in low-calorie diets. *Metabolism* 33:820–825.

Hoffer, L.J., Yang, R.D., Matthews, D.E., Bistrian, B.R., Bier, D.M., and Young, V.R. 1985. Effects of meal consumption on whole body leucine and alanine kinetics in young adult men. *Br. J. Nutr.* 53:31–38.

Hoffer, L.J., and Forse, R.A. 1990. Protein metabolic effects of a prolonged fast and hypocaloric refeeding. *Am. J. Physiol.* 258:E832–E840.

Hoffer, L.J., Taveroff, A., Robitaille, L., Hamadeh, M.J., and Mamer, O.A. 1997. Effects of leucine on whole body leucine, valine, and threonine metabolism in humans. *Am. J. Physiol.* 272:E1037–E1042.

Holloway, L., Butterfield, G.E., Hintz, R.L., Gesundheit, N., and Marcus, R. 1994. Effects of recombinant human growth hormone on metabolic indices, body composition, and bone turnover in healthy elderly women. *J. Clin. Endocrinol. Metab.* 79:470–479.

Horber, F.F., Scheidegger, J.R., Grunig, B.E., and Frey, F.J. 1985. Evidence that prednisone-induced myopathy is reversed by physical training. *J. Clin. Endocrinol. Metab.* 61:83–88.

Horber, F.F., and Haymond, M.W. 1990. Human growth hormone prevents the protein catabolic side effects of prednisone in humans. *J. Clin. Invest.* 86:265–272.

Hou, J.C.-S., Zhou, J.-N., Zhu, H.-W., Wu, J.-Z., Wu, J.-C., and Zhang, M.-W. 1986. Dynamic aspects of whole-body nitrogen metabolism in uremic patients on dietary therapy. *Nephron* 44:288–294.

Hsu, C.-J., Kimball, S.R., Antonetti, D.A., and Jefferson, L.S. 1992. Effects of insulin on total RNA, poly(A)$^+$ RNA, and mRNA in primary cultures of rat hepatocytes. *Am. J. Physiol.* 263:E1106–E1112.

Hunter, K.A., Ballmer, P.E., Anderson, S.E., Broom, J., Garlick, P.J., and McNurlan, M.A. 1995. Acute stimulation of albumin synthesis rate with oral meal feeding in healthy subjects measured with [*ring*-$^{2}H_{5}$]phenylalanine. *Clin. Sci.* 88:235–242.

Hussain, M.A., Schmitz, O., Mengel, A., Keller, A., Christiansen, C., Zapf, J. et al. 1993. Insulin-like growth factor I stimulates lipid oxidation, reduces protein oxidation, and enhances insulin sensitivity in humans. *J. Clin. Invest.* 92:2249–2256.

Huszar, G., Koivisto, V., Davis, E., and Felig, P. 1982. Urinary 3-methylhistidine excretion in juvenile-onset diabetics: evidence of increased protein catabolism in the absence of ketoacidosis. *Metabolism* 31:188–191.

Inculet, R.I., Stein, T.P., Peacock, J.L., Leskiw, M., Maher, M., Gorschboth, C.M. et al. 1987. Altered leucine metabolism in noncachectic sarcoma patients. *Cancer Res.* 47:4746–4749.

Ionasescu, V. 1975. Distinction between Duchenne and other muscular dystrophies by ribosomal protein synthesis. *J. Med. Genet.* 12:49–54.

Irving, C.S., Thomas, M.R., Malphus, E.W., Marks, L., Wong, W.W., Boutton, T.W. et al. 1986. Lysine and protein metabolism in young women. Subdivision based on the novel use of multiple stable isotopic labels. *J. Clin. Invest.* 77:1321–1331.

Itoh, R., and Suyama, Y. 1996. Sodium excretion in relation to calcium and hydroxyproline excretion in a healthy Japanese population. *Am. J. Clin. Nutr.* 63:735–740.

Jackson, A.A., Golden, M.H.N., Byfield, R., Jahoor, F., Royes, J., and Soutter, L. 1983. Whole-body protein turnover and nitrogen balance in young children at intakes of protein and energy in the region of maintenance. *Hum. Nutr. Clin. Nutr.* 37C:433–446.

Jackson, A.A. 1993. Chronic malnutrition: protein metabolism. *Proc. Nutr. Soc.* 52:1–10.

Jahoor, F., Desai, M., Herndon, D.N., and Wolfe, R.R. 1988. Dynamics of the protein metabolic response to burn injury. *Metabolism* 37:330–337.

Jahoor, F., Shangraw, R.E., Miyoshi, H., Wallfish, H., Herndon, D.N., and Wolfe, R.R. 1989. Role of insulin and glucose oxidation in mediating the protein catabolism of burns and sepsis. *Am. J. Physiol.* 257:E323–E331.

Jeejeebhoy, K.N., Anderson, G.H., Nakhooda, A.F., Greenberg, G.R., Sanderson, I., and Marliss, E.B. 1976. Metabolic studies in total parenteral nutrition with lipid in man. Comparison with glucose. *J. Clin. Invest.* 57:125–136.

Jeevanandam, M., Horowitz, G.D., Lowry, S.F., and Brennan, M.F. 1984. Cancer cachexia and protein metabolism. *Lancet* 1:1423–1426.

Jeevanandam, M., Brennan, M., Horowitz, M.F., Rose, D., Mihranian, M.H., Daly, J. et al. 1985. Tracer priming in human protein turnover studies with [^{15}N]glycine. *Biochem. Med.* 34:214–225.

Jeevanandam, M., Lowry, S.F., and Brennan, M.F. 1987. Effect of route of nutrient administration on whole-body protein kinetics in man. *Metabolism* 36:968–973.

Jennissen, H.P. 1995. Ubiquitin and the enigma of intracellular protein degradation. *Eur. J. Biochem.* 231:1–30.

Jensen, J., Christiansen, C., and Rodbro, P. 1986. Oestrogen-progestogen replacement therapy changes body composition in early post-menopausal women. *Maturitas* 8:209–216.

Jensen, M.D., Miles, J.M., Gerich, J.E., Cryer, P.E., and Haymond, M.W. 1988. Preservation of insulin effects on glucose production and proteolysis during fasting. *Am. J. Physiol.* 254:E700–E707.

Jensen, M.D., and Haymond, M.W. 1991. Protein metabolism in obesity: effects of body fat distribution and hyperinsulinemia on leucine turnover. *Am. J. Clin. Nutr.* 53:172–176.

Jepson, M.M., Pell, J.M., Bates, P.C., and Millward, D.J. 1986. The effects of endotoxaemia on protein metabolism in skeletal muscle and liver of fed and fasted rats. *Biochem. J.* 235:329–336.

Jepson, M.M., Bates, P.C., Broadbent, P., Pell, J.M., and Millward, D.J. 1988. Relationship between glutamine concentration and protein synthesis in rat skeletal muscle. *Am. J. Physiol.* 255:E166–E172.

Johannsson, G., Grimby, G., Sunnerhagen, K.S., and Bengtsson, B.-A. 1997. Two years of growth hormone (GH) treatment increase isometric and isokinetic muscle strength in GH-deficient adults. *J. Clin. Endocrinol. Metab.* 82:2877–2884.

Johansson, A.G., Lindh, E., Blum, W.F., Kollerup, G., Sorensen, O.H., and Ljunghall, S. 1996. Effects of growth hormone and insulin-like growth factor I in men with idiopathic osteoporosis. *J. Clin. Endocrinol. Metab.* 81:44–48.

Jones, M.O., Pierro, A., Garlick, P.J., McNurlan, M.A., Donnell, S.C., and Lloyd, D.A. 1995. Protein metabolism kinetics in neonates: effect of intravenous carbohydrate and fat. *J. Pediatr. Surg.* 30:458–462.

Jonsson, K.B., Ljunghall, S., Karlstrom, O., Johansson, A.G., Mallmin, H., and Ljunggren, O. 1993. Insulin-like growth factor I enhances the formation of type I collagen in hydrocortisone-treated human osteoblasts. *Biosci. Rep.* 13:297–302.

Jorgensen, J.O.L., Pedersen, S.A., Thuesen, L., Jorgensen, J., Ingemann-Hansen, T., Skakkebaek, N.E. et al. 1989. Beneficial effects of growth hormone treatment in GH-deficient adults. *Lancet* 1:1221–1225.

Jorgensen, J.O.L., Vahl, N., Hansen, T.B., Thuesen, L., Hagen, C., and Christiansen, J.S. 1996. Growth hormone versus placebo treatment for one year in growth hormone deficient adults: increase in exercise capacity and normalization of body composition. *Clin. Endocrinol.* 45:681–688.

Jurasinski, C.V., Kilpatrick, L., and Vary, T.C. 1995. Amrinone prevents muscle protein wasting during chronic sepsis. *Am. J. Physiol.* 268:E491–E500.

Kaiser, F.E., Silver, A.J., and Morley, J.E. 1991. The effect of recombinant human growth hormone on malnourished older individuals. *J. Am. Geriatr. Soc.* 39:235–240.

Kalhan, S.C., Rossi, K.Q., Gruca, L.L., Super, D.M., and Savin, S.M. 1998. Relation between transamination of branched-chain amino acids and urea synthesis: evidence from human pregnancy. *Am. J. Physiol.* 275:E423–E431.

Kaminski, H.J., and Ruff, R.L. 1994. Endocrine myopathies. In *Myology,* ed. A.G. Engel and C. Franzini-Armstrong, pp. 1726–1753. New York: McGraw-Hill.

Kanaley, J.A., Haymond, M.W., and Jensen, M.D. 1993. Effects of exercise and weight loss on leucine turnover in different types of obesity. *Am. J. Physiol.* 264:E687–E692.

Karinch, A.M., Kimball, S.R., Vary, T.C., and Jefferson, L.S. 1993. Regulation of eukaryotic initiation factor-2B activity in muscle of diabetic rats. *Am. J. Physiol.* 264:E101–E108.

Katz, J., and Wolfe, R.R. 1988. On the measurement of lactate turnover in humans. *Metabolism* 37:1078–1080.

Kayali, A.G., Young, V.R., and Goodman, M.N. 1987. Sensitivity of myofibrillar proteins to glucocorticoid-induced muscle proteolysis. *Am. J. Physiol.* 252:E621–E626.

Kaysen, G.A., and Schoenfeld, P.Y. 1984. Albumin homeostasis in patients undergoing continuous ambulatory peritoneal dialysis. *Kidney Int.* 25:107–114.

Kaysen, G.A., and Rathore, V. 1996. Derangements of protein metabolism in chronic renal failure. *Blood Purif.* 14:373–381.

Kettelhut, I.C., and Goldberg, A.L. 1988. Tumor necrosis factor can induce fever in rats without activating protein breakdown in muscle or lipolysis in adipose tissue. *J. Clin. Invest.* 81:1384–1389.

Keys, A., Brozek, J., Henschel, A., Mickelsen, O., and Taylor, H.L. 1950. *The Biology of Human Starvation.* Minneapolis: University of Minnesota Press

Khaleeli, A.A., Edwards, R.H.T., Gohil, K., McPhail, G., Rennie, M.J., Round, J. et al. 1983. Corticosteroid myopathy: a clinical and pathological study. *Clin. Endocrinol.* 18:155–166.

Kien, C.L., Young, V.R., Rohrbaugh, D.K., and Burke, J.F. 1978a. Increased rates of whole body protein synthesis and breakdown in children recovering from burns. *Ann. Surg.* 187:383–391.

Kien, C.L., Young, V.R., Rohrbaugh, D.K., and Burke, J.F. 1978b. Whole-body protein synthesis and breakdown rates in children before and after reconstructive surgery of the skin. *Metabolism* 27:27–34.

Kien, C.L., and Camitta, B.M. 1983. Increased whole-body protein turnover in sick children with newly diagnosed leukemia or lymphoma. *Cancer Res.* 43:5586–5592.

Kilberg, M.S., Hutson, R.G., and Laine, R.O. 1994. Amino acid-regulated gene expression in eukaryotic cells. *FASEB J.* 8:13–19.

Kimball, S.R., and Jefferson, L.S. 1988. Effect of diabetes on guanine nucleotide exchange factor activity in skeletal muscle and heart. *Biochem. Biophys. Res. Commun.* 156:706–711.

Kimball, S.R., Everson, W.V., Flaim, K.E., and Jefferson, L.S. 1989. Initiation of protein synthesis in a cell-free system prepared from rat hepatocytes. *Am. J. Physiol.* 256:C28–C34.

Kimball, S.R., Jurasinski, C.V., Lawrence, J.C.J., and Jefferson, L.S. 1997. Insulin stimulates protein synthesis in skeletal muscle by enhancing the association of eIF-4E and eIF-4G. *Am. J. Physiol.* 272:C754–C759.

Knapik, J., Meredith, C., Jones, B., Fielding, R., Young, V., and Evans, W. 1991. Leucine metabolism during fasting and exercise. *J. Appl. Physiol.* 70:43–47.

Knowlton, K., Kenyon, A.T., Sandiford, I., Lotwin, G., and Fricker, R. 1942. Comparative study of metabolic effects of estradiol benzoate and testosterone propionate in man. *J. Clin. Endocrinol.* 2:671–684.

Kohrt, W.M., Landt, M., and Birge, S.J. 1996. Serum leptin levels are reduced in response to exercise training, but not hormone replacement therapy, in older women. *J. Clin. Endocrinol. Metab.* 81:3980–3985.

Kohrt, W.M., Hasten, D.L., Pak, J.Y., Gischler, J., Ehsani, A., and Yarasheski, K.E. 1997. Glucose, lipid and amino acid metabolism in 78+ year old physically frail men and women. *FASEB J.* 11:A426(Abstract)

Kondrup, J., Nielsen, K., and Juul, A. 1997. Effect of long-term refeeding on protein metabolism in patients with cirrhosis of the liver. *Br. J. Nutr.* 77:197–212.

Kountz, W.B. 1951. Revitalization of tissue and nutrition in older individuals. *Ann. Intern. Med.* 35:1055–1067.

Kraenzlin, M.E., Keller, U., Keller, A., Thelin, A., Arnaud, M.J., and Stauffacher, W. 1989. Elevation of plasma epinephrine concentrations inhibits proteolysis and leucine oxidation in man via β-adrenergic mechanisms. *J. Clin. Invest.* 84:388–393.

Krishna, R.G., and Wold, F. 1993. Post-translational modification of proteins. *Adv. Enzymol.* 67:265–298.

Kristal, B.S., and Yu, B.P. 1992. An emerging hypothesis: synergistic induction of aging by free radicals and Maillard reactions. *J. Gerontol.* 47:B107–B114.

Kudo, Y., Iwashita, M., Iguchi, T., and Takeda, Y. 1996. The regulation of type-I collagen synthesis by insulin-like growth factor-I in human osteoblastic-like SaOS-2 cells. *Pflugers Arch.* 433:123–128.

Kushner, I. 1982. The phenomenon of the acute phase response. *Ann. N.Y. Acad. Sci.* 389:39–48.

Laager, R., Ninnis, R., and Keller, U. 1993. Comparison of the effects of recombinant human insulin-like growth factor-I and insulin on glucose and leucine kinetics in humans. *J. Clin. Invest.* 92:1903–1909.

Laidlaw, S.A., and Kopple, J.D. 1987. Newer concepts of the indispensable amino acids. *Am. J. Clin. Nutr.* 46:593–605.

Lamberts, S.W.J., and Birkenhager, J.C. 1976. Body composition in Cushing's disease. *J. Clin. Endocrinol. Metab.* 42:864–868.

Lamond, A.I., and Earnshaw, W.C. 1998. Structure and function in the nucleus. *Science* 280:547–553.

Lamont, L.S., Patel, D.G., and Kalhan, S.C. 1989. β-Adrenergic blockade alters whole-body leucine metabolism in humans. *J. Appl. Physiol.* 67:221–225.

Lamont, L.S., Patel, D.G., and Kalhan, S.C. 1990. Leucine kinetics in endurance-trained humans. *J. Appl. Physiol.* 69:1–6.

Landau, R.L., and Poulos, J.T. 1971. The metabolic influence of progestins. *Adv. Metab. Disorders* 5:119–147.

Larbaud, D., Debras, E., Taillandier, D., Samuels, S.E., Temparis, S., Champredon, C. et al. 1996. Euglycemic hyperinsulinemia and hyperaminoacidemia decrease skeletal muscle ubiquitin mRNA in goats. *Am. J. Physiol.* 271:E505–E512.

Lariviere, F., Moussalli, R., and Garrel, D.R. 1994. Increased leucine flux and leucine oxidation during the luteal phase of the menstrual cycle in women. *Am. J. Physiol.* 267:E422–E428.

LeBlanc, A., Rowe, R., Schneider, V., Evans, H., and Hedrick, T. 1995. Regional muscle loss after short duration spaceflight. *Aviat. Space Environ. Med.* 66:1151–1154.

LeBricon, T., Gugins, S., Cynober, L., and Baracos, V. 1995. Negative impact of cancer chemotherapy on protein metabolism in healthy and tumor-bearing rats. *Metabolism* 44:1340–1348.

Lecavalier, L., De Feo, P., and Haymond, M.W. 1991. Isolated hypoisoleucinemia impairs whole body but not hepatic protein synthesis in humans. *Am. J. Physiol.* 261:E578–E586.

Lee, H.-K., and Marzella, L. 1994. Regulation of intracellular protein degradation with special reference to lysosomes: role in cell physiology and pathology. *Int. Rev. Exp. Pathol.* 35:39–145.

Leger, J., Carel, C., Legrand, I., Paulsen, A., Hassan, M., and Czernichow, P. 1994. Magnetic resonance imaging evaluation of adipose tissue and muscle tissue mass in children with growth hormone (GH) deficiency, Turner's Syndrome, and intrauterine growth retardation during the first year of treatment with GH. *J. Clin. Endocrinol. Metab.* 78:904–909.

Leinskold, T., Permert, J., Olaison, G., and Larsson, J. 1995. Effect of postoperative insulin-like growth factor I supplementation on protein metabolism in humans. *Br. J. Surg.* 82:921–925.

Lemon, P.W.R., and Mullin, J.P. 1980. Effect of initial muscle glycogen levels on protein catabolism during exercise. *J. Appl. Physiol.* 48:624–629.

Lemon, P.W.R., Tarnopolsky, M.A., MacDougall, J.D., and Atkinson, S.A. 1992. Protein requirements and muscle mass/strength changes during intensive training in novice bodybuilders. *J. Appl. Physiol.* 73:767–775.

Lemon, P.W.R. 1996. Is increased dietary protein necessary or beneficial for individuals with a physically active lifestyle? *Nutr. Rev.* 54:S169–S175.

Lemon, P.W.R., Dolny, D.G., and Yarasheski, K.E. 1997. Moderate physical activity can increase dietary protein needs. *Can. J. Appl. Physiol.* 22:494–503.

Leonard, J.I., Leach, C.S., and Rambaut, P.C. 1983. Quantitation of tissue loss during prolonged space flight. *Am. J. Clin. Nutr.* 38:667–679.

Leverve, X., Guignier, M., Carpenter, F., Serre, J.C., and Caravel, J.P. 1984. Effect of parenteral nutrition on muscle amino acid output and 3-methylhistidine excretion in septic patients. *Metabolism* 33:471–477.

Lewallen, C.G., Rall, J.E., and Berman, M. 1959. Studies of iodoalbumin metabolism. II. The effects of thyroid hormone. *J. Clin. Invest.* 38:88–101.

Lewis, G.F., Uffelman, K.D., Szeto, L.W., and Steiner, G. 1993. Effects of acute hyperinsulinemia on VLDL triglyceride and VLDL ApoB production in normal weight and obese individuals. *Diabetes* 42:833–842.

Lewis, S.E.M., Kelly, F.J., and Goldspink, D.F. 1984. Pre- and post-natal growth and protein turnover in smooth muscle, heart and slow- and fast-twitch skeletal muscles of the rat. *Biochem. J.* 217:517–526.

Li, J.B., and Jefferson, L.S. 1978. Influence of amino acid availability on protein turnover in perfused skeletal muscle. *Biochim. Biophys. Acta* 544:351–359.

Lieberman, S.A., Butterfield, G.E., Harrison, D., and Hoffman, A.R. 1994. Anabolic effects of recombinant insulin-like growth factor-I in cachectic patients with the acquired immunodeficieny syndrome. *J. Clin. Endocrinol. Metab.* 78:404–410.

Lin, E., Goncalves, J.A., and Lowry, S.F. 1998. Efficacy of nutritional pharmacology in surgical patients. *Curr. Opin. Nutr. Metab. Care* 1:41–50.

Lin, F.D., Smith, T.K., and Bayley, H.S. 1988. A role for tryptophan in regulation of protein synthesis in porcine muscle. *J. Nutr.* 118:445–449.

Ling, P.R., Schwartz, J.H., and Bistrian, B.R. 1997. Mechanisms of host wasting induced by administration of cytokines in rats. *Am. J. Physiol.* 272:E333–E339.

Ljungqvist, O.H., Persson, M., Schimke, J., Ford, G.C., and Nair, K.S. 1996. Effect of meal-induced insulin on muscle protein synthesis: measurement using amino acyl tRNA. *Diabetes* 45:103A(Abstract)

Ljungqvist, O.H., Persson, M., Ford, G.C., and Nair, K.S. 1997. Functional heterogeneity of leucine pools in human skeletal muscle. *Am. J. Physiol.* 273:E564–E570.

Llovera, M., Garcia-Martinez, C., Agell, N., Lopez-Soriano, F., and Argiles, J.M. 1995. Muscle wasting associated with cancer cachexia is linked to an important activation of the ATP-dependent ubiquitin-mediated proteolysis. *Int. J. Cancer* 61:138–141.

Llovera, M., Garcia-Martinez, C., Agell, N., Lopez-Soriano, F., and Argiles, J.M. 1997. TNF can directly induce the expression of ubiquitin-dependent proteolytic system in rat soleus muscle. *Biochem. Biophys. Res. Commun.* 230:238–241.

Lo, H.-C., and Ney, D.M. 1996. GH and IGF-I differentially increase protein synthesis in skeletal muscle and jejunum of parenterally fed rats. *Am. J. Physiol.* 271:E872–E878.

Long, C.L., Schiller, W.R., Blakemore, W.S., Geiger, J.W., O'Dell, M., and Henderson, K. 1977. Muscle protein catabolism in the septic patient as measured by 3-methylhistidine excretion. *Am. J. Clin. Nutr.* 30:1349–1352.

Long, C.L., Birkhahn, R.H., Geiger, J.W., Betts, J.E., Schiller, W.R., and Blakemore, W.S. 1981. Urinary excretion of 3-methylhistidine: an assessment of muscle protein catabolism in adult normal subjects and during malnutrition, sepsis, and skeletal trauma. *Metabolism* 30:765–776.

Long, C.L., Dillard, D.R., Bodzin, J.H., Geiger, J.W., and Blakemore, W.S. 1988.

Validity of 3-methylhistidine excretion as an indicator of skeletal muscle protein breakdown in humans. *Metabolism* 37:844–849.

Long, C.L., Nelson, K.M., DiRienzo, D.B., Weis, J.K., Stahl, R.D., Broussard, T.D. et al. 1995. Glutamine supplementation of enteral nutrition: impact on whole body protein kinetics and glucose metabolism in critically ill patients. *J. Parent. Ent. Nutr.* 19:470–476.

Long, W.M., Chua, B.H.L., Munger, B.L., and Morgan, H.E. 1984. Effects of insulin on cardiac lysosomes and protein degradation. *Fed. Proc.* 43:1295–1300.

Lorite, M.J., Cariuk, P., and Tisdale, M.J. 1997. Induction of muscle protein degradation by a tumour factor. *Br. J. Cancer* 76:1035–1040.

Louard, R.J., Fryburg, D.A., Gelfand, R.A., and Barrett, E.J. 1992. Insulin sensitivity of protein and glucose metabolism in human forearm skeletal muscle. *J. Clin. Invest.* 90:2348–2354.

Louard, R.J., Bhushan, R., Gelfand, R.A., Barrett, E.J., and Sherwin, R.S. 1994. Glucocorticoids antagonize insulin's antiproteolytic action on skeletal muscle in humans. *J. Clin. Endocrinol. Metab.* 79:278–284.

Louard, R.J., Barrett, E.J., and Gelfand, R.A. 1990. Effect of infused branched-chain amino acids on muscle and whole-body amino acid metabolism in man. *Clin. Sci.* 79:457–466.

Louard, R.J., Barrett, E.J., and Gelfand, R.A. 1995. Overnight branched-chain amino acid infusion causes sustained suppression of muscle proteolysis. *Metabolism* 44:424–429.

Lovejoy, J.C., Smith, S.R., Bray, G.A., DeLany, J.P., Rood, J.C., Gouvier, D. et al. 1997. A paradigm of experimentally induced mild hyperthyroidism: effects on nitrogen balance, body composition, and energy expenditure in healthy young men. *J. Clin. Endocrinol. Metab.* 82:765–770.

Lowell, B.B., Ruderman, N.B., and Goodman, M.N. 1986. Evidence that lysosomes are not involved in the degradation of myofibrillar proteins in rat skeletal muscle. *Biochem. J.* 234:237–240.

Lucidi, P., Lauteri, M., Laureti, S., Celleno, R., Santoni, S., Volpi, E. et al. 1998. A dose-response study of growth hormone (GH) replacement on whole body protein and lipid kinetics in GH-deficient adults. *J. Clin. Endocrinol. Metab.* 83:353–357.

Lundeberg, S., Belfrage, M., Wernerman, J., von der Decken, A., Thunell, S., and Vinnars, E. 1991. Growth hormone improves muscle protein metabolism and whole body nitrogen economy in man during a hyponitrogenous diet. *Metabolism* 40:315–322.

Lundholm, K., and Schersten, T. 1975. Leucine incorporation into proteins and cathepsin-D activity in human skeletal muscles. The influence of the age of the subject. *Exp. Gerontol.* 10:155–159.

Lundholm, K., Bylund, A.-C., Holm, J., and Schersten, T. 1976. Skeletal muscle metabolism in patients with malignant tumor. *Eur. J. Cancer* 12:465–473.

Lundholm, K., Edstrom, S., Ekman, L., Karlberg, I., Bylund, A.-C., and Schersten, T. 1978. A comparative study of the influence of malignant tumor on host metabolism in mice and man. *Cancer* 42:453–461.

Lundholm, K., Edstrom, S., Ekman, L., Karlberg, I., Walker, P., and Schersten, T. 1981a. Protein degradation in human skeletal muscle tissue: the effect of insulin, leucine, amino acids and ions. *Clin. Sci.* 60:319–326.

Lundholm, K., Karlberg, I., Ekman, L., Edstrom, S., and Schersten, T. 1981b. Evaluation of anorexia as the cause of altered protein synthesis in skeletal muscles from nongrowing mice with sarcoma. *Cancer Res.* 41:1989–1996.

Lundholm, K., Bennegard, K., Eden, E., Svaninger, G., Emery, P.W., and Rennie, M.J. 1982. Efflux of 3-methylhistidine from the leg in cancer patients who experience weight loss. *Cancer Res.* 42:4807–4811.

Luzi, L., Castellino, P., Simonson, D.C., Petrides, A.S., and DeFronzo, R.A. 1990. Leucine metabolism in IDDM. Role of insulin and substrate availability. *Diabetes* 39:38–48.

Luzi, L., Petrides, A.S., and DeFronzo, R.A. 1993. Different sensitivity of glucose and amino acid metabolism to insulin in NIDDM. *Diabetes* 42:1868–1877.

Luzi, L., Castellino, P., and DeFronzo, R.A. 1996. Insulin and hyperaminoacidemia regulate by a different mechanism leucine turnover and oxidation in obesity. *Am. J. Physiol.* 270:E273–E281.

Macallan, D.C., McNurlan, M.A., Milne, E., Calder, A.G., Garlick, P.J., and Griffin, G.E. 1995. Whole-body protein turnover from leucine kinetics and the response to nutrition in human immunodeficiency virus infection. *Am. J. Clin. Nutr.* 61:818–826.

MacDougall, J.D., Gibala, M.J., Tarnopolsky, M.A., MacDonald, J.R., Interisano, S.A., and Yarasheski, K.E. 1995. The time course for elevated muscle protein synthesis following heavy resistance exercise. *Can. J. Appl. Physiol.* 20:480–486.

Macfie, J., Smith, R.C., and Hill, G.L. 1981. Glucose or fat as a nonprotein energy source? A controlled clinical trial in gastroenterological patients requiring intravenous nutrition. *Gastroenterology* 80:103–107.

MacGillivray, M.H. 1995. Disorders of growth and development. In *Endocrinology and Metabolism,* ed. P. Felig, J.D. Baxter, and L.A. Frohman, pp. 1619–1673. New York: McGraw-Hill

MacLennan, P.A., Brown, R.A., and Rennie, M.J. 1987. A positive relationship between protein synthetic rate and intracellular glutamine concentration in perfused rat skeletal muscle. *FEBS Let* 215:187–191.

MacLennan, P.A., Smith, K., Weryk, B., Watt, P.W., and Rennie, M.J. 1988. Inhibition of protein breakdown by glutamine in perfused rat skeletal muscle. *FEBS Let.* 237:133–136.

Manary, M.J., Brewster, D.R., Broadhead, R.L., Crowley, J.R., Fjeld, C.R., and Yarasheski, K.E. 1997. Protein metabolism in children with edematous malnutrition and acute lower respiratory infection. *Am. J. Clin. Nutr.* 65:1005–1010.

Mancini, D.M., Walter, G., Reichek, N., Lenkinski, R., McCully, K.K., Mullen, J.L. et al. 1992. Contribution of skeletal muscle atrophy to exercise intolerance and altered muscle metabolism in heart failure. *Circulation* 85:1364–1373.

Manson, J.M., Smith, R.J., and Wilmore, D.W. 1988. Growth hormone stimulates protein synthesis during hypocaloric parenteral nutrition. *Ann. Surg.* 208:136–142.

Mansoor, O., Beaufrere, B., Boirie, Y., Ralliere, C., Taillandier, D., Aurousseau, E. et al. 996. Increased mRNA levels for components of the lysosomal, Ca^{2+}-activated, and ATP-ubiquitin-dependent proteolytic pathways in skeletal muscle from head trauma patients. *Proc. Natl. Acad. Sci. USA* 93:2714–2718.

Mansoor, O., Cayol, M., Gachon, P., Boirie, Y., Schoeffler, P., Obled, C. et al. 1997. Albumin and fibrinogen syntheses increase while muscle protein synthesis decreases in head-injured patients. *Am. J. Physiol.* 273:E898–E902.

Marchesini, G., Zoli, M., Angiolini, A., Dondi, C., Bianchi, F.B., and Pisi, E. 1981. Muscle protein breakdown in liver cirrhosis and the role of altered carbohydrate metabolism. *Hepatology* 1:294–299.

Marchesini, G., Forlani, G., Zoli, M., Vannini, P., and Pisi, E. 1982. Muscle protein breakdown in uncontrolled diabetes as assessed by urinary 3-methylhistidine excretion. *Diabetologia* 23:456–458.

Marcus, R., Butterfield, G.E., Holloway, L., Gilliland, L., Baylink, D.J., Hintz, R.L. et al. 1990. Effects of short term administration of recombinant human growth hormone to elderly people. *J. Clin. Endocrinol. Metab.* 70:519–527.

Marliss, E.B., Murray, F.T., and Nakhooda, A.F. 1978. The metabolic response to hypocaloric protein diets in obese man. *J. Clin. Invest.* 62:468–479.

Maroni, B.J., Staffeld, C., Young, V.R., Manatunga, A., and Tom, K. 1997. Mechanisms permitting nephrotic patients to achieve nitrogen equilibrium with a protein-restricted diet. *J. Clin. Invest.* 99:2479–2487.

Martin, A.F. 1981. Turnover of cardiac troponin subunits. Kinetic evidence for a precursor pool of troponin-I. *J. Biol. Chem.* 256:964–968.

Martin, F.C., and Peters, T.J. 1985. Assessment *in vitro* and *in vivo* of muscle degradation in chronic skeletal muscle myopathy of alcoholism. *Clin. Sci.* 68:693–700.

Martin, W.H.I., Spina, R.J., Korte, E., Yarasheski, K.E., Angelopoulos, T.J., Nemeth, P.M. et al. 1991. Mechanisms of impaired exercise capacity in short duration experimental hyperthyroidism. *J. Clin. Invest.* 88:2047–2053.

Matthews, D.E., Bier, D.M., Rennie, M.J., Edwards, R.H.T., Halliday, D., Millward, D.J. et al. 1981. Regulation of leucine metabolism in man: a stable isotope study. *Science* 214:1129–1131.

Matthews, D.E., Schwarz, H.P., Yang, R.D., Motil, K.J., Young, V.R., and Bier, D.M. 1982. Relationship of plasma leucine and α-ketoisocaproate during a L-[1-^{13}C]leucine infusion in man: a method for measuring human intracellular leucine tracer enrichment. *Metabolism* 31:1105–1112.

Matthews, D.E., Pesola, G., and Campbell, R.G. 1990. Effect of epinephrine on amino acid and energy metabolism in humans. *Am. J. Physiol.* 258:E948–E956.

Mauras, N., Haymond, M.W., Darmaun, D., Vieira, N.E., Abrams, S.A., and Yergey, A.L. 1994. Calcium and protein kinetics in prepubertal boys. Positive effects of testosterone. *J. Clin. Invest.* 93:1014–1019.

Mauras, N. 1995a. Estrogens do not affect whole-body protein metabolism in the prepubertal female. *J. Clin. Endocrinol. Metab.* 80:2842–2845.

Mauras, N. 1995b. Combined recombinant human growth hormone and recombinant human insulin-like growth factor I: lack of synergy on whole body protein anabolism in normally fed subjects. *J. Clin. Endocrinol. Metab.* 80:2633–2637.

Mauras, N., and Beaufrere, B. 1995. Recombinant human insulin-like growth factor-I enhances whole body protein anabolism and significantly diminishes the protein catabolic effects of prednisone in humans without a diabetogenic effect. *J. Clin. Endocrinol. Metab.* 80:869–874.

Mauras, N., Doi, S.Q., and Shapiro, J.R. 1996. Recombinant human insulin-like growth factor I, recombinant human growth hormone, and sex steroids: effects on markers of bone turnover in humans. *J. Clin. Endocrinol. Metab.* 81:2222–2226.

Mauras, N., Horber, F.F., and Haymond, M.W. 1992. Low dose recombinant human insulin-like growth factor-I fails to affect protein anabolism but inhibits islet cell secretion in humans. *J. Clin. Endocrinol. Metab.* 75:1192–1197.

Mauras, N., Martha, P.M., Quarmby, V., and Haymond, M.W. 1997. rhIGF-I administration in humans: differential metabolic effects of bolus vs. continuous subcutaneous delivery. *Am. J. Physiol.* 272:E628–E633.

Mauras, N., Hayes, V., Welch, S., Rini, A., Helgeson, K., Dokler, M. et al. 1998a. Testosterone deficiency in young men: marked alterations in whole body protein kinetics, strength and adiposity. *J. Clin. Endocrinol. Metab.* 83:1886–1892.

Mauras, N., Welch, S., Rini, A., and Haymond, M.W. 1998b. Ovarian hyperandrogenism is associated with insulin resistance to both peripheral carbohydrate and whole-body protein metabolism in postpubertal young females: a metabolic study. *J. Clin. Endocrinol. Metab.* 83:1900–1905.

May, R.C., Kelly, R.A., and Mitch, W.E. 1986. Metabolic acidosis stimulates protein degradation in rat muscle by a glucocorticoid-dependent mechanism. *J. Clin. Invest.* 77:614–621.

Mays, P.K., McAnulty, R.J., and Laurent, G.J. 1991. Age-related changes in rates of protein synthesis and degradation in rat tissues. *Mech. Ageing Dev.* 59:229–241.

McClelland, I.S.M., Persaud, C., and Jackson, A.A. 1997. Urea kinetics in healthy women during normal pregnancy. *Br. J. Nutr.* 77:165–181.

McCullough, A.J., Mullen, K.D., Tavill, A.S., and Kalhan, S.C. 1992. In vivo differences between the turnover rates of leucine and leucine's ketoacid in stable cirrhosis. *Gastroenterology* 103:571–578.

McCullough, A.J., Mullen, K.D., and Kalhan, S.C. 1992. Body cell mass and leucine metabolism in cirrhosis. *Gastroenterology* 102:1325–1333.

McHardy, K.C., McNurlan, M.A., Milne, E., Calder, A.G., Fearns, L.M., Broom, J. et al. 1991. The effect of insulin suppression on postprandial nutrient metabolism: studies with infusion of somatostatin and insulin. *Eur. J. Clin. Nutr.* 45:515–526.

McKeran, R.O., Halliday, D., and Purkiss, P. 1977. Increased myofibrillar protein catabolism in Duchenne muscular dystrophy measured by 3-methylhistidine excretion in the urine. *J. Neurol. Neurosurg. Psychiatr.* 40:979–981.

McNulty, P.H., Louard, R.J., Deckelbaum, L.I., Zaret, B.L., and Young, L.H. 1995. Hyperinsulinemia inhibits myocardial protein degradation in patients with cardiovascular disease and insulin resistance. *Circulation* 92:2151–2156.

McNurlan, M.A., Pain, V.M., and Garlick, P.J. 1980. Conditions that alter rates of tissue protein synthesis *in vivo*. *Biochem. Soc. Trans.* 8:283–285.

McNurlan, M.A., Fern, E.B., and Garlick, P.J. 1982. Failure of leucine to stimulate protein synthesis *in vivo*. *Biochem. J.* 204:831–838.

McNurlan, M.A., McHardy, K.C., Broom, J., Milne, E., Fearns, L.M., Reeds, P.J. et al. 1987. The effect of indomethacin on the response of protein synthesis to feeding in rats and man. *Clin. Sci.* 73:69–75.

McNurlan, M.A., Essen, P., Heys, S.D., Buchan, V., Garlick, P.J., and Wernerman, J. 1991. Measurement of protein synthesis in human skeletal muscle: further investigation of the flooding technique. *Clin. Sci.* 81:557–564.

McNurlan, M.A., Essen, P., Milne, E., Vinnars, E., Garlick, P.J., and Wernerman, J. 1993. Temporal responses of protein synthesis in human skeletal muscle to feeding. *Br. J. Nutr.* 69:117–126.

McNurlan, M.A., Essen, P., Thorell, A., Calder, A.G., Anderson, S.E., Ljungqvist, O.H. et al. 1994a. Response of protein synthesis in human skeletal muscle to insulin: an investigation with L-[$^2H_5$]phenylalanine. *Am. J. Physiol.* 267:E102–E108.

McNurlan, M.A., Heys, S.D., Park, K.G.M., Broom, J., Brown, D.S., Eremin, O. et al. 1994b. Tumor and host tissue responses to branched-chain amino acid supplementation of patients with cancer. *Clin. Sci.* 86:339–345.

McNurlan, M.A., Sandgren, A., Hunter, K., Essen, P., Garlick, P.J., and Wernerman, J. 1996. Protein synthesis rates of skeletal muscle, lymphocytes, and albumin with stress hormone infusion in healthy man. *Metabolism* 45:1388–1394.

McNurlan, M.A., Garlick, P.J., Steigbigel, R.T., DeCristofaro, K.A., Frost, R.A., Lang, C.H. et al. 1997. Responsiveness of muscle protein synthesis to growth hormone

administration in HIV-infected individuals declines with severity of disease. *J. Clin. Invest.* 100:2125–2132.
McNurlan, M.A., Garlick, P.J., Frost, R.A., DeCristofaro, K.A., Lang, C.H., Steigbigel, R.T. et al. 1998. Albumin synthesis and bone collagen formation in human immunodeficiency virus-positive subjects: differential effects of growth hormone administration. *J. Clin. Endocrinol. Metab.* 83:3050–3055.
McPherron, A.C., and Lee, S.-J. 1997a. Regulation of skeletal muscle mass in mice by a new TGF-β superfamily member. *Nature* 387:83–90.
McPherron, A.C., and Lee, S.-J. 1997b. Double muscling in cattle due to mutations in the myostatin gene. *Proc. Natl. Acad. Sci. USA* 94:12457–12461.
McSherry, E., and Morris, R.C. 1978. Attainment and maintenance of normal stature with alkali therapy in infants and children with classic renal tubular acidosis. *J. Clin. Invest.* 61:509–527.
Medina, R., Wing, S.S., and Goldberg, A.L. 1995. Increase in levels of polyubiquitin and proteasome mRNA in skeletal muscle during starvation and denervation atrophy. *Biochem. J.* 307:631–637.
Meier, P.R., Peterson, R.G., Bonds, D.R., Meschia, G., and Battaglia, F.C. 1981. Rates of protein synthesis and turnover in fetal life. *Am. J. Physiol.* 240:E320–E324.
Melville, S., McNurlan, M.A., McHardy, K.C., Broom, J., Milne, E., Calder, A.G. et al. 1989. The role of degradation in the acute control of protein balance in adult man: failure of feeding to stimulate protein synthesis as assessed by L-[1-^{13}C]leucine infusion. *Metabolism* 38:248–255.
Melville, S., McNurlan, M.A., Calder, A.G., and Garlick, P.J. 1990. Increased protein turnover despite normal energy metabolism and responses to feeding in patients with lung cancer. *Cancer Res.* 50:1125–1131.
Meredith, C.N., Zackin, M.J., Frontera, W.R., and Evans, W.J. 1989. Dietary protein requirements and body protein metabolism in endurance-trained men. *J. Appl. Physiol.* 66:2850–2856.
Meredith, C.N., Frontera, W.R., O'Reilly, K.P., and Evans, W.J. 1992. Body composition in elderly men: effect of dietary modification during strength training. *J. Am. Geriatr. Soc.* 40:155–162.
Merrick, W.C. 1992. Mechanism and regulation of eukaryotic protein synthesis. *Microbiol. Rev.* 56:291–315.
Merry, B.J., Lewis, S.E.M., and Goldspink, D.F. 1992. The influence of age and chronic restricted feeding on protein synthesis in the small intestine of the rat. *Exp. Gerontol.* 27:191–200.
Miles, J.M., Nissen, S.L., Rizza, R.A., Gerich, J.E., and Haymond, M.W. 1983. Failure of infused β-hydroxybutyrate to decrease proteolysis in man. *Diabetes* 32:197–205.
Miles, J.M., Nissen, S.L., Gerich, J.E., and Haymond, M.W. 1984. Effects of epinephrine infusion on leucine and alanine kinetics in humans. *Am. J. Physiol.* 247:E166–E172.
Miller, W.L., and Tyrell, J.B. 1995. The adrenal cortex. In *Endocrinology and Metabolism,* ed. P. Felig, J.D. Baxter, and L.A. Frohman, pp. 555–711. New York: McGraw-Hill.
Millikan, W.J.J., Henderson, J.M., Galloway, J.R., Warren, W.D., Matthews, D.E., McGhee et al. 1985. In vivo measurement of leucine metabolism with stable isotopes in normal subjects and in those with cirrhosis fed conventional and branched-chain amino acid-enriched diets. *Surgery* 98:405–412.
Millward, D.J. 1970. Protein turnover in skeletal muscle. II. The effect of starvation and a protein-free diet on the synthesis and catabolism of skeletal muscle proteins in comparison to liver. *Clin. Sci.* 39:591–603.

Millward, D.J., Davies, C.T.M., Halliday, D., Wolman, S.L., Matthews, D.E., and Rennie, M.J. 1982. Effect of exercise on protein metabolism in humans as explored with stable isotopes. *Fed. Proc.* 41:2686–2691.

Millward, D.J., Bowtell, J.L., Pacy, P., and Rennie, M.J. 1994. Physical activity, protein metabolism and protein requirements. *Proc. Nutr. Soc.* 53:223–240.

Millward, D.J., Fereday, A., Gibson, N., and Pacy, P.J. 1997. Aging, protein requirements, and protein turnover. *Am. J. Clin. Nutr.* 66:774–786.

Mitch, W.E., Medina, R., Grieber, S., May, R.C., England, B.K., Price, S.R. et al. 1994. Metabolic acidosis stimulates muscle protein degradation by activating the adenosine triphosphate-dependent pathway involving ubiquitin and proteasomes. *J. Clin. Invest.* 93:2127–2133.

Mitch, W.E., and Goldberg, A.L. 1996. Mechanisms of muscle wasting. The role of the ubiquitin-proteasome pathway. *N. Engl. J. Med.* 335:1897–1905.

Mitch, W.E., Walser, M., and Sapir, D.G. 1981. Nitrogen sparing induced by leucine compared with that induced by its keto analogue, α-ketoisocaproate, in fasting obese man. *J. Clin. Invest.* 67:553–562.

Moller-Loswick, A.-C., Zachrisson, H., Hyltander, A., Korner, U., Matthews, D.E., and Lundholm, K. 1994. Insulin selectively attenuates breakdown of nonmyofibrillar proteins in peripheral tissues of normal men. *Am. J. Physiol.* 266:E645–E652.

Monnier, V.M., Sell, D.R., Nagaraj, R.H., and Miyata, S. 1991. Mechanisms of protection against damage mediated by the Maillard reaction in aging. *Gerontology* 37:152–165.

Morais, J.A., Gougeon, R., Pencharz, P.B., Jones, P.J.H., Ross, R., and Marliss, E.B. 1997. Whole-body protein turnover in the healthy elderly. *Am. J. Clin. Nutr.* 66:880–889.

Morales, A.J., Nolan, J.J., Nelson, J.C., and Yen, S.S.C. 1994. Effects of replacement dose of dehydroepiandrosterone in men and women of advancing age. *J. Clin. Endocrinol. Metab.* 78:1360–1367.

Morkin, E., Yazaki, Y., Katagiri, T., and Laraia, P.J. 1973. Comparison of the synthesis of the light and heavy chains of adult skeletal myosin. *Biochim. Biophys. Acta* 324:420–429.

Morlese, J.F., Forrester, T., and Jahoor, F. 1998. Acute-phase protein response to infection in severe malnutrition. *Am. J. Physiol.* 275:E112–E117.

Morrison, W.L., Gibson, J.N.A., and Rennie, M.J. 1988. Skeletal muscle and whole body protein turnover in cardiac cachexia: influence of branched-chain amino acid administration. *Eur. J. Clin. Invest.* 18:648–654.

Morrison, W.L., Gibson, J.N.A., Jung, R.T., and Rennie, M.J. 1988. Skeletal muscle and whole body protein turnover in thyroid disease. *Eur. J. Clin. Invest.* 18:62–68.

Morrison, W.L., Bouchier, I.A.D., Gibson, J.N.A., and Rennie, M.J. 1990. Skeletal muscle and whole-body protein turnover in cirrhosis. *Clin. Sci.* 78:613–619.

Mortimore, G.E., Poso, A.R., Kadowaki, M., and Wert, J.J.J. 1987. Multiphasic control of hepatic protein degradation by regulatory amino acids. General features and hormonal modulation. *J. Biol. Chem.* 262:16322–16327.

Mortimore, G.E., and Poso, A.R. 1988. Amino acid control of intracellular protein degradation. *Meth. Enzymol.* 166:461–476.

Mortola, J.F., and Yen, S.S.C. 1990. The effects of oral dehydroepiandrosterone on endocrine-metabolic parameters in postmenopausal women. *J. Clin. Endocrinol. Metab.* 71:696–704.

Motil, K.J., Bier, D.M., Matthews, D.E., Burke, J.F., and Young, V.R. 1981a. Whole body leucine and lysine metabolism studied with [1-^{13}C]leucine and [α-^{15}N]lysine: response in healthy young men given excess energy intake. *Metabolism* 30:783–791.

Motil, K.J., Matthews, D.E., Bier, D.M., Burke, J.F., Munro, H.N., and Young, V.R. 1981b. Whole-body leucine and lysine metabolism: response to dietary protein intake in young men. *Am. J. Physiol.* 240:E712–E721.

Motil, K.J., Montandon, C.M., Hachey, D.L., Boutton, T.W., Klein, P.D., and Garza, C. 1989. Whole-body protein metabolism in lactating and nonlacating women. *J. Appl. Physiol.* 66:370–376.

Motil, K.J., Montandon, C.M., Thotathuchery, M., and Garza, C. 1990. Dietary protein and nitrogen balance in lactating and nonlactating women. *Am. J. Clin. Nutr.* 51:378–384.

Motil, K.J., Opekun, A.R., Montandon, C.M., Berthold, H.K., Davis, T.A., Klein, P.D. et al. 1994a. Leucine oxidation changes rapidly after dietary protein intake is altered in adult women but lysine flux is unchanged as is lysine incorporation into VLDL-apolipoprotein B-100. *J. Nutr.* 124:41–51.

Motil, K.J., Thotathuchery, M., Montandon, C.M., Hachey, D.L., Boutton, T.W., Klein, P.D. et al. 1994b. Insulin, cortisol and thyroid hormones modulate maternal protein status and milk production and composition in humans. *J. Nutr.* 124:1248–1257.

Motil, K.J., Davis, T.A., Montandon, C.M., Wong, W.W., Klein, P.D., and Reeds, P.J. 1996. Whole-body protein turnover in the fed state is reduced in response to dietary protein restriction in lactating women. *Am. J. Clin. Nutr.* 64:32–39.

Motil, K.J., Sheng, H.-P., Kertz, B.L., Montandon, C.M., and Ellis, K.J. 1998. Lean body mass of well-nourished women is preserved during lactation. *Am. J. Clin. Nutr.* 67:292–300.

Mullen, K.D., Denne, S.C., McCullough, A.J., Savin, S.M., Bruno, D., Tavill, A.S. et al. 1986. Leucine metabolism in stable cirrhosis. *Hepatology* 6:622–630.

Mulligan, K., Grunfeld, C., Hellerstein, M.K., Neese, R.A., and Schambelan, M. 1993. Anabolic effects of recombinant human growth hormone in patients with wasting associated with human immunodefiency virus infection. *J. Clin. Endocrinol. Metab.* 77:956–962.

Mulligan, K., and Bloch, A.S. 1998. Energy expenditure and protein metabolism in human immunodeficiency virus infection and cancer cachexia. *Semin. Oncol.* 25:82–91.

Mulligan, K., Tai, V.W., and Schambelan, M. 1998. Effects of chronic growth hormone treatment on energy intake and resting energy metabolism in patients with human immunodeficiency virus-associated wasting—a Clinical Research Center study. *J. Clin. Endocrinol. Metab.* 83:1542–1547.

Munro, H.N. 1964a. Historical introduction: the origin and growth of our present concepts of protein metabolism. In *Mammalian Protein Metabolism,* ed. H.N. Munro and J.B. Allison, pp. 1–29. New York: Academic Press.

Munro, H.N. 1964b. General aspects of the regulation of protein metabolism by diet and by hormones. In *Mammalian Protein Metabolism,* ed. H.N. Munro and J.B. Allison, pp. 381–481. New York: Academic Press.

Munro, H.N., and Fleck, A. 1969. Analysis of tissues and body fluids for nitrogenous constitutents. In *Mammalian Protein Metabolism,* ed. H.N. Munro and J.B. Allison, pp. 423–525. New York: Academic Press

Naeye, R.L., and Roode, P.R. 1970. The sizes and numbers of cells in visceral organs in human obesity. *Am. J. Clin. Pathol.* 54:251–253.

Nagasaka, S., Sugimoto, H., Nakamura, T., Kusaka, I., Fujisawa, G., Sakuma, N. et al. 1997. Antithyroid therapy improves bony manifestations and bone metabolic markers in patients with Graves' thyrotoxicosis. *Clin. Endocrinol.* 47:215–221.

Nair, K.S., Garrow, J.S., Ford, G.C., Mahler, R.F., and Halliday, D. 1983. Effect of poor diabetic control and obesity on whole body protein metabolism in man. *Diabetologia* 25:400–403.

Nair, K.S., Ford, G.C., and Halliday, D. 1987. Effect of intravenous insulin treatment on in vivo whole body leucine kinetics and oxygen consumption in insulin-deprived type I diabetic patients. *Metabolism* 36:491–495.

Nair, K.S., Halliday, D., Ford, G.C., Heels, S., and Garrow, J.S. 1987a. Failure of carbohydrate to spare leucine oxidation in obese subjects. *Int. J. Obesity* 11:537–544.

Nair, K.S., Halliday, D., Matthews, D.E., and Welle, S.L. 1987b. Hyperglucagonemia during insulin deficiency accelerates protein catabolism. *Am. J. Physiol.* 253:E208–E213.

Nair, K.S., Woolf, P.D., Welle, S.L., and Matthews, D.E. 1987c. Leucine, glucose, and energy metabolism after 3 days of fasting in healthy human subjects. *Am. J. Clin. Nutr.* 46:557–562.

Nair, K.S., Halliday, D., and Griggs, R.C. 1988. Leucine incorporation into mixed skeletal muscle protein in humans. *Am. J. Physiol.* 254:E208–E213.

Nair, K.S., Welle, S.L., Halliday, D., and Campbell, R.G. 1988. Effect of β-hydroxybutyrate on whole-body leucine kinetics and fractional mixed skeletal muscle protein synthesis in humans. *J. Clin. Invest.* 82:198–205.

Nair, K.S., Halliday, D., Ford, G.C., and Garrow, J.S. 1989. Effect of triiodothyronine on leucine kinetics, metabolic rate, glucose concentration, and insulin secretion rate during two weeks of fasting in obese women. *Int. J. Obesity* 13:487–496.

Nair, K.S., Schwartz, R.G., and Welle, S. 1992. Leucine as a regulator of whole body and skeletal muscle protein metabolism in humans. *Am. J. Physiol.* 263:E928–E934.

Nair, K.S., Ford, G.C., Ekberg, K., Fernqvist-Forbes, E., and Wahren, J. 1995. Protein dynamics in whole body and in splanchnic and leg tissues in type I diabetic patients. *J. Clin. Invest.* 95:2926–2937.

Nakshabendi, I.M., Obeidat, W., Russell, R.I., Downie, S., Smith, K., and Rennie, M.J. 1995. Gut mucosal protein synthesis measured using intravenous and intragastric delivery of stable tracer amino acids. *Am. J. Physiol.* 269:E996–E999.

Negrutskii, B.S., and Deutscher, M.P. 1991. Channeling of aminoacyl-tRNA for protein synthesis *in vivo*. *Proc. Natl. Acad. Sci. USA* 88:4991–4995.

Nestler, J.E., Barlascini, C.O., Clore, J.N., and Blackard, W.G. 1988. Dehydroepiandrosterone reduces serum low density lipoprotein levels and body fat but does not alter insulin sensitivity in normal men. *J. Clin. Endocrinol. Metab.* 66:57–61.

Newman, E., Heslin, M.J., Wolf, R.F., Pisters, P.W.T., and Brennan, M.F. 1992. The effect of insulin on glucose and protein metabolism in the forearm of cancer patients. *Surg. Oncol.* 1:257–267.

Newman, E., Heslin, M.J., Wolf, R.F., Pisters, P.W.T., and Brennan, M.F. 1994. The effect of systemic hyperinsulinemia with concomitant amino acid infusion on skeletal muscle protein turnover in the human forearm. *Metabolism* 43:70–78.

Newsholme, E.A., and Crabtree, B. 1976. Substrate cycles in metabolic regulation and in heat generation. *Biochem. Soc. Symp.* 41:61–109.

Nordenstrom, J., Askanazi, J., Elwyn, D.H., Martin, P., Carpenter, Y.A., Robin, A.P. et al. 1983. Nitrogen balance during total parenteral nutrition. Glucose vs. fat. *Ann. Surg.* 197:27–33.

Norton, J.A., Stein, T.P., and Brennan, M.F. 1981. Whole body protein synthesis and turnover in normal man and malnourished patients with and without known cancer. *Ann. Surg.* 194:123–128.

O'Brien, C. 1994. Missing link in insulin's path to protein production. *Science* 266:542–543.
O'Keefe, S.J.D., Sender, P.M., and James, W.P.T. 1974. "Catabolic" loss of body nitrogen in response to surgery. *Lancet* 2:1035–1038.
O'Keefe, S.J.D., Abraham, R., El-Zayadi, A., Marshall, W., Davis, M., and Williams, R. 1981. Increased plasma tyrosine concentrations in patients with cirrhosis and fulminant hepatic failure associated with increased plasma tyrosine flux and reduced hepatic oxidation capacity. *Gastroenterology* 81:1017–1024.
O'Keefe, S.J.D., Ogden, J., Ramjee, G., and Rund, J. 1990. Contribution of elevated protein turnover and anorexia to cachexia in patients with hepatocellular carcinoma. *Cancer Res.* 50:1226–1230.
O'Sullivan, A.J., Kelly, J.J., Hoffman, D.M., Freund, J., and Ho, K.K.Y. 1994. Body composition and energy expenditure in acromegaly. *J. Clin. Endocrinol. Metab.* 78:381–386.
Oddoye, E.A., and Margen, S. 1979. Nitrogen balance studies in humans: long-term effect of high nitrogen intake on nitrogen accretion. *J. Nutr.* 109:363–377.
Oehri, M., Ninnis, R., Girard, J., Frey, F.J., and Keller, U. 1996. Effects of growth hormone and IGF-I on glucocorticoid-induced protein catabolism in humans. *Am. J. Physiol.* 270:E552–E558.
Olufemi, O.S., Humes, P., Whittaker, P.G., Read, M.A., Lind, T., and Halliday, D. 1990. Albumin synthetic rate: a comparison of arginine and alpha-ketoisocaproate precursor methods using stable isotope methods. *Eur. J. Clin. Nutr.* 44:351–361.
Olufemi, O.S., Whittaker, P.G., Halliday, D., and Lind, T. 1991. Albumin metabolism in fasted subjects during late pregnancy. *Clin. Sci.* 81:161–168.
Olufemi, O.S., Whittaker, P.G., and Lind, T. 1991. Glycine and urea metabolism during normal and diabetic pregnancies. *Proc. Nutr. Soc.* 50:200A(Abstract)
Osella, G., Terzolo, M., Reimondo, G., Piovesan, A., Pia, A., Termine, A. et al. 1997. Serum markers of bone and collagen turnover in patients with Cushing's syndrome and in subjects with adrenal incidentalomas. *J. Clin. Endocrinol. Metab.* 82:3303–3307.
Owen, O.E., Felig, P., Morgan, A.P., Wahren, J., and Cahill, G.F. 1969. Liver and kidney metabolism during prolonged starvation. *J. Clin. Invest.* 48:574–583.
Owen, O.E., Mozzoli, M.A., Smalley, K.J., Kavle, E.C., and D'Alessio, D.A. 1992. Oxidative and nonoxidative nutrient disposal in lean and obese men after mixed meals. *Am. J. Clin. Nutr.* 55:630–636.
Owen, O.E., Smalley, K.J., D'Alessio, D.A., Mozzoli, M.A., and Dawson, E.K. 1998. Protein, fat, and carbohydrate requirements during starvation: anaplerosis and cataplerosis. *Am. J. Clin. Nutr.* 68:12–34.
Ozawa, E., Yoshida, M., Suzuki, A., Mizuno, Y., Hagiwara, Y., and Noguchi, S. 1995. Dystrophin-associated proteins in muscular dystrophy. *Hum. Mol. Genet.* 4:1711–1716.
Paans, A.M.J., Pruim, J., van Waarde, A., Willemsen, A.T.M., and Vaalburg, W. 1996. Radiolabelled tyrosine for the measurement of protein synthesis rate in vivo by positron emission tomography. *Balliere's Clin. Endocrinol. Metab.* 10:497–510.
Pacy, P.J., and Halliday, D. 1989. Muscle protein synthesis in steriod-induced proximal myopathy: a case report. *Muscle Nerve* 12:378–381.
Pacy, P.J., Nair, K.S., Ford, G.C., and Halliday, D. 1989. Failure of insulin infusion to stimulate fractional muscle protein synthesis in type I diabetic patients. Anabolic effect of insulin and decreased proteolysis. *Diabetes* 38:618–624.

Pacy, P.J., Cheng, K.N., Ford, G.C., and Halliday, D. 1990. Influence of glucagon on protein and leucine metabolism: a study in fasting man with induced insulin resistance. *Br. J. Surg.* 77:791–794.

Pacy, P.J., Thompson, G.N., and Halliday, D. 1991. Measurement of whole-body protein turnover in insulin-dependent (type 1) diabetic patients during insulin withdrawal and infusion: comparison of [^{13}C]leucine and [$^{2}H_5$]phenylalanine methodologies. *Clin. Sci.* 80:345–352.

Pacy, P.J., Preedy, V.R., Peters, T.J., Read, M., and Halliday, D. 1991. The effect of chronic alcohol ingestion on whole body and muscle protein synthesis—a stable isotope study. *Alcohol Alcohol* 26:505–513.

Pacy, P.J., Price, G.M., Halliday, D., Quevedo, M.R., and Millward, D.J. 1994. Nitrogen homeostasis in man: the diurnal responses of protein synthesis and degradation and amino acid oxidation to diets with increasing protein intakes. *Clin. Sci.* 86:103–118.

Pacy, P.J., Read, M., and Halliday, D. 1990. Influence of insulin on albumin and non-albumin protein fractional synthetic rates in post-absorptive type I diabetic patients. *Eur. J. Clin. Nutr.* 44:343–349.

Palmer, R.M. 1990. Prostaglandins and the control of muscle protein synthesis and degradation. *Prostaglandins Leukotrienes and Essential Fatty Acids* 39:95–104.

Pannemans, D.L.E., Halliday, D., and Westerterp, K.R. 1995. Whole-body protein turnover in elderly men and women: responses to two protein intakes. *Am. J. Clin. Nutr.* 61:33–38.

Pannemans, D.L.E., Halliday, D., Westerterp, K.R., and Kester, A.D.M. 1995. Effect of variable protein intake on whole-body protein turnover in young men and women. *Am. J. Clin. Nutr.* 61:69–74.

Pannemans, D.L.E., Wagenmakers, A.J.M., Westerterp, K.R., Schaafsma, G., and Halliday, D. 1997. The effect of an increase of protein intake on whole-body protein turnover in elderly women is tracer dependent. *J. Nutr.* 127:1788–1794.

Parhofer, K.G., Barrett, P.H.R., Bier, D.M., and Schonfeld, G. 1991. Determination of kinetic parameters of apolipoprotein B metabolism using amino acids labeled with stable isotopes. *J. Lipid Res.* 32:1311–1323.

Pasquali, R., Baraldi, G., Biso, P., Piazzi, S., Patrono, D., Capelli, M. et al. 1984. Effect of "physiological" doses of triiodothyronine replacement on the hormonal and metabolic adaptation to short-term semistarvation and to low-calorie diet in obese patients. *Clin. Endocrinol.* 21:357–367.

Pasquali, R., Casimirri, F., Melchionda, N., Grossi, G., Bortoluzzi, L., Morselli Labate, A.M. et al. 1992. Effects of chronic administration of ephedrine during very-low-calorie diets on energy expenditure, protein metabolism and hormone levels in obese subjects. *Clin. Sci.* 82:85–92.

Pasquali, R., Casimirri, F., and Melchionda, N. 1987. Protein metabolism in obese patients during very low-calorie mixed diets containing different amounts of proteins and carbohydrates. *Metabolism* 36:1141–1148.

Patterson, B.W., Nguyen, T., Pierre, E., Herndon, D.N., and Wolfe, R.R. 1997a. Urea and protein metabolism in burned children: effect of dietary protein intake. *Metabolism* 46:573–578.

Patterson, B.W., Zhang, X.-J., Chen, Y.P., Klein, S., and Wolfe, R.R. 1997b. Measurement of very low stable isotope enrichments by gas chromatography/mass spectrometry: application to measurement of muscle protein synthesis. *Metabolism* 46:943–948.

Patti, M.-E., Brambilla, E., Luzi, L., Landaker, E.J., and Kahn, C.R. 1998. Bidirectional modulation of insulin action by amino acids. *J. Clin. Invest.* 101:1519–1529.

Pawan, G.L.S., and Semple, S.J.G. 1983. Effect of 3-hydroxybutyrate in obese subjects on very-low-energy diets and during therapeutic starvation. *Lancet* 1:15–17.
Pearce, G., Tabensky, D.A., Delmas, P.D., Baker, H.W.G., and Seeman, E. 1998. Corticosteroid-induced bone loss in men. *J. Clin. Endocrinol. Metab.* 83:801–806.
Peavy, D.E., Taylor, J.M., and Jefferson, L.S. 1981. Protein synthesis in perfused rat liver following thyroidectomey and hormone treatment. *Am. J. Physiol.* 240:E18–E23.
Pell, J.M., and Bates, P.C. 1992. Differential actions of growth hormone and insulin-like growth factor-I on tissue protein metabolism in dwarf mice. *Endocrinol.* 130:1942–1950.
Perriello, G., Nurjhan, N., Stumvoll, M., Bucci, A., Welle, S., Dailey, G. et al. 1997. Regulation of gluconeogenesis by glutamine in normal postabsorptive humans. *Am. J. Physiol.* 272:E437–E445.
Persani, L., Preziati, D., Matthews, C.H., Sartorio, A., Chatterjee, V.K.K., and Beck-Peccoz, P. 1997. Serum levels of carboxyterminal cross-linked telopeptide of type I collagen (ICTP) in the differential diagnosis of the syndromes of inappropriate secretion of TSH. *Clin. Endocrinol.* 47:207–214.
Peters, J.-M. 1994. Proteasomes: protein degradation machines of the cell. *Trends Biochem. Sci.* 19:377–382.
Petersen, S.R., Holaday, N.J., and Jeevanandam, M. 1994. Enhancement of protein synthesis efficiency in parenterally fed trauma victims by adjuvant recombinant human growth hormone. *J. Trauma* 36:726–733.
Petersen, S.R., Jeevanandam, M., Shahbazian, L.M., and Holaday, N.J. 1997. Reprioritization of liver protein synthesis resulting from recombinant human growth hormone supplementation in parenterally fed trauma patients: the effect of growth hormone on the acute-phase response. *J. Trauma* 42:987–996.
Petersson, B., von der Decken, A., Vinnars, E., and Wernerman, J. 1994. Long-term effects of postoperative total parenteral nutrition supplemented with glycylglutamine on subjective fatigue and muscle protein synthesis. *Br. J. Surg.* 81:1520–1523.
Petersson, B., Hultman, E., Andersson, K., and Wernerman, J. 1995. Human skeletal muscle protein: effect of malnutrition, elective surgery and total parenteral nutrition. *Clin. Sci.* 88:479–484.
Petrides, A.S., Luzi, L., Reuben, A., Riely, C., and DeFronzo, R.A. 1991. Effect of insulin and plasma amino acid concentration on leucine metabolism in cirrhosis. *Hepatology* 14:432–441.
Petrides, A.S., Luzi, L., and DeFronzo, R.A. 1994. Time-dependent regulation by insulin of leucine metabolism in young healthy adults. *Am. J. Physiol.* 267:E361–E368.
Philips, A.V., Timchenko, L.T., and Cooper, T.A. 1998. Disruption of splicing regulated by a CUG-binding protein in myotonic dystrophy. *Science* 280:737–741.
Phillips, S.M., Atkinson, S.A., Tarnopolsky, M.A., and MacDougall, J.D. 1993. Gender differences in leucine kinetics and nitrogen balance in endurance athletes. *J. Appl. Physiol.* 75:2134–2141.
Phillips, S.M., Tipton, K.D., Aarsland, A.A., Wolf, S.E., and Wolfe, R.R. 1997. Mixed muscle protein synthesis and breakdown after resistance exercise in humans. *Am. J. Physiol.* 273:E99–E107.
Phillips, S.M., Ferrando, A.A., Tipton, K.D., and Wolfe, R.R. 1998. Resistance training reduces muscle protein turnover. *FASEB J.* 12:A653(Abstract)
Piatti, P.M., Monti, L.D., Magni, F., Fermo, I., Baruffaldi, L., Nasser, R. et al. 1994. Hypocaloric high-protein diet improves glucose oxidation and spares lean body mass: comparison to hypocaloric high-carbohydrate diet. *Metabolism* 43:1481–1487.

Picou, D., and Taylor-Roberts, T. 1969. The measurement of total protein synthesis and catabolism and nitrogen turnover in infants in different nutritional states and receiving different amount of dietary protein. *Cli. Sci.* 36:283–296.

Pisters, P.W.T., and Pearlstone, D.B. 1993. Protein and amino acid metabolism in cancer cachexia: investigative techniques and therapeutic interventions. *Crit. Rev. Clin. Lab. Sci.* 30:223–272.

Pivarnik, J.M., Hickson, J.F., and Wolinsky, I. 1989. Urinary 3-methylhistidine excretion increases with repeated weight training exercise. *Med. Sci. Sports Exerc.* 21:283–287.

Poehlman, E.T., Toth, M.J., and Gardner, A.W. 1995. Changes in energy balance and body composition at menopause: a controlled longitudinal study. *Ann. Intern. Med.* 123:673–675.

Poindexter, B.B., Karn, C.A., Ahlrichs, J.A., Wang, J., Leitch, C.A., Liechty, E.A. et al. 1997. Amino acids suppress proteolysis independent of insulin throughout the neonatal period. *Am. J. Physiol.* 272:E592–E599.

Polin, R.A., Yoder, M.C., Douglas, S.D., McNelis, W., Nissim, I., and Yudkoff, M. 1989. Fibronectin turnover in the premature neonate measured with [^{15}N]glycine. *Am. J. Clin. Nutr.* 49:314–319.

Powis, M.R., Smith, K., Rennie, M.J., Halliday, D., and Pierro, A. 1998. Effect of major abdominal operations on energy and protein metabolism in infants and children. *J. Pediatr. Surg.* 33:49–53.

Preedy, V.R., McNurlan, M.A., and Garlick, P.J. 1983. Protein synthesis in skin and bone of the young rat. *Br. J. Nutr.* 49:517–523.

Preedy, V.R., and Garlick, P.J. 1985. The effect of glucagon administration on protein synthesis in skeletal muscles, heart and liver *in vivo*. *Biochem. J.* 228:575–581.

Preedy, V.R., and Garlick, P.J. 1988. Inhibition of protein synthesis by glucagon in different rat muscles and protein fractions *in vivo* and in the perfused rat hemicorpus. *Biochem. J.* 251:727–732.

Preedy, V.R., Paska, L., Sugden, P.H., Schofield, P.S., and Sugden, M.C. 1988. The effects of surgical stress and short-term fasting on protein synthesis *in vivo* in diverse tissues of the mature rat. *Biochem. J.* 250:179–188.

Preedy, V.R., Peters, T.J., Patel, V.B., and Miell, J.P. 1994. Chronic alcoholic myopathy: transcription and translational alterations. *FASEB J.* 8:1146–1151.

Preedy, V.R., Patel, V.B., Why, H.J.F., Corbett, J.M., Dunn, M.J., and Richardson, P.J. 1996. Alcohol and the heart: biochemical alterations. *Cardivasc. Res.* 31:139–147.

Preedy, V.R., Macallan, D.C., Griffin, G.E., Cook, E.B., Palmer, T.N., and Peters, T.J. 1997. Total contractile protein contents and gene expression in skeletal muscle in response to chronic ethanol consumption in the rat. *Alcohol* 14:545–549.

Preston, T., Fearon, K.C.H., McMillan, D.C., Winstanley, F.P., Slater, C., Shenkin, A. et al. 1995. Effect of ibuprofen on the acute-phase response and protein metabolism in patients with cancer and weight loss. *Br. J. Surg.* 82:229–234.

Prestwood, K.M., Pilbeam, C.C., Burleson, J.A., Woodiel, F.N., Delmas, P.D., Deftos, L.J. et al. 1994. The short term effects of conjugated estrogen on bone turnover in older women. *J. Clin. Endocrinol. Metab.* 79:366–371.

Price, G.M., Halliday, D., Pacy, P.J., Quevedo, M.R., and Millward, D.J. 1994a. Nitrogen homeostasis in man: influence of protein intake on the amplitude of diurnal cycling of body nitrogen. *Clin. Sci.* 86:91–102.

Price, S.R., England, B.K., Bailey, J.L., Van Vreede, K., and Mitch, W.E. 1994b. Acidosis and glucocorticoids concomitantly increase ubiquitin and proteasome subunit mRNAs in rat muscle. *Am. J. Physiol.* 267:C955–C960.

Price, S.R., Bailey, J.L., Wang, X., Jurkovitz, C., England, B.K., Ding, X. et al. 1996.

Muscle wasting in insulinopenic rats results from activation of the ATP-dependent, ubiquitin-proteasome proteolytic pathway by a mechanism including gene transcription. *J. Clin. Invest.* 98:1703–1708.
Prummel, M.F., Wiersinga, W.M., Lips, P., Sanders, G.T.B., and Sauerwein, H.P. 1991. The course of biochemical parameters of bone turnover during treatment with corticosteroids. *J. Clin. Endocrinol. Metab.* 72:382–386.
Pussell, B.A., Peake, P.W., Brown, M.A., and Charlesworth, J.A. 1985. Human fibronectin metabolism. *J. Clin. Invest.* 76:143–148.
Quevedo, M.R., Price, G.M., Halliday, D., Pacy, P.J., and Millward, D.J. 1994. Nitrogen homeostasis in man: diurnal changes in nitrogen excretion, leucine oxidation and whole body leucine kinetics during a reduction from a high to a moderate protein intake. *Clin. Sci.* 86:185–193.
Quinn, L.S., Haugk, K.L., and Grabstein, K.H. 1995. Interleukin-15: a novel anabolic cytokine for skeletal muscle. *Endocrinology* 136:3669–3672.
Radha, E., and Bessman, S.P. 1983. Effect of exercise on protein degradation: 3-methylhistidine and creatinine excretion. *Biochem. Med.* 29:96–100.
Rall, L.C., Rosen, C.J., Dolnikowski, G., Hartman, W.J., Lundgren, N., Abad, L.W. et al. 1996. Protein metabolism in rheumatoid arthritis and aging. *Arthritis Metab.* 39:1115–1124.
Ralliere, C., Tauveron, I., Taillandier, D., Guy, L., Boiteux, J.-P., Giraud, B. et al. 1997. Glucocorticoids do not regulate the expression of proteolytic genes in skeletal muscle from Cushing's syndrome patients. *J. Clin. Endocrinol. Metab.* 82:3161–3164.
Randall, V.A. 1994. Role of 5α-reductase in health and disease. *Balliere's Clin. Endocrinol. Metab.* 8:405–431.
Rannels, S.R., Rannels, D.E., Pegg, A.E., and Jefferson, L.S. 1978. Glucocorticoid effects on peptide-chain initiation in skeletal muscle and heart. *Am. J. Physiol.* 235:E134–E139.
Rao, B.S.N., and Nagabhushan, V.S. 1973. Urinary excretion of 3-methyl histidine in children suffering from protein-calorie malnutrition. *Life Sci.* 12:205–210.
Rasch, P.J., and Pierson, W.R. 1962. Effect of a protein dietary supplement on muscular strength and hypertrophy. *Am. J. Clin. Nutr.* 11:530–532.
Rathmacher, J.A., Flakoll, P.J., and Nissen, S.L. 1995. A compartmental model of 3-methylhistidine metabolism in humans. *Am. J. Physiol.* 269:E193–E198.
Reaich, D., Channon, S.M., Scrimgeour, C.M., and Goodship, T.H.J. 1992. Ammonium chloride-induced acidosis increases protein breakdown and amino acid oxidation in humans. *Am. J. Physiol.* 263:E735–E739.
Reaich, D., Channon, S.M., Scrimgeour, C.M., Daley, S.E., Wilkinson, R., and Goodship, T.H.J. 1993. Correction of acidosis in humans with CRF decreases protein degradation and amino acid oxidation. *Am. J. Physiol.* 265:E230–E235.
Redpath, N.T., and Proud, C.G. 1994. Molecular meachanisms in the control of translation by hormones and growth factors. *Biochim. Biophys. Acta* 1220:147–162.
Reeds, P.J., and Palmer, R.M. 1983. The possible involvement of prostaglandin $F_{2\alpha}$ in the stimulation of muscle protein synthesis by insulin. *Biochem. Biophys. Res. Commun.* 116:1084–1090.
Reeds, P.J., Hay, S.M., Glennie, R.T., Mackie, W.S., and Garlick, P.J. 1985. The effect of indomethacin on the stimulation of protein synthesis by insulin in young post-absorptive rats. *Biochem. J.* 227:255–261.
Reeds, P.J., Hachey, D.L., Patterson, B.W., Motil, K.J., and Klein, P.D. 1992. VLDL apolipoprotein B-100, a potential indicator of the isotopic labeling of the hepatic protein synthetic precursor pool in humans: studies with multiple stable isotopically labeled amino acids. *J. Nutr.* 122:457–466.

Reid, I.R., Wattie, D.J., Evans, M.C., and Stapleton, J.P. 1996. Testosterone therapy in glucocorticoid-treated men. *Arch. Intern. Med.* 156:1173–1177.

Reilly, M.E., Mantle, D., Richardson, P.J., Salisbury, J., Jones, J., Peters, T.J. et al. 1997. Studies on the time-course of ethanol's acute effects on skeletal muscle protein synthesis: comparison with acute changes in proteolytic activity. *Alcoholism: Clin. Exp. Res.* 21:792–798.

Reizman, H. 1997. The ins and outs of protein translocation. *Science* 278:1728–1729.

Rennie, M.J., Edwards, R.H.T., Davies, C.T.M., Krywawych, S., Halliday, D., Waterlow, J.C. et al. 1980. Protein and amino acid turnover during and after exercise. *Biochem. Soc. Trans.* 8:499–501.

Rennie, M.J., Edwards, R.H.T., Krywawych, S., Davies, C.T.M., Halliday, D., Waterlow, J.C. et al. 1981. Effect of exercise on protein turnover in man. *Clin. Sci.* 61:627–639.

Rennie, M.J., Edwards, R.H.T., Halliday, D., Matthews, D.E., Wolman, S.L., and Millward, D.J. 1982. Muscle protein synthesis measured by stable isotope techniques in man: the effects of feeding and fasting. *Clin. Sci.* 63:519–523.

Rennie, M.J., and Millward, D.J. 1983. 3-Methylhistidine excretion and the urinary 3-methylhistidine/creatinine ratio are poor indicators of skeletal muscle protein breakdown. *Clin. Sci.* 65:217–225.

Rennie, M.J., Bennegard, K., Eden, E., Emery, P.W., and Lundholm, K. 1984. Urinary excretion and efflux from the leg of 3-methylhistidine before and after major surgical operation. *Metabolism* 33:250–256.

Rennie, M.J., Smith, K., and Watt, P.W. 1994. Measurement of human tissue protein synthesis: an optimal approach. *Am. J. Physiol.* 266:E298–E307.

Richardson, D.P., Wayler, A.H., Scrimshaw, N.S., and Young, V.R. 1979. Quantitative effect of an isoenergetic exchange of fat for carbohydrate on dietary protein utilization in healthy young men. *Am. J. Clin. Nutr.* 32:2217–2226.

Ricketts, W.G., Birchenall-Sparks, M.C., Hardwick, J.P., and Richardson, A. 1985. Effect of age and dietary restriction on protein synthesis by isolated kidney cells. *J. Cell. Physiol.* 125:492–498.

Rifai, Z., Welle, S., Moxley, R.T. III, Lorenson, M., and Griggs, R.C. 1995. Effect of prednisone on protein metabolism in Duchenne dystrophy. *Am. J. Physiol.* 268:E67–E74.

Risteli, J., and Risteli, L. 1997. Assays of type I procollagen domains and collagen fragments: problems to be solved and future trends. *Scand. J. Clin. Lab. Invest.* 57(Suppl.227):105–113.

Robert, J.J., Beaufrere, B., Koziet, J., Desjeux, J.F., Bier, D.M., Young, V.R. et al. 1985. Whole body de novo amino acid synthesis in type I (insulin-dependent) diabetes studied with stable isotope-labeled leucine, alanine, and glycine. *Diabetes* 34: 67–73.

Robinson, S., Coldham, N., Gelding, S.V., Murphy, C., Beard, R.W., Halliday, D. et al. 1992. Leucine flux is increased whilst glucose turnover is normal, in pregnancy complicated by gestational diabetes mellitus. *Diabetologia,* 35(Suppl 2):A177(Abstract)

Robinson, S., and Pendergast, C.H. 1996. Protein metabolism in pregnancy. *Balliere's Clin. Endocrinol. Metab.* 10:571–587.

Robinson, S.M., Jaccard, C., Persaud, C., Jackson, A.A., Jequier, E., and Schutz, Y. 1990. Protein turnover and thermogenesis in response to high-protein and high-carbohydrate feeding in men. *Am. J. Clin. Nutr.* 52:72–80.

Rodier, M., Richard, J.L., Bringer, J., Cavalie, G., Bellet, H., and Mirouze, J. 1984. Thyroid status and muscle protein breakdown as assessed by urinary 3-methylhistidine excretion: study in thyrotoxic patients before and after treatment. *Metabolism* 33:97–100.

Rooyackers, O.E., Adey, D.B., Ades, P.A., and Nair, K.S. 1996. Effect of age on *in vivo* rates of mitochondrial protein synthesis in human skeletal muscle. *Proc. Natl. Acad. Sci. USA* 93:15364–15369.

Rooyackers, O.E., and Nair, K.S. 1997. Hormonal regulation of human muscle protein metabolism. *Ann. Rev. Nutr.* 17:457–485.

Rothschild, M.A., Oratz, M., Zimmon, D., Schreiber, S.S., Weiner, I., and Van Caneghem, A. 1969. Albumin synthesis in cirrhotic subjects with ascites studied with carbonate-^{14}C. *J. Clin. Invest.* 48:344–350.

Roubenoff, R., Roubenoff, R.A., Ward, L.M., and Stevens, M.B. 1990. Catabolic effects of high-dose corticosteroids persist despite therapeutic benefit in rheumatoid arthritis. *Am. J. Clin. Nutr.* 52:1113–1117.

Roubenoff, R., Roubenoff, R.A., Cannon, J.G., Kehayias, J.J., Zhuang, H., Dawson-Hughes, B. et al. 1994. Rheumatoid cachexia: cytokine-driven hypermetabolism accompanying reduced body cell mass in chronic inflammation. *J. Clin. Invest.* 93:2379–2386.

Rucker, R.B., and McGee, C. 1993. Chemical modifications of proteins in vivo: selected examples important to cellular regulation. *J. Nutr.* 123:977–990.

Rudman, D., Feller, A.G., Nagraj, H.N., Gergans, G.A., Lalitha, P.Y., Goldberg, A.F. et al. 1990. Effects of human growth hormone in men over 60 years old. *N. Engl. J. Med.* 323:1–6.

Russell, J.D., Mira, M., Allen, B.J., Stewart, P.M., Vizzard, J., Arthur, B. et al. 1994. Protein repletion and treatment in anorexia nervosa. *Am. J. Clin. Nutr.* 59:98–102.

Russell-Jones, D.L., Weissberger, A.J., Bowes, S.B., Kelly, J.M., Thomason, M., Umpleby, A.M. et al. 1993. The effects of growth hormone on protein metabolism in adult growth hormone deficient patients. *Clin. Endocrinol.* 38:427–431.

Russell-Jones, D.L., Umpleby, A.M., Hennessy, T.R., Bowes, S.B., Shojaee-Moradie, F., Hopkins, K.D. et al. 1994. Use of a leucine clamp to demonstrate that IGF-I actively stimulates protein synthesis in normal humans. *Am. J. Physiol.* 267:E591–E598.

Russell-Jones, D.L., Bowes, S.B., Rees, S.E., Jackson, N.C., Weissberger, A.J., Hovorka, R. et al. 1998. Effect of growth hormone on postprandial protein metabolism in growth hormone-deficient adults. *Am. J. Physiol.* 274:E1050–E1056.

Saido, T.C., Sorimachi, H., and Suzuki, K. 1994. Calpain: new perspectives in molecular diversity and physiological-pathological involvement. *FASEB J.* 8:814–822.

Sakurai, Y., Zhang, X.-J., and Wolfe, R.R. 1994. Effect of tumor necrosis factor on substrate and amino acid kinetics in conscious dogs. *Am. J. Physiol.* 266:E936–E945.

Sakurai, Y., Aarsland, A., Herndon, D.N., Chinkes, D.L., Pierre, E., Nguyen, T.T. et al. 1995. Stimulation of muscle protein synthesis by long-term insulin infusion in severely burned patients. *Ann. Surg.* 222:283–297.

Salbe, A.D., Kotler, D.P., Soave, R., Matthews, D.E., Wang, J., Ma, R.M. et al. 1995. Protein turnover and resting metabolic rate are normal in asymptomatic, HIV-infected men. *J. Invest. Med.* 43:370A(Abstract)

Sapir, D.G., Walser, M., Moyer, E.D., Rosenshein, N.B., Stewart, P.M., Moreadith, C. et al. 1983. Effects of α-ketoisocaproate and of leucine on nitrogen metabolism in postoperative patients. *Lancet* 1:1010–1013.

Sax, H.C., Talamini, M.A., Hasselgren, P.-O., Rosenblum, L., Ogle, C.K., and Fischer,

J.E. 1988. Increased synthesis of secreted hepatic proteins during abdominal sepsis. *J. Surg. Res.* 44:109–116.
Schiefermeier, M., Ratheiser, K.M., Zauner, C., Roth, E., Eichler, H.G., and Matthews, D.E. 1997. Epinephrine does not impair utilization of exogenous amino acids in humans. *Am. J. Clin. Nutr.* 65:1765–1773.
Schoenheimer, R.S., Rutner, S., and Rittenberg, D. 1939. Studies in protein metabolism; metabolic activity of body proteins investigated with L-leucine containing 2 isotopes. *J. Biol. Chem.* 130:730–732.
Schonheyder, F., Heilskov, N.S.C., and Olesen, K. 1954. Isotopic studies on the mechanism of negative nitrogen balance produced by immobilization. *Scand. J. Clin. Lab. Invest.* 6:178–188.
Schultz, R.M., and Liebman, M.N. 1997. Proteins I: composition and structure. In *Textbook of Biochemistry with Clinical Correlations,* ed. T.M. Devlin, pp. 23–83. New York: Wiley-Liss.
Schwenk, W.F., and Haymond, M.W. 1987. Effects of leucine, isoleucine, or threonine infusion on leucine metabolism in humans. *Am. J. Physiol.* 253:E428–E434.
Scrimshaw, N.S., Hussein, M.A., Murray, E., Rand, W.M., and Young, V.R. 1972. Protein requirements of man: variations in obligatory urinary and fecal nitrogen losses in young men. *J. Nutr.* 102:1595–1604.
Seene, T., and Alev, K. 1985. Effect of glucocorticoids on the turnover rate of actin and myosin heavy and light chains on different types of skeletal muscle fibres. *J. Steroid Biochem.* 22:767–771.
Seglen, P.O., and Bohley, P. 1992. Autophagy and other vacuolar protein degradation mechanisms. *Experientia* 48:158–172.
Selberg, O., Suttmann, U., Melzer, A., Deicher, H., Muller, M.-J., Henkel, E. et al. 1995. Effect of increased protein intake and nutritional status on whole-body protein metabolism of AIDS patients with weight loss. *Metabolism* 44:1159–1165.
Shamoon, H., Jacob, R., and Sherwin, R.S. 1980. Epinephrine-induced hypoaminoacidemia in normal and diabetic human subjects. Effect of beta blockade. *Diabetes* 29:875–881.
Shanbhogue, R.L.K., Bistrian, B.R., Lakshman, K., Crosby, L., Swenson, S., Wagner, D. et al. 1987. Whole body leucine, phenylalanine, and tyrosine kinetics in end-stage liver disease before and after hepatic transplantation. *Metabolism* 36:1047–1053.
Shangraw, R.E., Stuart, C.A., Prince, M.J., Peters, E.J., and Wolfe, R.R. 1988. Insulin responsiveness of protein metabolism in vivo following bedrest in humans. *Am. J. Physiol.* 255:E548–E558.
Shaw, J.H.F., Wildbore, M., and Wolfe, R.R. 1987. Whole body protein kinetics in severely septic patients. *Ann. Surg.* 205:288–294.
Shaw, J.H.F., Humberstone, D.A., Douglas, R.G., and Koea, J. 1991. Leucine kinetics in patients with benign disease, non-weight-losing cancer, and cancer cachexia: studies at the whole-body and tissue level and the response to nutritional support. *Surgery* 109:37–50.
Sherwin, R.S., Hendler, R., and Felig, P. 1975. Effect of ketone infusions on amino acid and nitrogen metabolism in man. *J. Clin. Invest.* 55:1382–1390.
Sherwin, R.S. 1978. Effect of starvation on the turnover and metabolic response to leucine. *J. Clin. Invest.* 61:1471–1481.
Silk, D.B.A., and Green, C.J. 1998. Perioperative nutrition: parenteral versus enteral. *Curr. Opin. Nutr. Metab. Care* 1:21–27.
Simmons, P.S., Miles, J.M., Gerich, J.E., and Haymond, M.W. 1984. Increased pro-

teolysis. An effect of increases in cortisol within the physiologic range. *J. Clin. Invest.* 73:412–420.

Sjolin, J., Stjernstrom, H., Friman, G., Larsson, J., and Wahren, J. 1990. Total and net muscle protein breakdown in infection determined by amino acid effluxes. *Am. J. Physiol.* 258:E856–E863.

Smith, K., Barua, J.M., Watt, P.W., Scrimgeour, C.M., and Rennie, M.J. 1992. Flooding with L-[1-^{13}C]leucine stimulates human muscle protein incorporation of continuously infused L-[1-^{13}C]valine. *Am. J. Physiol.* 262:E372–E376.

Smith, K., Downie, S., Barua, J.M., Watt, P.W., Scrimgeour, C.M., and Rennie, M.J. 1994. Effect of a flooding dose of leucine in stimulating incorporation of constantly infused valine into albumin. *Am. J. Physiol.* 266:E640–E644.

Smith, K., Reynolds, N., Downie, S., Patel, A., and Rennie, M.J. 1998a. Effects of flooding amino acids on incorporation of labeled amino acids into human muscle protein. *Am. J. Physiol.* 275:E73–E78.

Smith, K.L., and Tisdale, M.J. 1993. Increased protein degradation and decreased protein synthesis in skeletal muscle during cancer cachexia. *Br. J. Cancer* 67:680–685.

Smith, S.M., Nillen, J.L., LeBlanc, A., Lipton, A., Demers, L.M., Lane, H.W. et al. 1998b. Collagen cross-link excretion during space flight and bed rest. *J. Clin. Endocrinol. Metab.* 83:3584–3591.

Snyder, D.K., Clemmons, D.R., and Underwood, L.E. 1988. Treatment of obese, diet-restricted subjects with growth hormone for 11 weeks: effects on anabolism, lipolysis, and body composition. *J. Clin. Endocrinol. Metab.* 67:54–61.

Soares, M.J., Piers, L.S., Shetty, P.S., Robinson, S., Jackson, A.A., and Waterlow, J.C. 1991. Basal metabolic rate, body composition and whole-body protein turnover in Indian men with differing nutritional status. *Clin. Sci.* 81:419–425.

Soares, M.J., Piers, L.S., Shetty, P.S., Jackson, A.A., and Waterlow, J.C. 1994. Whole body protein turnover in chronically undernourished individuals. *Clin. Sci.* 86:441–446.

Solini, A., Bonora, E., Bonadonna, R., Castellino, P., and DeFronzo, R.A. 1997. Protein metabolism in human obesity: relationship with glucose and lipid metabolism and with visceral adipose tissue. *J. Clin. Endocrinol. Metab.* 82:2552–2558.

Sommer, T., and Wolf, D.H. 1997. Endoplasmic reticulum degradation: reverse protein flow of no return. *FASEB J.* 11:1227–1233.

Sonntag, W.E., Lenham, J.E., and Ingram, R.L. 1992. Effects of aging and dietary restriction on tissue protein synthesis: relationship to plasma insulin-like growth factor-1. *J. Gerontol.* 47:B159–B163.

Stadtman, E.R. 1988. Protein modification in aging. *J. Gerontol.* 43:B112–B120.

Stadtman, E.R. 1992. Protein oxidation and aging. *Science* 257:1220–1224.

Stadtman, E.R. 1995. Role of oxidized amino acids in protein breakdown and stability. *Methods Enzymol* 258:379–393.

Starnes, H.F., Warren, R.S., Jeevanandam, M., Gabrilove, J.L., Larchian, W., Oettgen, H.F. et al. 1988. Tumor necrosis factor and the acute metabolic response to tissue injury in man. *J. Clin. Invest.* 82:1321–1325.

Starnes, H.F., Warren, R.S.J., and Brennan, M.F. 1987. Protein synthesis in hepatocytes isolated from patients with gastrointestinal malignancy. *J. Clin. Invest.* 80:1384–1390.

Staten, M.A., Matthews, D.E., and Bier, D.M. 1986. Leucine metabolism in type II diabetes mellitus. *Diabetes* 35:1249–1253.

Stein, T.P., Oram-Smith, J.C., Wallace, H.W., and Leskiw, M.J. 1976. The effect of trauma on protein synthesis. *J. Surg. Res.* 21:201–203.

Stein, T.P., Mullen, J.L., Oram-Smith, J.C., Rosato, E.F., Wallace, H.W., and Hargrove, W.C.I. 1978. Relative rates of tumor, normal gut, liver, and fibrinogen protein synthesis in man. *Am. J. Physiol.* 234:E648–E652.

Stein, T.P., Leskiw, M.J., Buzby, G.P., Giandomenico, A.L., Wallace, H.W., and Mullen, J.L. 1980. Measurement of protein synthesis rates with [^{15}N]glycine. *Am. J. Physiol.* 239:E294–E300.

Stein, T.P., Nutinsky, C., Condoluci, D., Schluter, M.D., and Leskiw, M.J. 1990. Protein and energy substrate metabolism in AIDS patients. *Metabolism* 39:876–881.

Stein, T.P., Rumpler, W.V., Leskiw, M.J., Schluter, M.D., Staples, R., and Bodwell, C.E. 1991. Effect of reduced dietary intake on energy expenditure, protein turnover, and glucose cycling in man. *Metabolism* 40:478–483.

Stein, T.P., Leskiw, M.J., and Schluter, M.D. 1993. Effect of spaceflight on human protein metabolism. *Am. J. Physiol.* 264:E824–E828.

Stein, T.P., Leskiw, M.J., and Schluter, M.D. 1996. Diet and nitrogen metabolism during spaceflight on the shuttle. *J. Appl. Physiol.* 81:82–97.

Stein, T.P., and Schluter, M.D. 1997. Human skeletal muscle protein breakdown during spaceflight. *Am. J. Physiol.* 272:E688–E695.

Stiegler, H., Rett, K., Wicklmayr, M., and Menhert, H. 1989. Metabolic effects of prostaglandin E_1 on human skeletal muscle with special regard to the amino acid metabolism. *Vasa (Suppl)* 28:14–18.

Stroud, M.A., Jackson, A.A., and Waterlow, J.C. 1996. Protein turnover rates of two human subjects during an unassisted crossing of Antartica. *Br. J. Nutr.* 76:165–174.

Stryer, L. 1995. *Biochemistry.* New York: W.H. Freeman.

Stuart, C.A., Shangraw, R.E., Peters, E.J., and Wolfe, R.R. 1990. Effect of dietary protein on bed-rest-related changes in whole-body-protein synthesis. *Am. J. Clin. Nutr.* 52:509–514.

Svanberg, E., Moller-Loswick, A.-C., Matthews, D.E., Korner, U., Andersson, M., and Lundholm, K. 1996. Effects of amino acids on synthesis and degradation of skeletal muscle proteins in humans. *Am. J. Physiol.* 271:E718–E724.

Swart, G.R., van den Berg, J.W.O., Wattimena, J.L.D., Rietveld, T., van Vuure, J.K., and Frenkel, M. 1988. Elevated protein requirements in cirrhosis of the liver investigated by whole body protein turnover studies. *Clin. Sci.* 75:101–107.

Tappy, L., Owen, O.E., and Boden, G. 1988. Effect of hyperinsulinemia on urea pool size and substrate oxidation rates. *Diabetes* 37:1212–1216.

Tarnopolsky, M.A., MacDougall, J.D., and Atkinson, S.A. 1988. Influence of protein intake and training status on nitrogen balance and lean body mass. *J. Appl. Physiol.* 64:187–193.

Tarnopolsky, L.J., MacDougall, J.D., Atkinson, S.A., Tarnopolsky, M.A., and Sutton, J.R. 1990. Gender differences in substrate for endurance exercise. *J. Appl. Physiol.* 68:302–308.

Tarnopolsky, M.A., Atkinson, S.A., MacDougall, J.D., Senor, B.B., Lemon, P.W.R., and Schwarcz, H. 1991. Whole body leucine metabolism during and after resistance exercise in fed humans. *Med. Sci. Sports Exerc.* 23:326–333.

Tarnopolsky, M.A., Atkinson, S.A., MacDougall, J.D., Chesley, A., Phillips, S., and Schwarcz, H.P. 1992. Evaluation of protein requirements for trained strength athletes. *J. Appl. Physiol.* 73:1986–1995.

Tauveron, I., Larbaud, D., Champredon, C., Debras, E., Tesseraud, S., Bayle, G. et al.

1994. Effect of hyperinsulinemia and hyperaminoacidemia on muscle and liver protein synthesis in lactating goats. *Am. J. Physiol.* 267:E877–E885.

Tauveron, I., Charrier, S., Champredon, C., Bonnet, Y., Berry, C., Bayle, G. et al. 1995. Response of leucine metabolism to hyperinsulinemia under amino acid replacement in experimental hyperthyroidism. *Am. J. Physiol.* 269:E499–E507.

Taveroff, A., Lapin, H., and Hoffer, L.J. 1994. Mechanism governing short-term fed-state adaptation to dietary protein restriction. *Metabolism* 43:320–327.

Tavill, A.S., Craigie, A., and Rosenoer, V.M. 1968. The measurement of the synthetic rate of albumin in man. *Clin. Sci.* 34:1–28.

Tawa, N.E.J., Kettelhut, I.C., and Goldberg, A.L. 1992. Dietary protein deficiency reduces lysosomal and nonlysosomal ATP-dependent proteolysis in muscle. *Am. J. Physiol.* 263:E326–E334.

Tawa, N.E.J., Odessey, R., and Goldberg, A.F. 1997. Inhibitors of the proteasome reduce the accelerated proteolysis in atrophying rat skeletal muscles. *J. Clin. Invest.* 100:197–203.

Temparis, S., Asensi, M., Taillandier, D., Aurousseau, E., Larbaud, D., Obled, A. et al. 1994. Increased ATP-ubiquitin-dependent proteolysis in skeletal muscles of tumor-bearing rats. *Cancer Res.* 54:5568–5573.

Tessari, P., Nissen, S.L., Miles, J.M., and Haymond, M.W. 1986a. Inverse relationship of leucine flux and oxidation to free fatty acid availability in vivo. *J. Clin. Invest.* 77:575–581.

Tessari, P., Trevisan, R., Inchiostro, S., Biolo, G., Nosadini, R., De Kreutzenberg, S.V. et al. 1986b. Dose-response curves of effects of insulin on leucine kinetics in humans. *Am. J. Physiol.* 251:E334–E342.

Tessari, P., Inchiostro, S., Biolo, G., Trevisan, R., Fantin, G., Marescotti, M.C. et al. 1987. Differential effects of hyperinsulinemia and hyperaminoacidemia on leucine-carbon metabolism in vivo. Evidence for distinct mechanisms in regulation of net amino acid deposition. *J. Clin. Invest.* 79:1062–1069.

Tessari, P., Inchiostro, S., Biolo, G., Marescotti, M.C., Fantin, G., Boscarato, M.T. et al. 1989. Leucine kinetics and the effects of hyperinsulinemia in patients with Cushing's syndrome. *J. Clin. Endocrinol. Metab.* 68:256–262.

Tessari, P., Biolo, G., Inchiostro, S., Sacca, L., Nosadini, R., Boscarato, M.T. et al. 1990. Effects of insulin on whole body and forearm leucine and KIC metabolism in type 1 diabetes. *Am. J. Physiol.* 259:E96–E103.

Tessari, P., Inchiostro, S., Biolo, G., Vincenti, E., Sabadin, L., and Vettore, M. 1991. Effects of acute systemic hyperinsulinemia on forearm muscle proteolysis in healthy man. *J. Clin. Invest.* 88:27–33.

Tessari, P., Biolo, G., Inchiostro, S., Orlando, R., Vettore, M., and Sergi, G. 1993. Leucine and phenylalanine kinetics in compensated liver cirrhosis: effects of insulin. *Gastroenterology* 104:1712–1721.

Tessari, P., Biolo, G., Bruttomesso, D., Inchiostro, S., Panebianco, G., Vedovato, M. et al. 1994a. Effects of metformin treatment on whole-body and splanchnic amino acid turnover in mild type 2 diabetes. *J. Clin. Endocrinol. Metab.* 79:1553–1560.

Tessari, P., Inchiostro, S., Barazzoni, R., Zanetti, M., Orlando, R., Biolo, G. et al. 1994b. Fasting and postprandial phenylalanine and leucine kinetics in liver cirrhosis. *Am. J. Physiol.* 267:E140–E149.

Tessari, P., Barazzoni, R., Zanetti, M., Kiwanuka, E., and Tiengo, A. 1996a. The role of substrates in the regulation of protein metabolism. *Balliere's Clin. Endocrinol. Metab.* 10:511–532.

Tessari, P., Barazzoni, R., Zanetti, M., Vettore, M., Normand, S., Bruttomesso, D. et al. 1996b. Protein degradation and synthesis measured with multiple amino acid tracers in vivo. *Am. J. Physiol.* 271:E733–E741.

Tessari, P., Garibotto, G., Inchiostro, S., Robaudo, C., Saffiotti, S., Vettore, M. et al. 1996c. Kidney, splanchnic, and leg protein turnover in humans. Insight from leucine and phenylalanine kinetics. *J. Clin. Invest.* 98:1481–1492.

Tessari, P., Zanetti, M., Barazzoni, R., Biolo, G., Orlando, R., Vettore, M. et al. 1996d. Response of phenylalanine and leucine kinetics to branched chain-enriched amino acids and insulin in patients with cirrhosis. *Gastroenterology* 111:127–137.

Tessari, P., Zanetti, M., Barazzoni, R., Vettore, M., and Michielan, F. 1996e. Mechanisms of postprandial protein accretion in human skeletal muscle. Insight from leucine and phenylalanine forearm kinetics. *J. Clin. Invest.* 98:1361–1372.

Tessari, P., Iori, E., Vettore, M., Zanetti, M., Kiwanuka, E., Davanzo, G. et al. 1997. Evidence for acute stimulation of fibrinogen production by glucagon in humans. *Diabetes* 46:1368–1371.

Thompson, C., Blumenstock, F.A., Saba, T.M., Feustel, P.J., Kaplan, J.E., Fortune, J.B., Hough, L., and Gray, V. 1989. Plasma fibronectin synthesis in normal and injured humans as determined by stable isotope incorporation. *J. Clin. Invest.* 84:1226–1235.

Thompson, G.N., Pacy, P.J., Merritt, H., Ford, G.C., Read, M.A., Cheng, K.N., and Halliday, D. 1989. Rapid measurement of whole body and forearm protein turnover using a [$^2H_5$]phenylalanine model. *Am. J. Physiol.* 256:E631–E639.

Thompson, G.N., Bresson, J.L., Pacy, P.J., Bonnefont, J.P., Walter, J.H., Leonard, J.V. et al. 1990. Protein and leucine metabolism in maple syrup urine disease. *Am. J. Physiol.* 258:E654–E660.

Thompson, G.N., and Halliday, D. 1992. Protein turnover in pregnancy. *Eur. J. Clin. Invest.* 46:411–417.

Thompson, J.L., Butterfield, G.E., Marcus, R., Hintz, R.L., Van Loan, M., Ghiron, L. et al. 1995. The effects of recombinant human insulin-like growth factor-I and growth hormone on body composition in elderly women. *J. Clin. Endocrinol. Metab.* 80:1845–1852.

Thornton, C., Welle, S., Griggs, R.C., and Abraham, G.N. 1996. Human IgG production in vivo. Determination of synthetic rate by nonradioactive tracer incorporation. *J. Immunol.* 157:950–955.

Thornton, C.A., Griggs, R.C., Welle, S., Forbes, G.B., and Moxley, R.T. 1993. Recombinant human growth hormone (rhGH) treatment increases lean body mass in patients with myotonic dystrophy. *Neurology* 43:A280(Abstract)

Tipton, K.D., Ferrando, A.A., Williams, B.D., and Wolfe, R.R. 1996. Muscle protein metabolism in female swimmers after a combination of resistance and endurance exercise. *J. Appl. Physiol.* 81:2034–2038.

Tischler, M.E., Desautels, M., and Goldberg, A.L. 1982. Does leucine, leucyl-tRNA, or some metabolite of leucine regulate protein synthesis and degradation in skeletal and cardiac muscle? *J. Biol. Chem.* 257:1613–1621.

Tjader, I., Essen, P., Thorne, A., Garlick, P.J., Wernerman, J., and McNurlan, M.A. 1996. Muscle protein synthesis rate decreases 24 hours after abdominal surgery irrespective of total parenteral nutrition. *J. Parent. Ent. Nutr.* 20:135–138.

Todorov, P., Cariuk, P., McDevitt, T., Coles, B., Fearon, K., and Tisdale, M. 1996. Characterization of a cancer cachectic factor. *Nature* 379:739–742.

Toffolo, G., Foster, D.M., and Cobelli, C. 1993. Estimation of protein fractional synthetic rate from tracer data. *Am. J. Physiol.* 264:E128–E135.

Tom, K., Young, V.R., Chapman, T., Masud, T., Akpele, L., and Maroni, B.J. 1995. Long-term adaptive responses to dietary protein restriction in chronic renal failure. *Am. J. Physiol.* 268:E668–E677.

Tomas, F.M., Ballard, F.J., and Pope, L.M. 1979. Age-dependent changes in the rate of myofibrillar protein degradation in humans as assessed by 3-methylhistidine and creatinine excretion. *Clin. Sci.* 56:341–346.

Tomkins, J.K., Collins, S.P., Baker, W.D., and Kidman, A.D. 1982. In vitro skeletal muscle protein and RNA incorporation in neuromuscular disease. *J. Neurol. Sci.* 54: 59–68.

Tomkins, A.M., Garlick, P.J., Schofield, W.N., and Waterlow, J.C. 1983. The combined effects of infection and malnutrition on protein metabolism in children. *Clin. Sci.* 65:313–324.

Toth, M.J., Gottlieb, S.S., Fisher, M.L., and Poehlman, E.T. 1997. Skeletal muscle atrophy and peak oxygen consumption in heart failure. *Am. J. Cardiol.* 79:1267–1269.

Tsalikian, E., and Lim, V.S. 1989. L-Triodothyronine at slightly over physiologic dose increases leucine flux, which suggests an increase in protein degradation in normal subjects. *J. Lab. Clin. Med.* 114:171–175.

Tsujinaka, T., Fujita, J., Ebisui, C., Yano, M., Kominami, E., Suzuki, K., Tanaka, K. et al. 1996. Interleukin 6 receptor antibody inhibits muscle atrophy and modulates proteolytic systems in interleukin 6 transgenic mice. *J. Clin. Invest.* 97:244–249.

Turkalj, I., Keller, U., Ninnis, R., Vosmeer, S., and Stauffacher, W. 1992. Effect of increasing doses of recombinant human insulin-like growth factor-I on glucose, lipid, and leucine metabolism in man. *J. Clin. Endocrinol. Metab.* 75:1186–1191.

Uauy, R., Winterer, J.C., Bilmazes, C., Haverberg, L.N., Scrimshaw, N.S., Munro, H.N. et al. 1978. The changing pattern of whole body protein metabolism in aging humans. *J. Gerontol.* 33:663–671.

Umpleby, A.M., Boroujerdi, M.A., Brown, P.M., Carson, E.R., and Sonksen, P.H. 1986. The effect of metabolic control on leucine metabolism in Type 1 (insulin-dependent) diabetic patients. *Diabetologia* 29:131–141.

Umpleby, A.M., and Russell-Jones, D.L. 1996. The hormonal control of protein metabolism. *Balliere's Clin. Endocrinol. Metab.* 10:551–570.

Urban, R.J., Bodenburg, Y.H., Gilkison, C., Foxworth, J., Coggan, A.R., Wolfe, R.R. et al. 1995. Testosterone administration to elderly men increases skeletal muscle strength and protein synthesis. *Am. J. Physiol.* 269:E820–E826.

Usiskin, K.S., Butterworth, S., Clore, J.N., Arad, Y., Ginsberg, H.N., Blackard, W.G. et al. 1990. Lack of effect of dehydroepiandrosterone in obese men. *Int. J. Obesity* 14:457–463.

Vaisman, N., Clarke, R., Rossi, M., Goldberg, E., Zello, G.A., and Pencharz, P.B. 1992. Protein turnover and resting energy expenditure in patients with undernutrition and chronic lung disease. *Am. J. Clin. Nutr.* 55:63–69.

Vaisman, N., Zadik, Z., Akivias, A., Voet, H., Katz, I., Yair, S. et al. 1994. Changes in body composition, resting energy expenditure, and thermic effect of food in short children on growth hormone therapy. *Metabolism* 43:1543–1548.

Valk, N.K., Lely, A.J.V.D., De Herder, W.W., Lindemans, J., and Lamberts, S.W. 1994. The effects of human growth hormone (GH) administration in GH-deficient adults: a 20-day metabolic ward study. *J. Clin. Endocrinol. Metab.* 79:1070–1076.

Van Goudoever, J.B., Sulkers, E.J., Halliday, D., Degenhart, H.J., Carnielli, V.P., Wattimena, J.L.D. et al. 1995. Whole-body protein turnover in preterm appropriate for gestational age and small for gestational age infants: comparison of [^{15}N]glycine and [1-^{13}C]leucine administered simultaneously. *Pediatr. Res.* 37:381–388.

Varcoe, R., Halliday, D., Carson, E.R., Richards, P., and Tavill, A.S. 1975. Efficiency of utilization of urea nitrogen for albumin synthesis by chronically uraemic and normal man. *Clin. Sci.* 48:379–390.

Vary, T.C., and Kimball, S.R. 1992. Regulation of hepatic protein synthesis in chronic inflammation and sepsis. *Am. J. Physiol.* 262:C445–C452.

Vazquez, J.A., Morse, E.L., and Adibi, S.A. 1985. Effect of dietary fat, carbohydrate, and protein on branched-chain amino acid catabolism during caloric restriction. *J. Clin. Invest.* 76:737–743.

Vazquez, J.A., and Adibi, S.A. 1992. Protein sparing during treatment of obesity: ketogenic versus nonketogenic very low calorie diet. *Metabolism* 41:406–414.

Vignati, L., Finley, R.J., Hagg, S., and Aoki, T.T. 1978. Protein conservation during prolonged fast: a function of triiodothyronine levels. *Trans. Assoc. Am. Physicians* 91:169–179.

Vlachopapadopoulou, E., Zachwieja, J.J., Gertner, J.M., Manzione, D., Bier, D.M., Matthews, D.E. et al. 1995. Metabolic and clinical response to recombinant human insulin-like growth factor I in myotonic dystrophy—a clinical research center study. *J. Clin. Endocrinol. Metab.* 80:3715–3723.

Vogiatzi, M.G., Nair, K.S., Beckett, P.R., and Copeland, K.C. 1997. Insulin does not stimulate protein synthesis acutely in prepubertal children with insulin-dependent diabetes mellitus. *J. Clin. Endocrinol. Metab.* 82:4083–4087.

Volpi, E., Lucidi, P., Cruciani, G., Monacchia, F., Reboldi, G., Brunetti, P. et al. 1997. Nicotinamide counteracts alcohol-induced impairment of hepatic protein metabolism in humans. *J. Nutr.* 127:2199–2204.

Volpi, E., Lucidi, P., Cruciani, G., Monacchia, F., Santoni, S., Reboldi, G. et al. 1998. Moderate and large doses of ethanol differentially affect hepatic protein metabolism in humans. *J. Nutr.* 128:198–203.

Vom Dahl, S., Hallbrucker, C., Lang, F., Gerok, W., and Haussinger, D. 1991. Regulation of liver cell volume and proteolysis by glucagon and insulin. *Biochem. J.* 278:771–777.

Von Allmen, D., Hasselgren, P.-O., Higashiguchi, T., Frederick, J., Zamir, O., and Fischer, J.E. 1992. Increased intestinal protein synthesis during sepsis and following the administration of tumour necrosis factor α or interleukin-1α. *Biochem. J.* 286:585–589.

Walberg, J.L., Leidy, M.K., Sturgill, D.J., Hinkle, D.E., Ritchey, S.J., and Sebolt, D.R. 1988. Macronutrient content of a hypoenergy diet affects nitrogen retention and muscle function in weightlifters. *Int. J. Sports Med.* 9:261–266.

Walker, M., Shmueli, E., Daley, S.E., Cooper, B.G., and Alberti, K.G.M.M. 1993. Do nonesterified fatty acids regulate skeletal muscle protein turnover in humans? *Am. J. Physiol.* 265:E357–E361.

Walsh, C.H., Soler, N.G., James, H., Harvey, T.C., Thomas, B.J., Fremlin, J.H. et al. 1976. Studies in whole body potassium and whole body nitrogen in newly diagnosed diabetics. *Q. J. Med.* 45:295–301.

Walton, K.W., Scott, P.J., Dykes, P.W., and Davies, J.W.L. 1965. The significance of alterations in serum lipids in thyroid dysfunction. II. Alterations of the metabolism and turnover of ^{131}I-low-density lipoproteins in hypothyroidism and thyrotoxicosis. *Clin. Sci.* 29:217–238.

Wang, C., Eyre, D.R., Clark, R., Kleinberg, D., Newman, C., Iranmanesh, A. et al. 1996. Sublingual testosterone replacement improves muscle mass and strength, decreases bone resorption, and increases bone formation markers in hypogonadal men—a Clinical Research Center Study. *J. Clin. Endocrinol. Metab.* 81:3654–3662.

Wannemacher, R.W., Dinterman, R.E., Pekarek, R.S., Bartelloni, P.J., and Beisel, W.R. 1975. Urinary amino acid excretion during experimentally induced sandfly fever in man. *Am. J. Clin. Nutr.* 28:110–118.

Ward, H.C., Johnson, A.W., Halliday, D., and Sim, A.J.W. 1985. Elevated rates of whole body protein metabolism in patients with disseminated malignancy in the immediate postoperative period. *Br. J. Surg.* 72:983–986.

Ward, W.F. 1988. Enhancement by food restriction of liver protein synthesis in the aging Fischer 344 rat. *J. Gerontol.* 43:B50–B53.

Wasa, M., Bode, B.P., Abcouwer, S.F., Collins, C.L., Tanabe, K.K., and Souba, W.W. 1996. Glutamine as a regulator of DNA and protein biosynthesis in human solid tumor cell lines. *Ann. Surg.* 224:189–197.

Waterhouse, C., and Mason, J. 1981. Leucine metabolism in patients with malignant disease. *Cancer* 48:939–944.

Waterlow, J.C., Garlick, P.J., and Millward, D.J. 1978. *Protein turnover in mammalian tissues and in the whole body.* Amsterdam: North-Holland

Waterlow, J.C., Golden, M.H.N., and Garlick, P.J. 1978. Protein turnover in man measured with ^{15}N: comparison of end products and dose regimes. *Am. J. Physiol.* 235:E165–E174.

Waterlow, J.C. 1984. Protein turnover with special reference to man. *Quart. J. Exp. Physiol.* 69:409–438.

Watt, P.W., Lindsay, Y., Scrimgeour, C.M., Chien, P.A.F., Gibson, J.N.A., Taylor, D.J. et al. 1991. Isolation of aminoacyl-tRNA and its labeling with stable-isotope tracers: Use in studies of human tissue protein synthesis. *Proc. Natl. Acad. Sci. USA* 88:5892–5896.

Watt, P.W., Corbett, M.E., and Rennie, M.J. 1992. Stimulation of protein synthesis in pig skeletal muscle by infusion of amino acids during constant insulin availability. *Am. J. Physiol.* 263:E453–E460.

Weindruch, R., and Sohal, R.S. 1997. Caloric intake and aging. *N. Engl. J. Med.* 337:986–994.

Welle, S., Matthews, D.E., Campbell, R.G., and Nair, K.S. 1989. Stimulation of protein turnover by carbohydrate overfeeding in men. *Am. J. Physiol.* 257:E413–E417.

Welle, S., and Nair, K.S. 1990a. Failure of glyburide and insulin treatment to decrease leucine flux in obese type II diabetic patients. *Int. J. Obesity* 14:701–710.

Welle, S., and Nair, K.S. 1990b. Relationship of resting metabolic rate to body composition and protein turnover. *Am. J. Physiol.* 258:E990–E998.

Welle, S., Jozefowicz, R., and Statt, M. 1990. Failure of dehydroepiandrosterone to influence energy and protein metabolism in humans. *J. Clin. Endocrinol. Metab.* 71:1259–1264.

Welle, S., Barnard, R.R., Statt, M., and Amatruda, J.M. 1992a. Increased protein turnover in obese women. *Metabolism* 41:1028–1034.

Welle, S., Forbes, G.B., Statt, M., Barnard, R.R., and Amatruda, J.M. 1992b. Energy expenditure under free-living conditions in normal-weight and overweight women. *Am. J. Clin. Nutr.* 55:14–21.

Welle, S., Jozefowicz, R., Forbes, G., and Griggs, R.C. 1992c. Effect of testosterone on metabolic rate and body composition in normal men and men with muscular dystrophy. *J. Clin. Endocrinol. Metab.* 74:332–335.

Welle, S., Thornton, C., Jozefowicz, R., and Statt, M. 1993. Myofibrillar protein synthesis in young and old men. *Am. J. Physiol.* 264:E693–E698.

Welle, S., Statt, M., Barnard, R., and Amatruda, J.M. 1994a. Differential effect of insulin on whole-body proteolysis and glucose metabolism in normal-weight, obese, and reduced-obese women. *Metabolism* 43:441–445.

Welle, S., Thornton, C., Statt, M., and McHenry, B. 1994b. Postprandial myofibrillar and whole-body protein synthesis in young and old human subjects. *Am. J. Physiol.* 267:E599–E604.

Welle, S., Thornton, C., and Statt, M. 1995. Myofibrillar protein synthesis in young and old human subjects after three months of resistance training. *Am. J. Physiol.* 268:E422–E427.

Welle, S., Thornton, C., Statt, M., and McHenry, B. 1996a. Growth hormone increases muscle mass and strength but does not rejuvenate myofibrillar protein synthesis in healthy subjects over 60 years old. *J. Clin. Endocrinol. Metab.* 81:3239–3243.

Welle, S., Thornton, C., Totterman, S., and Forbes, G.B. 1996b. Utility of creatinine excretion in body-composition studies of healthy men and women older than 60 y. *Am. J. Clin. Nutr.* 63:151–156.

Welle, S., and Thornton, C.A. 1998. High-protein meals do not enhance myofibrillar synthesis after resistance exercise in 62-to 75-yr-old men and women. *Am. J. Physiol.* 274:E677–E683.

Wernerman, J., von der Decken, A., and Vinnars, E. 1985. Size distribution of ribosomes in biopsy specimens of human skeletal muscle during starvation. *Metabolism* 34:665–669.

Wernerman, J., von der Decken, A., and Vinnars, E. 1986. Polyribosome concentration in human skeletal muscle after starvation and parenteral or enteral refeeding. *Metabolism* 35:447–451.

Wernerman, J., Hammarqvist, F., von der Decken, A., and Vinnars, E. 1987. Ornithine-alpha-ketoglutarate improves skeletal muscle protein synthesis as assessed by ribosome analysis and nitrogen use after surgery. *Ann. Surg.* 206:674–678.

Wernerman, J., Botta, D., Hammarqvist, F., Thunell, S., von der Decken, A., and Vinnars, E. 1989. Stress hormones given to healthy volunteers alter the concentration and configuration of ribosomes in skeletal muscle, reflecting changes in protein synthesis. *Clin. Sci.* 77:611–616.

Wheatley, D.N., Grisolia, S., and Hernandez-Yago, J. 1982. Significance of the rapid degradation of newly synthesized proteins in mammalian cells: a working hypothesis. *J. Theor. Biol.* 98:283–300.

Whicher, J.T., Bell, A.M., Martin, M.F.R., Marshall, L.A., and Dieppe, P.A. 1984. Prostaglandins cause an increase in serum acute-phase proteins in man, which is diminished in systemic sclerosis. *Clin. Sci.* 66:165–171.

Whitney, E.N., and Cataldo, C.B. 1983. *Understanding Normal and Clinical Nutrition.* St. Paul: West Publishing Co.

Wicklmayr, M., Rett, K., Schwiegelshohn, B., Wolfram, G., Hailer, S., and Dietze, G. 1987. Inhibition of muscular amino acid release by lipid infusion in man. *Eur. J. Clin. Invest.* 17:301–305.

Wigmore, S.J., Fearon, K.C., Maingay, J.P., Lai, P.B.S., and Ross, J.A. 1997. Interleukin-8 can mediate acute-phase protein production by isolated human hepatocytes. *Am. J. Physiol.* 273:E720–E726.

Wigmore, S.J., Fearon, K.C., and Ross, J.A. 1997. Modulation of human hepatocyte acute phase protein production *in vitro* by n-3 and n-6 polyunsaturated fatty acids. *Ann. Surg.* 225:103–111.

Wilkinson, K.D. 1997. Regulation of ubiquitin-dependent processes by deubiquitinating enzymes. *FASEB J.* 11:1245–1256.

Wilkinson, P., and Mendenhall, C.L. 1963. Serum albumin turnover in normal subjects and patients with cirrhosis measured by ^{131}-I-labelled human albumin. *Clin. Sci.* 25:281–292.

Williams, B.D., Wolfe, R.R., Bracy, D.P., and Wasserman, D.H. 1996. Gut proteolysis contributes essential amino acids during exercise. *Am. J. Physiol.* 270:E85–E90.

Williamson, D.H., Farrell, R., Kerr, A., and Smith, R. 1977. Muscle-protein catabolism after injury in man, as measured by urinary excretion of 3-methylhistidine. *Clin. Sci. Mol. Med.* 52:527–533.

Wilmore, D.W., Goodwin, C.W., Aulick, L.H., Powanda, M.C., Mason, A.D., and Pruitt, B.A. 1980. Effect of injury and infection on visceral metabolism and circulation. *Ann. Surg.* 192:491–500.

Wilson, J.H.P., and Lamberts, S.W.J. 1981. The effect of triiodothyronine on weight loss and nitrogen balance of obese patients on a very-low-calorie liquid-formula diet. *Int. J. Obesity* 5:279–282.

Wing, S.S., Haas, A.L., and Goldberg, A.L. 1995. Increase in ubiquitin-protein conjugates concomitant with the increase in proteolysis in rat skeletal muscle during starvation and atrophy denervation. *Biochem. J.* 307:639–645.

Wing, S.S., and Bedard, N. 1996. Insulin-like growth factor I stimulates degradation of an mRNA transcript encoding the 14 kDa ubiquitin-conjugating enzyme. *Biochem. J.* 319:455–461.

Winterer, J.C., Steffee, W.P., Davy, W., Perera, A., Uauy, R., Scrimshaw, N.S. et al. 1976. Whole body protein turnover in aging man. *Exp. Gerontol.* 11:79–87.

Wolf, R.F., Heslin, M.J., Newman, E., Pearlstone, D.B., Gonenne, A., and Brennan, M. 1992a. Growth hormone and insulin combine to improve whole-body and skeletal muscle protein kinetics. *Surgery* 112:284–292.

Wolf, R.F., Pearlstone, D.B., Newman, E., Heslin, M.J., Gonenne, A., Burt, M.E. et al. 1992b. Growth hormone and insulin reverse net whole body and skeletal muscle protein catabolism in cancer patients. *Ann. Surg.* 216:280–290.

Wolfe, B.M., Culebras, J.M., Aoki, T.T., O'Connor, N.E., Finley, R.J., Kaczowka, A. et al. 1979. The effects of glucagon on protein metabolism in normal man. *Surgery* 86:248–257.

Wolfe, R.R., Wolfe, M.H., Nadel, E.R., and Shaw, J.H.F. 1984. Isotopic determination of amino acid-urea interactions in exercise in humans. *J. Appl. Physiol.* 56:221–229.

Wolfe, R.R. 1992. *Radioactive and Stable Isotope Tracers in Biomedicine: Principles and Practice of Kinetic Analysis.* New York: Wiley-Liss.

Wolman, S.L., Sheppard, H., Fern, M., and Waterlow, J.C. 1985. The effect of tri-iodothyronine (T_3) on protein turnover and metaboic rate. *Int. J. Obesity* 9:459–463.

Wolthers, O.D., Hansen, M., Juul, A., Nielsen, H.K., and Pedersen, S. 1997. Knemometry, urine cortisol excretion, and measures of the insulin-like growth factor axis and collagen turnover in children treated with inhaled glucocorticosteroids. *Pediatr. Res.* 41:44–50.

Wu, G., and Thompson, J.R. 1990. The effect of glutamine on protein turnover in chick skeletal muscle *in vitro*. *Biochem. J.* 265:593–598.

Yang, R.D., Matthews, D.E., Bier, D.M., Wen, Z.M., and Young, V.R. 1986. Response of alanine metabolism in humans to manipulation of dietary protein and energy intakes. *Am. J. Physiol.* 250:E39–E46.

Yang, R.D., Mack, G.W., Wolfe, R.R., and Nadel, E.R. 1998. Albumin synthesis after intense intermittent exercise in human subjects. *J. Appl. Physiol.* 84:584–592.

Yarasheski, K.E., Campbell, J.A., Smith, K., Rennie, M.J., Holloszy, J.O., and Bier, D.M. 1992. Effect of growth hormone and resistance exercise on muscle growth in young men. *Am. J. Physiol.* 262:E261–E267.

Yarasheski, K.E., Zachwieja, J.J., Angelopoulos, T.J., and Bier, D.M. 1993. Short-term growth hormone treatment does not increase muscle protein synthesis in experienced weight lifters. *J. Appl. Physiol.* 74:3073–3076.

Yarasheski, K.E., Zachwieja, J.J., Campbell, J.A., and Bier, D.M. 1995. Effect of growth hormone and resistance exercise on muscle growth and strength in older men. *Am. J. Physiol.* 268:E268–E276.

Yarasheski, K.E., Pak, J.Y., Crowley, J.R., and Hasten, D.L. 1998. Fractional synthesis rates of mixed muscle protein, myosin heavy chain, and actin in young and elderly men and women. *FASEB J.* 12:A430(Abstract)

Yarasheski, K.E., Zachwieja, J.J., and Bier, D.M. 1993. Acute effects of resistance exercise on muscle protein synthesis rate in young and elderly men and women. *Am. J. Physiol.* 265:E210–E214.

Yates, L.D., and Greaser, M.L. 1983. Quantitative determination of myosin and actin in rabbit skeletal muscle. *J. Mol. Biol.* 168:123–141.

Young, L.H., McNulty, P.H., Morgan, C., Deckelbaum, L.I., Zaret, B.L., and Barrett, E.J. 1991. Myocardial protein turnover in patients with coronary artery disease. effect of branched chain amino acid infusion. *J. Clin. Invest.* 87:554–560.

Young, L.H., Stirewalt, W., McNulty, P.H., Revkin, J.H., and Barrett, E.J. 1994. Effect of insulin on rat heart and skeletal muscle phenylalanyl-tRNA labeling and protein synthesis in vivo. *Am. J. Physiol.* 267:E337–E342.

Young, V.R., Steffee, W.P., Pencharz, P.B., Winterer, J.C., and Scrimshaw, N.S. 1975. Total human body protein synthesis in relation to protein requirements at various ages. *Nature* 253:192–194.

Young, V.R., and Munro, H.N. 1978. N$^{\tau}$-Methylhistidine (3-methylhistidine) and muscle protein turnover: an overview. *Fed. Proc.* 37:2291–2300.

Young, V.R., Bier, D.M., and Pellett, P.L. 1989. A theoretical basis for increasing current estimates of the amino acid requirements in adult man, with experimental support. *Am. J. Clin. Nutr.* 50:80–92.

Young, V.R., and Marchini, J.S. 1990. Mechanisms and nutritional significance of metabolic responses to altered intakes of protein and amino acids, with reference to nutritional adaptation in humans. *Am. J. Clin. Nutr.* 51:270–289.

Yu, Y.M., Yang, R.D., Matthews, D.E., Burke, J.F., Bier, D.M., and Young, V.R. 1985. Quantitative aspects of glycine and alanine nitrogen metabolism in postabsorptive young men: effects of level of nitrogen and dispensable amino acid intake. *J. Nutr.* 115:399–410.

Yudkoff, M., Nissim, I., McNellis, W., and Polin, R. 1987. Albumin synthesis in premature infants: determination of turnover with [^{15}N]glycine. *Pediatr. Res.* 21:49–53.

Zachwieja, J.J., Bier, D.M., and Yarasheski, K.E. 1994. Growth hormone administration in older adults: effects on albumin synthesis. *Am. J. Physiol.* 266:E840–E844.

Zanze, M., Souberbielle, J.C., Kindermans, C., Rossignol, C., and Garabedian, M. 1997. Procollagen propeptide and pyridinium cross-links as markers of type I col-

lagen turnover: sex- and age-related changes in healthy children. *J. Clin. Endocrinol. Metab.* 82:2971–2977.
Zeman, R.J., Kameyama, T., Matsumoto, K., Bernstein, P., and Etlinger, J.D. 1985. Regulation of protein degradation in muscle by calcium. *J. Biol. Chem.* 260:13619–13624.
Zhang, X.-J., Chinkes, D., Doyle, D.J., and Wolfe, R.R. 1998. Metabolism of skin and muscle protein is regulated differently in response to nutrition. *Am. J. Physiol.* 274:E484–E492.
Zhou, X., and Thompson, J.R. 1997. Regulation of protein turnover by glutamine in heat-shocked skeletal myotubes. *Biochim. Biophys. Acta* 1357:234–242.
Zillikens, M.C., van den Berg, J.W.O., Wattimena, J.L.D., Rietveld, T., and Swart, G.R. 1993. Nocturnal oral glucose supplementation. The effects on protein metabolism in cirrhotic patients and in healthy controls. *J. Hepatol.* 17:377–383.
Zimmer, D.M., Golichowski, A.M., Karn, C.A., Brechtel, G., Baron, A.D., and Denne, S.C. 1996. Glucose and amino acid turnover in untreated gestational diabetes. *Diabetes Care* 19:591–596.
Zoli, M., Marchesini, G., Dondi, C., Bianchi, G.P., and Pisi, E. 1982. Myofibrillar protein catabolic rates in cirrhotic patients with and without muscle wasting. *Clin. Sci.* 62:683–686.

Index